An Introduction to the Geometry and Topology of Fluid Flows

NATO Science Series

A Series presenting the results of scientific meetings supported under the NATO Science Programme.

The Series is published by IOS Press, Amsterdam, and Kluwer Academic Publishers in conjunction with the NATO Scientific Affairs Division

Sub-Series

I. Life and Behavioural Sciences	IOS Press
II. Mathematics, Physics and Chemistry	Kluwer Academic Publishers
III. Computer and Systems Science	IOS Press
IV. Earth and Environmental Sciences	Kluwer Academic Publishers
V. Science and Technology Policy	IOS Press

The NATO Science Series continues the series of books published formerly as the NATO ASI Series.

The NATO Science Programme offers support for collaboration in civil science between scientists of countries of the Euro-Atlantic Partnership Council. The types of scientific meeting generally supported are “Advanced Study Institutes” and “Advanced Research Workshops”, although other types of meeting are supported from time to time. The NATO Science Series collects together the results of these meetings. The meetings are co-organized bij scientists from NATO countries and scientists from NATO's Partner countries – countries of the CIS and Central and Eastern Europe.

Advanced Study Institutes are high-level tutorial courses offering in-depth study of latest advances in a field.
Advanced Research Workshops are expert meetings aimed at critical assessment of a field, and identification of directions for future action.

As a consequence of the restructuring of the NATO Science Programme in 1999, the NATO Science Series has been re-organized and there are currently five sub-series as noted above. Please consult the following web sites for information on previous volumes published in the Series as well as details of earlier sub-series.

http://www.nato.int/science
http://www.wkap.nl
http://www.iospress.nl
http://www.wtv-books.de/nato-pco.htm

Series II: Mathematics, Physics and Chemistry – Vol. 47

An Introduction to the Geometry and Topology of Fluid Flows

edited by

Renzo L. Ricca
Mathematics Department,
University College London,
London, United Kingdom

Kluwer Academic Publishers

Dordrecht / Boston / London

Published in cooperation with NATO Scientific Affairs Division

Proceedings of the NATO Advanced Study Institute on
Pedagogical Workshop on Geometry and Topology of Fluid Flows
Cambridge, United Kingdom
11 – 22 September 2000

A C.I.P. Catalogue record for this book is available from the Library of Congress.

ISBN 1-4020-0206-8

Published by Kluwer Academic Publishers,
P.O. Box 17, 3300 AA Dordrecht, The Netherlands.

Sold and distributed in North, Central and South America
by Kluwer Academic Publishers,
101 Philip Drive, Norwell, MA 02061, U.S.A.

In all other countries, sold and distributed
by Kluwer Academic Publishers,
P.O. Box 322, 3300 AH Dordrecht, The Netherlands.

Printed on acid-free paper

Printed in the Netherlands.

Any idea?

CONTENTS

PREFACE

In editing this collection of lectures I kept in mind the spirit with which Hassan Aref, Tsutomu Kambe, Richard Pelz and myselforganized the programme on Geometry and Topology of Fluid Flows held at the Isaac Newton Institute for Mathematical Sciences Cambridge, UK, September-December, 2000: to promote and fosterinterchange of knowledge and information between mathematicians, pure and applied, and physicists on modern applications of geometric and topological techniques in fluid mechanics and magnetohydrodynamics. I therefore asked the authors to contribute with papers that could reflect this spirit. I hope we achieved our goal with a volume that will provide a valuable addition to your library. If there is any merit in this our thanks must go to the contributors. Mistakes, misprints and sloppy presentations of material, however, should be entirely ascribed to this editor. I would like to thank my friends and co-organisers of the GTFF programme, the Director of the Isaac Newton Institute, Professor H.K. Moffatt, and the dedicated staff,for their continuing support and help towards the organisation of that successful programme. I would also like to acknowledge financial support from NATO through the Advanced Study Institute held in Cambridge.

Renzo L. Ricca

Henry Keith Moffatt

Photograph: R.L. Ricca — Cetraro (Italy), July 2001

I. EIGHT PROBLEMS FOR THE XXI CENTURY

SOME REMARKS ON TOPOLOGICAL FLUID MECHANICS

H. KEITH MOFFATT
Isaac Newton Institute for Mathematical Sciences
20 Clarkson Road, Cambridge, CB3 0EH, UK
E-mail: *hkm2@newton.cam.ac.uk*

Abstract. Some fluid dynamical problems having a topological flavour are briefly reviewed, and some further problems having at least a topological starting point are posed.

1. Introduction

Topological Fluid Mechanics is primarily concerned with structures within a flow field which retain some coherence over a significant period of time. Under circumstances that may be described as 'ideal' relative to the type of structure considered, this 'significant period of time' is infinite; but insofar as circumstances are never ideal in reality, we must be equally concerned with the manner in which structural (or topological) properties of a flow may change with time (generally under the influence of some diffusive process).

These statements suffer from a degree of imprecision that can be removed only through consideration of particular problems. The purpose of this brief paper is to set out a number of such problems, all of which have at least a starting point that can be described as topological, and most of which are unsolved. There is no shortage of challenging problems of this type for which a combination of analytical, computational and experimental (ACE) techniques will be required if real progress is to be made.

2. Particle Paths

For any given flow field $\mathbf{u}(\mathbf{x}, t)$, whether in a finite or infinite domain, the path $\mathbf{X} = \mathbf{X}(\mathbf{x}, t)$ of the particle initially at position $\mathbf{x}$ is determined by the

R.L. Ricca (ed.), An Introduction to the Geometry and Topology of Fluid Flows, 3–10.

dynamical system

$$\frac{d\mathbf{X}}{dt} = \mathbf{u}(\mathbf{X}, t), \quad \mathbf{X}(\mathbf{x}, 0) = \mathbf{x}. \tag{1}$$

We shall restrict attention to incompressible flows for which $\nabla \cdot \mathbf{u} = 0$. Even with this restriction it is only in the simplest circumstances that the system (1) is integrable; for example, for steady two-dimensional flow with streamfunction $\psi(x, y)$, the particle paths coincide with the streamlines $\psi =$ cst. , and the distance travelled by a fluid particle along a streamline is obtained by integrating along the streamline.

As shown by Aref [1], the introduction of a time-periodic ingredient in a two-dimensional flow field, is sufficient to induce chaos in the particle paths, this chaos appearing near homoclinic or heteroclinic trajectories of the time-averaged velocity field $\bar{\mathbf{u}}(\mathbf{x})$. The particular case studied was that of the 'blinking vortices' for which this time-averaged flow has a single homoclinic (saddle) point; but there seems little doubt that the associated appearance of chaos is indeed a generic phenomenon. A passive scalar field $\theta(\mathbf{x}, t)$ subjected to advection in such a flow field suffers a very rapid stretching of its iso-scalar curves $\theta =$ cst. and an associated rapid decrease of scale to the level at which molecular diffusion becomes operative in smoothing out field variation [17].

In three dimensions, even steady flows $\mathbf{u}(\mathbf{x})$ have associated particle paths (coinciding with streamlines) that exhibit chaos. This behaviour was first identified for space-periodic flows, specifically the 'ABC-flow'

$$\mathbf{u} = (C \cos kz + B \sin ky\,,\ A \cos kx + C \sin kz\,,\ B \cos ky + A \sin kx)\,, \tag{2}$$

by Arnold [2] and Hénon [9], the analysis of this flow being developed later by Dombre *et al.* [7].

Chaos can likewise occur for three-dimensional steady flows in a bounded domain. The case of quadratic flows of the form

$$u_i(\mathbf{x}) = a_i + b_{ij} x_j + c_{ijk} x_j x_k \tag{3}$$

satisfying $\nabla \cdot \mathbf{u} = 0$ and $\mathbf{n} \cdot \mathbf{u} = 0$ on $|\mathbf{x}| = 1$ was treated by Bajer & Moffatt [3]. The vorticity field $\boldsymbol{\omega} = \nabla \times \mathbf{u}$ in such flows is linear in $\mathbf{x}$ and so satisfies $\nabla^2 \boldsymbol{\omega} = 0$. Thus, the flows are Stokes flows in the sphere $|\mathbf{x}| < 1$, driven by a prescribed tangential velocity on the surface $|\mathbf{x}| = 1$. These flows in general exhibit regions in which the particle paths are chaotic. This behaviour may be detected visually through computation of Poincaré sections; more formally, the chaotic regions may be defined as the set of initial positions $\mathbf{x}$ for which the Liapunov exponent

$$\lambda = \lim_{t\to\infty} \frac{1}{t} \ln |\boldsymbol{\xi}| \tag{4}$$

is positive; here $\boldsymbol{\xi}$ is an infinitesimal material line element originally at $\mathbf{x}$; λ represents the asymptotic rate of stretching of this material line element. Note that if $\mathbf{x}_1$, $\mathbf{x}_2$ lie on the same streamline, then $\lambda(\mathbf{x}_1) = \lambda(\mathbf{x}_2)$. In regions of 'regularity' of the flow, $|\boldsymbol{\xi}|$ grows merely linearly with time, and $\lambda(\mathbf{x}) = 0$.

A quantity of key importance in relation to the stirring efficiency of a flow is the volume fraction μ of the fluid domain in which $\lambda(\mathbf{x}) > 0$. This volume fraction is a fundamental structural parameter; it is topological in the sense that it is invariant under continuous volume-preserving deformations of the flow field (which deform the regions of chaos without change of volume). We may thus pose a first problem of topological character:

XXI Century Problem 1 *For a given bounded domain $\mathcal{D}$ and a given continuous flow field $\mathbf{u}(\mathbf{x})$ $(\mathbf{x} \in \mathcal{D})$, satisfying $\nabla \cdot \mathbf{u} = 0$ in $\mathcal{D}$ and $\mathbf{n} \cdot \mathbf{u} = 0$ on $\partial\mathcal{D}$, to determine the volume fraction μ of $\mathcal{D}$ for which $\lambda(\mathbf{x}) > 0$, i.e. for which the flow has chaotic particle paths.*

This problem has obvious generalisations for space-periodic flows and for time-periodic two-dimensional flows.

3. Scalar Field Problems

Let $\theta(\mathbf{x}, t)$ be a scalar field that is convected by a continuous velocity field $\mathbf{u}$, and suppose for the moment that molecular diffusion is negligible. Then θ is constant for each fluid particle, i.e.

$$\frac{D\theta}{Dt} \equiv \frac{\partial \theta}{\partial t} + \mathbf{u} \cdot \nabla\theta = 0\,. \tag{5}$$

To be specific, suppose that the flow domain $\mathcal{D}$ is in $\mathbb{R}^3$, and that $\mathbf{u}\cdot\mathbf{n} = 0$, $\theta =$ cst. on $\partial\mathcal{D}$. Equation (5) of course implies that the surfaces $\theta =$ cst. are transported with the flow. Their topology is therefore conserved. How is this topology to be described?

A start has been made [14] through consideration of the saddle points S_i of the field θ and the homoclinic iso-scalar surfaces Σ_i through these S_i. Together with $\partial\mathcal{D}$, the Σ_i divide $\mathcal{D}$ into a number of subdomains $\mathcal{D}_\alpha$ $(\alpha = 0, 1, 2, \ldots)$ where $\mathcal{D}_0$ is the subdomain that is bounded externally by $\partial\mathcal{D}$, and each $\mathcal{D}_\alpha$ $(\alpha = 1, 2, \ldots)$ is bounded by parts of one or (at most) two of the homoclinic surfaces. The volume of each $\mathcal{D}_\alpha$ is conserved under (5) as is the topology of the surfaces Σ_i.

Knowledge of the relative configurations of the Σ_i and of the volumes of the $\mathcal{D}_\alpha$ is an important first step in classifying the possible topologies of the θ-field. This is not all however. Within each $\mathcal{D}_\alpha$ are a family of surfaces $\theta =$ cst. and we may define a *signature function* $V_\alpha(\theta)$ in $\mathcal{D}_\alpha$ with the property that $(dV_\alpha/d\theta)\delta\theta$ is the volume between surfaces labelled $\theta, \theta + \delta\theta$; the function $V_\alpha(\theta)$ is then defined up to a constant C_α which may be chosen so that the signature function varies continuously in moving from one subdomain to another. The set of signature functions $\{V_\alpha(\theta)\}$ thus defined is clearly invariant under the evolution (5) and is therefore a topological property of the field θ.

There are now two interesting directions that merit investigation. We indicate these in the form of problems:

XXI Century Problem 2

Suppose now that molecular diffusivity κ is included, so that (5) is replaced by

$$\frac{D\theta}{Dt} = \frac{\partial\theta}{\partial t} + \mathbf{u} \cdot \nabla\theta = \kappa\nabla^2\theta \,. \tag{6}$$

What transitions in the topology of the set of homoclinic surfaces $\{\Sigma_i\}$ are possible as a result of this diffusion; and how does the set of signature functions $\{V_\alpha(\theta)\}$ evolve, particularly during such a change of topology?

XXI Century Problem 3

Suppose that the velocity field $\mathbf{u}$ is itself 'driven' by inhomogeneity of the θ-field, according to some well-defined dynamical prescription (e.g. θ could represent temperature variation in a gravity field, the flow being driven by the buoyancy force in the Boussinesq approximation (see, for example, [4]). The problem is to examine the evolution of the θ-field in the neighbourhood of its saddle-points, to determine whether singularities of $\nabla\theta$ can develop, and to examine the influence of weak molecular diffusivity κ in controlling the approach to such singularities.

4. Vector Field Problems

Each of the above scalar field problems has a counterpart in the context of a transported vector field, such as the magnetic field $\mathbf{B}(\mathbf{x}, t)$ in a conducting fluid. This field is divergence-free, i.e. $\nabla \cdot \mathbf{B} = 0$, and satisfies the induction equation

$$\frac{\partial \mathbf{B}}{\partial t} = \nabla \times (\mathbf{u} \times \mathbf{B}) + \eta\nabla^2\mathbf{B} \,, \tag{7}$$

the counterpart of (6). In the diffusionless (perfectly conducting) limit $\eta = 0$, (7) implies that $\mathbf{B}$-lines are 'frozen' in the fluid, the flux of $\mathbf{B}$

through any closed material curve C being conserved. This has the important consequence that any linkage between **B**-lines is conserved – clearly a result with topological content. The simplest measure of net linkage of the field inside any Lagrangian (i.e. material) surface S on which $\mathbf{n} \cdot \mathbf{B} = 0$ is the magnetic helicity

$$\mathcal{H}_M = \int_V \mathbf{A} \cdot \mathbf{B} \, dV \tag{8}$$

where V is the volume inside S, and $\mathbf{A}$ is a vector potential for $\mathbf{B}$, i.e. $\mathbf{B} = \nabla \times \mathbf{A}$. The integral (8) is gauge-independent, but it is usual to choose the gauge of $\mathbf{A}$ so that $\nabla \cdot \mathbf{A} = 0$.

XXI Century Problem 4 *Consider two linked unknotted flux tubes, each carrying flux Φ, the field within each tube being untwisted, so that the helicity $\mathcal{H}_M$ is $2\Phi^2$ [11]; we assume here that the linkage is right-handed. Suppose that the fluid motion brings the tubes into close proximity and that weak diffusion ($\eta > 0$) causes reconnexion of* **B***-lines in such a way that the two tubes become a single tube carrying flux Φ. In this process, the helicity (or at least some proportion of it) may survive through the appearance of internal twist in the resultant tube [15]. The problem is to determine precisely what is the total field helicity after reconnexion, the whole process being governed by equation (7).*

In the above problem, if the field **B** is sufficiently weak, then presumably it may be treated as dynamically passive, the velocity **u** being then independently prescribable. More realistically, however, the Lorentz force $\mathbf{j} \times \mathbf{B}$ where $\mathbf{j} = \nabla \times \mathbf{B}$, plays an important part in the reconnexion process. This is particularly the case when η is very weak since then very strong stretching of field lines occurs in conjunction with the flow that brings sections of the two initial tubes into close proximity. This leads to

XXI Century Problem 5 *Consider two flux tubes oblique to each other, carrying the same flux Φ, and driven towards each other by an imposed strain field; it is required to analyse the process of reconnexion, its dependence on Φ (via the Lorentz force), and again to determine how much twist is created in the reconnected tubes.*

Just as for the scalar field problem, there are circumstances in which the velocity field **u** is entirely driven by the Lorentz force distribution. If the fluid is viscous but perfectly conducting, then the field energy converts to kinetic energy which is dissipated by viscosity, and during this process the field topology is conserved. There is however an outstanding problem in relation to this 'magnetic relaxation' scenario that remains open:

XXI Century Problem 6 *Consider a smooth localised magnetic field of finite and nonzero magnetic helicity in a viscous, perfectly conducting, incompressible fluid initially at rest. It is known [12] that the kinetic energy tends to zero as $t \to \infty$. It is required to prove that $|\mathbf{u}(\mathbf{x},t)| \to 0$ at all points $\mathbf{x}$ as $t \to \infty$.*

This is almost certainly true, since otherwise the appearance of singularities (of implausible form in a viscous fluid) is implied; a proof should not be impossibly difficult.

If the initial field is confined to a single knotted flux tube of volume V and carrying flux Φ, and with internal twist such that the helicity is $\mathcal{H}_M = h\Phi^2$, then, on dimensional grounds the 'relaxed' state has minimal magnetic energy M^E given by

$$M^E = m(h)\Phi^2 V^{-1/3} \tag{9}$$

where $m(h)$ is a positive dimensionless function of the dimensionless twist parameter h, which depends only on the form of the knot. Particular interest attaches to value of h (for a given knot K) for which $m(h)$ is minimal, $m_{\min}$ say. It is to be expected that $m_{\min}$ increases with increasing knot complexity.

XXI Century Problem 7 *Determine $m_{\min}$ for knots of minimum crossing number 3, 4, 5, ...*

This problem presents a considerable computational challenge. A start has been made for torus knots by Chui & Moffatt [6].

5. The Finite-Time Singularity Problem

It is but a small step from the above 'magnetically active' problems to the Euler and Navier-Stokes problems that lie at the heart of fluid mechanics. We simply replace $\mathbf{B}$ in (7) by the vorticity field $\boldsymbol{\omega}$, and we take $\mathbf{u}$ to be the inverse curl of $\boldsymbol{\omega}$:

$$\mathbf{u} = (\mathrm{curl})^{-1}\boldsymbol{\omega}\,, \quad \nabla\cdot\mathbf{u} = 0\,; \tag{10}$$

and of course we replace η by kinematic viscosity ν.

The simplicity of the functional relationship (10) between $\mathbf{u}$ and $\boldsymbol{\omega}$ might suggest that, for example, the problem of viscous vortex tube reconnexion should be no more difficult than the problem of magnetic flux tube reconnexion (with Lorentz forces included). This however is just wishful thinking! Vortex tube reconnexion has attracted much study, both analytical [5] and computational [10], [18] (see also this volume), and yet we are still in the dark as regards the details of the process. In particular, we do not know whether the vorticity field within a zone of reconnexion remains finite for all t, or conversely whether a singularity of vorticity may develop within a finite time. The computational work cited above provides quite strong evidence for the appearance of finite-time singularities for Euler evolution ($\nu = 0$), whereas analytical studies based on the Leray similarity transformation [16], [19] point to the non-existence of finite-time singularities when $\nu > 0$. I have argued [13] that if finite-time singularities appear under Euler evolution, then the same type of singularity should appear when ν is positive but sufficiently small, i.e. that weak viscosity may not be able to prevent the formation of finite-time singularities. This suggests a problem that may provide a helpful stepping-stone towards the central ($1m) finite-time singularity problem as posed by the Clay Institute [8].

XXI Century Problem 8 *Suppose that there exists a smooth velocity field* $\mathbf{u}_0(\mathbf{x})$ *of finite energy in a bounded domain* $\mathcal{D}$ *such that, under Euler evolution starting from this initial condition, a singularity of* $\boldsymbol{\omega}(\mathbf{x}, t)$ *appears at some finite time* t^*. *Prove that, for* $0 < \nu < \nu_c$ *where* ν_c *is small, and with the same initial condition,* $\boldsymbol{\omega}(\mathbf{x}, t)$ *still becomes singular at finite time; or conversely, prove that* $\boldsymbol{\omega}(\mathbf{x}, t)$ *remains smooth for all* t.

References

1. Aref, H. (1984) Stirring by chaotic advection. *J. Fluid Mech.* **143**, 1–21.
2. Arnold, V.I. (1965) Sur la topologie des écoulements stationnaires des fluides parfaits. *C.R. Acad. Sci. Paris* **261**, 17–20.
3. Bajer, K. & Moffatt, H.K. (1990) On a class of steady confined Stokes flows with chaotic streamlines. *J. Fluid Mech.* **212**, 337–363.
4. Batchelor, G.K., Canuto, V.M. & Chasnov, J.R. (1992) Homogeneous buoyancy-generated turbulence. *J. Fluid Mech.* **235**, 349–378.
5. Beale, J.T., Kato, T. & Majda, A.J. (1989) Remarks on the breakdown of smooth solutions for the 3-D Euler equations. *Commun. Math. Phys.* **94**, 61–66.
6. Chui, A. & Moffatt, H.K. (1995) The energy and helicity of knotted magnetic flux tubes. *Proc. Roy. Soc. Lond. A* **451**, 609–629.
7. Dombre, T., Frisch, U., Greene, J.M., Hénon, M., Mehr, A. & Soward, A.M. (1986) Chaotic streamlines in the ABC flows. *J. Fluid Mech.* **167**, 353–391.
8. Fefferman, C. (2000) Existence and smoothness of the Navier-Stokes equations. `www.claymath.org/prizeproblems/navierstokes.htm` .
9. Hénon, M. (1966) Sur la topologie des lignes de courant dans un cas particulier. *C.R. Acad. Sci. Paris A* **262**, 312–314.
10. Kerr, R.M. (1993) Evidence for a singularity in the three-dimensional Euler equations. *Phys. Fluids* **6**, 1725–1746.
11. Moffatt, H.K. (1969) The degree of knottedness of tangled vortex lines. *J. Fluid Mech.* **35**, 117–129.
12. Moffatt, H.K. (1985) Magnetostatic equilibrium and analogous Euler flows of arbitrarily complex topology. Part 1. Fundamentals. *J. Fluid Mech.* **159**, 359–378.

13. Moffatt H.K. (2000) The interaction of skewed vortex pairs: a model for blow-up of the Navier-Stokes equations. *J. Fluid Mech.* **409**, 51–68.
14. Moffatt, H.K. (2001) The topology of scalar fields in 2D and 3D turbulence. In *Geometry and Statistics of Turbulence* (ed. T. Kambe *et al.*), pp. 13–22. Kluwer.
15. Moffatt, H.K. & Ricca, R.L. (1992) Helicity and the Călugăreanu invariant. *Proc. R. Soc. Lond. A* **439**, 411–429.
16. Nečas, J., Růžička, M. & Šverák, V. (1996) On Leray's self-similar solutions of the Navier-Stokes equations. *Acta Math.* **176**, 283–294.
17. Ottino, J.M. (1989) *The Kinematics of Mixing: Stretching, Chaos and Transport.* Cambridge University Press.
18. Pelz, R.B. (1997) Locally self-similar, finite-time collapse in a high-symmetry vortex filament model. *Phys. Rev. E* **55**, 1617–1626.
19. Tsai, T.-P. (1998) On Leray's self-similar solutions of the Navier-Stokes equations satisfying local energy estimates. *Arch. Rat. Mech. Anal.* **143**, 29–51.

II. MATHEMATICS BACKGROUND

DIFFERENTIAL GEOMETRY OF CURVES AND SURFACES

RÉMI LANGEVIN
Laboratoire de Topologie, Université de Bourgogne
9 Avenue Alain Savary, B.P. 47 870
21078 Dijon Cedex, France

E-mail: *langevin@u-bourgogne.fr*

Abstract. The goal of this article is to present the relation between some differential formulas, like the Gauss integral for a link, or the integral of the Gaussian curvature on a surface, and topological invariants like the linking number or the Euler characteristic.

1. Plane Curves

The curves we will consider should, if they are not polygonal, be smooth enough (of class at least $\mathcal{C}^2$; we will assume they are of class $\mathcal{C}^\infty$).

1.1. EQUATION AND OSCULATING CIRCLE

Locally, at a neighbouhood of the point m, the curve C has the equation $y = f(x)$ where x is an Euclidean coordinate on the line generated by T_mC and y is an Euclidean coordinate on the line $L(m)$ normal to the curve at m. We can chose the Euclidean coordinates x, y such that $f(0) = 0$ and $f'(0) = 0$. We can also chose the coordinate y in order that the orientation of the frame defined by the two axis (Ox, Oy) is positive. Let us write

$$f = \frac{1}{2}k \cdot x^2 + \text{ higher order terms .} \tag{1}$$

A circle tangent to C at m has the equation $(y - R)^2 + x^2 = R^2$, $R \neq 0$. The points of this circle near m have the equation $y = f_R(x) = \frac{1}{2R}x^2+$ higher order terms. The difference $f - f_R$ is $\frac{1}{2}[k - \frac{1}{R}]$.

Among all circles tangent to C at m the one such that the difference $f - f_R$ is of order higher than two is called the osculating circle at m. The

R.L. Ricca (ed.), An Introduction to the Geometry and Topology of Fluid Flows, 13–33.

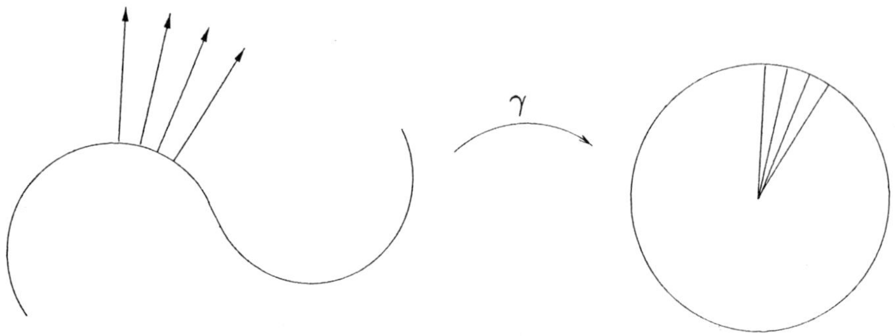

Figure 1. The Gauss map.

absolute value of the curvature $|k(m)| = 1/R$. The number $k(m)$ has a sign which depends on the orientation of the line $L(m)$ defined by the choice of the y-coordinate; it is positive if the line points towards the osculating circle.

1.2. THE GAUSS MAP

An essential tool in the study of hypersurfaces of $\mathbb{R}^n$, and first of planar curves, is the Gauss map γ, which associates to each point m of an oriented curve C its oriented normal $N(m) = R_{\pi/2}(T(m)) \in S^1$, where $T(m)$ is the oriented unit tangent at m to the curve (see Figure 1):

$$\gamma : C \to S^1 \ . \tag{2}$$

The degree of this Gauss map is called the *turning number* of the oriented curve C.

Proposition 1.1 *The Jacobian $d\gamma/dx$ of γ at a point $m \in C$, where the curve is locally parametrised by the arc-length of its tangent line at m_0, orienting the unit circle as usual, is the number $k(m)$ we defined above. It is called the* **curvature** *of C at that point.*

Corollary 1.2

$$t(C) = \int_C k(m)dm \ . \tag{3}$$

The proof of this corollary is very similar to the proof of the Theorem 1.4 below, using the number $n(z) = \sum_{m\in\gamma^{-1}(z)} \epsilon(m)$, where $\epsilon(m)$ is the sign of the Jacobian of the Gauss map at m. The degree theory (see [10]) implies that this number does not depend on the regular value z of the Gauss map we choose. So we shall skip it.

Notice that we could have used any parametrisation $c(t)$ of the curve C near m_0 such that $c(0) = m_0$ and $c'(0) = +1$. Notice also that we can define a Gauss map with value in the projective set $\mathbb{P}_1$ of line directions in $\mathbb{R}^2$, forgetting the orientation of $N(m)$. The tangent map $T : C \mapsto S^1$ mapping a point $m \in C$ to the oriented unit tangent to C at m, has the same Jacobian $k(m)$.

Remark 1.3 The projection of C on the oriented normal, using the previous parametrisation $y = f(x)$ of the curve near m_0, is just $y = f(x)$. Let $m_0 \in C$ be a non-critical point of the Gauss map. Then the point m_0 is a nondegenerate critical point of the orthogonal projection of C on the oriented line $L(m_0)$.

Proof. We shall prove the proposition and the remark above at the same time. Using our local coordinates, the normal $N(m)$ can be written:

$$N(m) = \frac{\text{grad}\,(F)}{|\text{grad}\,(F)|}\ , \ \text{where } F(x,y) = y - f(x)\ . \tag{4}$$

Then the vector $dN(m) \in \mathbb{R}^2$ is:

$$dN(m) = \frac{1}{|\text{grad}\,(F)|}\begin{pmatrix} -f'' \\ 0 \end{pmatrix} + \text{grad}\,(F) \cdot \frac{d}{dx}\left(\frac{1}{|\text{grad}\,(F)|}\right)\ .$$

At m_0, we have

$$\text{grad}\,(F) \cdot \frac{d}{dx}\left(\frac{1}{|\text{grad}\,(F)|}\right) = 0\ ,$$

therefore: $dN(m_0) = \begin{pmatrix} -f''(0) \\ 0 \end{pmatrix}$; this implies, taking into acount the orientation of the unit circle at the point $\begin{pmatrix} 0 \\ 1 \end{pmatrix}$, that $k(m_0) = d\gamma(m_0) = f''(0)$. This Jacobian $d\gamma(m_0)$ is zero exactly when the quadratic part (in x) of f is zero.

QED

For a direction L, denote by $|\mu(C, L)|$ the number of critical points of the orthogonal projection of C on L. This function $|\mu|$ is defined on the space $\mathbb{P}_1$ of directions in $\mathbb{R}^2$ as defined above.

The change of variable theorem implies then that there exists a neighbourhood v of m such that

$$\int_v |k(m)|dm = \int_{\mathbb{P}_1} |\mu(v, L)|\, dL\ . \tag{5}$$

The result holds globally on C.

Theorem 1.4

$$\int_C |k(m)| dm = \int_{\mathbb{P}_1} |\mu(C, L)| \, dL \, . \tag{6}$$

Proof. The proof relies on Sard's theorem. The set Σ of critical values of γ is of zero measure. Its inverse image $\gamma^{-1}(\Sigma)$ is the union of critical points of γ, where $k = 0$, and of non-critical points of γ with image in σ, the latter forming a set of measure zero. The complement of $\gamma^{-1}(\Sigma)$ is an enumerable union of open sets of C. Discarding at most an enumerable set of points if necessary, we get an enumerable union $\bigcup_i U_i$ of open sets of C where the restriction of γ is a diffeomorphism on its image. Using the change of variable theorem and summing on i we get:

$$\int_C |k(m)| \, dm = \int_{\bigcup_i (U_i)} |k(m)| \, dm = \sum_i (\gamma U_i = \int_{\mathbb{P}_1} |\mu(C, L)| \, dL \, . \tag{7}$$

(A similar result, with a similar proof, is true for surfaces in $\mathbb{R}^3$.)

QED

We can also count "most" of the critical points with a sign. Assign to the non degenerate critical points of the orthogonal projection of the oriented curve C on the oriented line L^+ the sign $\epsilon(m) = (-1)^{\text{index}(m)}$. When the two unit vectors contained in L are non degenerate values of the Gauss map, we can orient the line L at each point m, such that $\gamma(m) \subset L$, using the normal $N(m)$ to define $\epsilon(m)$. Thus, we get:

$$\mu(C, L) = \sum_{\gamma(m) \in L} \epsilon(m) \, . \tag{8}$$

A classical theorem for embedded closed planar curve states that:

Theorem 1.5

$$\left| \int_C k(m) \, dm \right| = 2\pi \, . \tag{9}$$

It is a consequence of the following fact that we shall explain below:

$$\mu(C, L) = 2 \cdot \text{degree} \, (\text{Gauss map}) = \pm 2 \, , \tag{10}$$

when C is a simple closed curve, and when $\mu(C, L)$ makes sense. As a corollary we get the inequality:

$$\int_C |k(m)| \, dm \geq 2\pi \, . \tag{11}$$

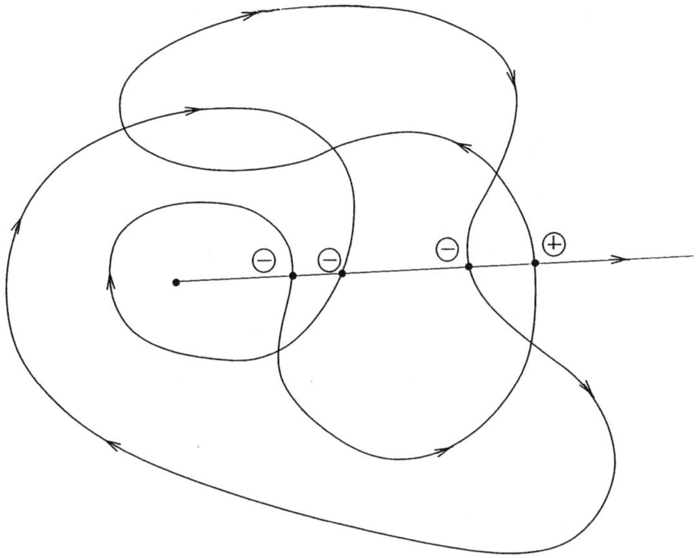

Figure 2. Algebraic intersection of a ray and an oriented curve.

This inequality is still true for closed curves embedded in $\mathbb{R}^3$; it was proved by Fenchel in 1929. If the closed curve $C \subset \mathbb{R}^3$ is knotted, a stronger inequality, proved independently by Fary, Fenchel and Milnor, is true:

$$\int_C |k(m)| \, dm > 4\pi \ . \tag{12}$$

1.3. THE INDEX OF A POINT RELATIVE TO A CLOSED CURVE

The closed curve C is now immersed in the plane. Let $m \notin C$. A ray D^+ with origin in m may intersect transversally the curve. In each transverse intersection point $m \in D^+ \cap C$, we define a sign $\epsilon(m) = \pm 1$. $\epsilon(m) = +1$ if the frame (u, T) is positive, where u is the unit vector contained in the ray D^+, and T is the unit tangent vector to C, defined by the orientation; $\epsilon(m) = -1$, if the orientation of the frame is negative (see Figure 2). If the ray hits a double point, we have to take care of the two strands of C. Summing up, this defines the *algebraic intersection number* of D^+ and C by:

$$D^+ \bullet C = \sum_{x \in C \cap D^+} \epsilon(x) \tag{13}$$

Proposition 1.6 (Index of m with respect to C) *For almost all ray D^+ with origin at $m \notin C$, the algebraic intersection number $D^+ \bullet C$ is defined.*

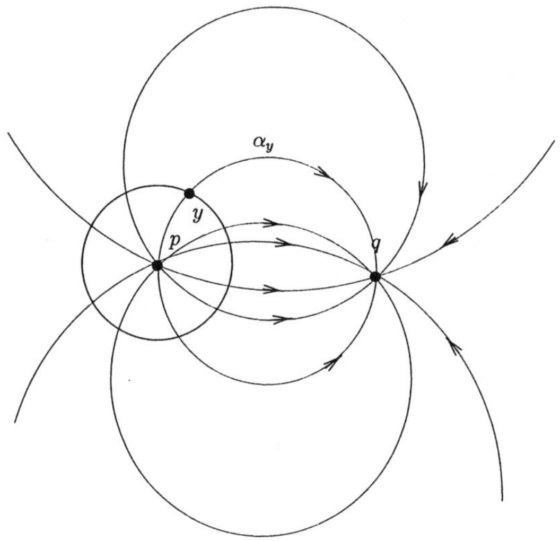

Figure 3. Linking of a pair of points and a curve.

Moreover this number depends only on the oriented curve C and m. We call this number, the index of m with respect to C and denote it by $\mathcal{I}_C(m)$.

Proof. To prove that the index $\mathcal{I}_C(m)$ is defined for almost all rays, let us consider the map $\Phi : C \to S^1$ defined by $\phi(x) = \frac{x-m}{|x-m|}$. Since the map is smooth enough, the measure of its critical values is zero. Since the critical values of Φ are exactly the unit vectors contained in rays which are not transversal to C, this proves the first part of the proposition. The proof of the second part amounts to observe that the number $D^+ \bullet C$ is precisely the degree of the map Φ.

QED

Remark 1.7 We can give an elementary proof the the previous proposition when C is a closed polygon. All rays not parallel to a side intersect C transversally. It is now enough to observe the possible "accidents" that ocur when the ray D^+ turns.

Remark 1.8 The index can also be defined using complex analysis, see Theorem 10.10 in [14]:

$$\mathcal{I}_C(m) = \frac{1}{2\pi i}\int_C \frac{c'(t)}{c(t)-m}\,dt\ . \tag{14}$$

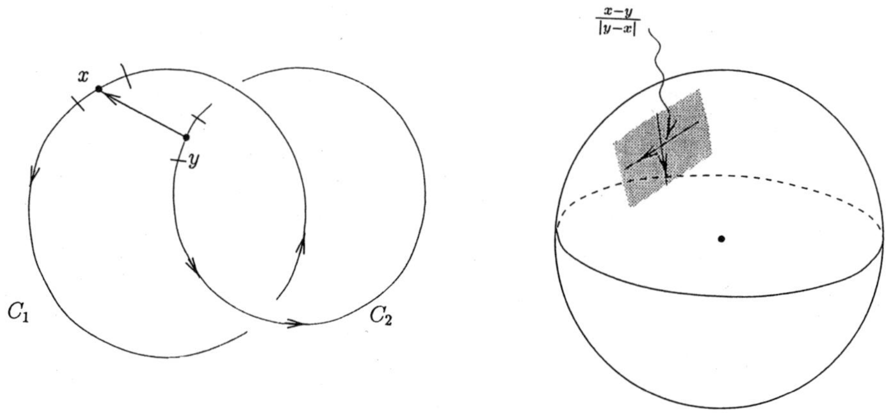

Figure 4. The linking number.

1.4. LINKING OF A PAIR OF POINTS AND AN ORIENTED CURVE

A slight generalisation of the previous construction can be made using two distinct points p, q with masses -1 and $+1$, respectively. We can interpret such a pair as a sum of two Dirac masses $-\delta_p + \delta_p$. Each circle of the family of circles passing through p and q can be cutted into two arcs that we orient from p to q. Using a small circle γ centered at p (or a small circle of the family with limit points p and q close to p), we can label these arcs α_y by the points $y \in \gamma$ (see Figure 3).

Proposition 1.9 (Linking number of $-\delta_p + \delta_p$ with C) *For almost all arcs α_y the algebraic intersection number $\alpha_y \bullet C$ is defined. Moreover this number depends only on the pair $-\delta_p + \delta_p$ and the oriented curve C. We call this number, the linking number of $-\delta_p + \delta_q$ and C and denote it by $\mathcal{L}(-\delta_p + \delta_q, C)$.*

Proof. The proof is very similar to the proof of the existence of the index, considering the map:

$$\Phi : C \to \gamma; \quad \Phi(x) \mapsto y \in \gamma \quad \text{such that } x \in \alpha_y \, . \tag{15}$$

QED

2. Curves in $\mathbb{R}^3$

For topological properties of closed curves embedded in $\mathbb{R}^3$ (knots and links, in particular) see the articles by Weber and by Kauffman collected in this volume. Here we want to consider the properties of a number associated with two oriented curves: the linking number. Then we show some differentiable properties of curves related to the idea of linking.

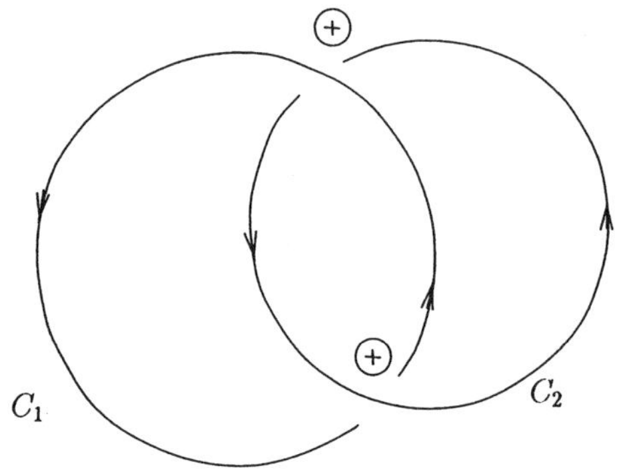

Figure 5. Counting signs on a projection.

2.1. THE LINKING NUMBER

Consider two disjoint smooth oriented closed curves in $\mathbb{R}^3$: C_1 and C_2. In general, we shall call the union of disjoint embedded curves in $\mathbb{R}^3$ a *link* and often use the letter $\mathcal{L}$. We can define a map from the torus $C_1 \times C_2$ to the unit sphere by: $\Phi(x, y) = \frac{x-y}{|y-x|}$ (Figure 4).

Definition 2.1 *The linking number of C_1 and C_2 is the degree of the map Φ.*

Note that the orientation of the two curves is used for the definition of the linking number. In the paper by Gauss [9] it appears as an integral:

$$L(C_1, C_2) = \frac{1}{4\pi} \int_{C_1 \times C_2} \frac{\det\left(\frac{dx}{ds}, \frac{dy}{ds}, \frac{x-y}{|x-y|}\right)}{|x-y|^2} . \tag{16}$$

One recognizes in the integrand the Jacobian of the map Φ, proving therefore that the value of the integral is an integer number. The integral form shows also that the linking number depends continuously on the link formed by C_1 and C_2. It will change continuously when deforming the two components, not allowing intersections: as the value is an integer, the linking number is constant when such deformations are performed.

We can project a link $\mathcal{L}$ on a plane. In general (that is for a set of full measure in the set $G(3, 2)$ of planes in $\mathbb{R}^3$) the projection $p(\mathcal{L})$ is the union of two oriented immersed planar curves which meet transversely. In each point of tranverse intersection or *crossing* we can add one piece of information: that is which curve passes below (or above) the other.

This is usually done by drawing the curve which passes below with an interruption (as in Figure 5). We can associate to each crossing a number $+1$ or -1 defined by using this information and the orientation of the frame

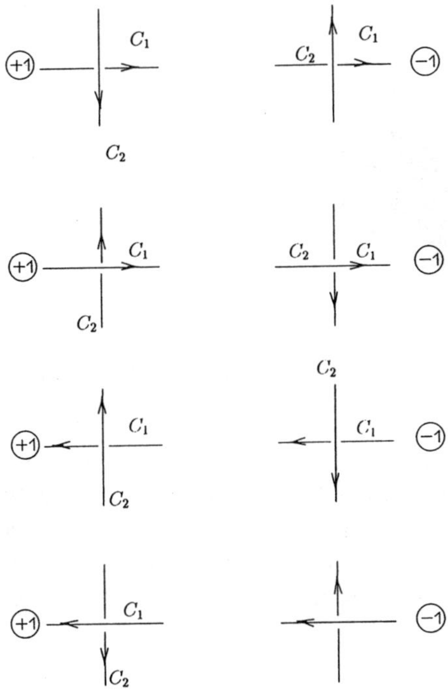

Figure 6. Sign $\epsilon(\kappa)$ of a crossing κ.

defined by the tangent vectors to $p(C_1)$ and $p(C_2)$ at the crossing. A crossing κ correspond to a pair $(x, y) \in C_1 \times C_2$ such that $\Phi(x, y) =$ North or South. Observe that when, on the projection, the intersection is transverse, then at the point (x, y) the map Φ is a local diffeomorphism. The sign $\epsilon(\kappa)$ that we have associated with each configuration of the figure by "counting signs on a projection" is precisely the sign of the Jacobian of Φ, that is $+1$ if Φ preserves the orientation, -1 if ϕ reverses the orientation.

This interpretation of the sign $\epsilon(\kappa)$ (Figure 6) proves that:

Proposition 2.2 *The linking number $L(C_1, C_2)$ satisfies:*

$$L(C_1, C_2) = \frac{1}{2} \sum \epsilon(\kappa) \ . \tag{17}$$

Topologists often use another definition of the linking number.

Theorem 2.3 *Let M be and orientable surface with boundary the oriented curve C_1; suppose also that the surface M is transverse to C_2. Then the algebraic intersection number $M \bullet C_2$ does not depend on the choice of M. Moreover*

$$L(C_1, C_2) = M \bullet C_2 \ . \tag{18}$$

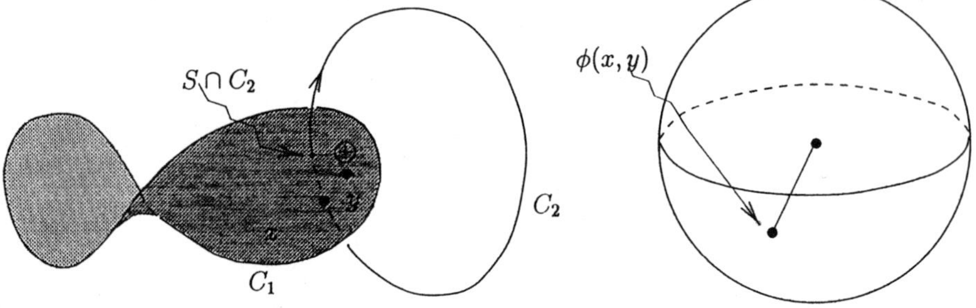

Figure 7. Sign of a crossing with the Seifert surface and $\tilde{\Phi}|_{\Sigma_m}$.

Likewise the algebraic intersection number of two planar curves, the algebraic intersection number between an oriented surface and an oriented curve is defined comparing at each intersection point the orientation of the ambient space and the orientation of a frame made of a positive frame of $T_m M$ followed by the oriented tangent to C_2 at m. If they coincide, $\epsilon(m) = +1$, if not $\epsilon(m) = -1$. Then:

$$M \bullet C = \sum_{m \in M \cap C} \epsilon(m) \, . \tag{19}$$

The proof of the last theorem relies on an important property of the degree of a map: the degree of the restriction of a map to the boundary of a manifold is zero [10].

So we shall construct a map $\tilde{\Phi}$ from a manifold N to S^2 such that:

- one component of the boundary of N is $C_1 \times C_2$;
- the restriction of $\tilde{\phi}$ to $C_1 \times C_2$ is the map $\Phi(x,y) = \frac{y-x}{|y-x|}$.

Let us take $N = M \times C_2 \setminus \bigcup_{m \in M \cap C_2} B(m, \eta)$, where η is small enough and $\tilde{\Phi}(x,y) = \frac{y-x}{|y-x|}$. The map $\tilde{\Phi}$ is well defined because we had removed a neighbourhood of the point of the intersection $M \cap C_2$ where $\frac{y-x}{|y-x|}$ is not defined. One component of the boundary of N is $C_1 \times C_2$. The rest of ∂N is a union of spheres Σ_m in $M \times C_2$ endowed with the "boundary orientation". This orientation is the opposite of the "usual" one, that is the orientation of Σ_m as boundary of the ball $B(m, \eta)$. Computing the degree of the restriction of $\tilde{\Phi}$ to $C_1 \times C_2$, we get the linking number. We need to see that degree$(\tilde{\Phi}|_{\Sigma_m})$ gives the intersection sign of C_2 with the chosen Seifert surface bounded by C_1 (a Seifert surface is an embedded, orientable surface bounded by the curve; this surface always exists and it is not unique; see

Figure 7). Then, by using the nullity of the degree on an oriented boundary, we get:

$$L(C_1, C_2) = \text{degree}\,(\Phi) = -\text{degree}\,(\tilde{\phi}|_{\cup_{m\in(M\cap C_2)}\Sigma_m}) \ . \qquad (20)$$

Since orientable surfaces bounded by a given curve always exist, the previous theorem provides another satisfying definition of linking number. When C_1 is a trivial knot (i.e. an unknotted closed curve), $L(C_1, C_2)$ is the number of times, counted with a sign, C_2 intersects an oriented embedded disc of boundary C_1. As the complement (in $S^3 = \mathbb{R}^3 \cup \{\infty\}$) of C_1 is a thick torus $S^1 \times D^2$, the linking number can be seen as the number of turns made by C_2 in this thick torus.

2.2. CURVATURE AND TORSION

We defined the curvature $k(m) \in \mathbb{R}$ of a plane curve, using a choice of a normal vector depending on the orientation of the plane. It is not possible to proceed in the same way when the curve C is in $\mathbb{R}^3$. Still we can define a positive number $|k(m)|$ at each point $m \in C$.

Once more we will use a map with value in the unit sphere. The orientation of the curve C determines the choice of a unit tangent vector $T(x)$ at each point $x \in C$. This defines a map: $T : C \to S^2$, $x \mapsto T(x)$. The ratio of infinitesimal lengths $d\tau$ on $T(C)$ and ds on C, $|k| = |\frac{d\tau}{ds}|$, is the curvature of C at x. At a point $m \in C$, where the curvature is different from zero, one can define a unit frame: the *Frenet frame* (see Figure 8). The first vector of this frame is $T(m)$ (or T). The second is $N = \frac{dT}{ds}$ (we omit reference to m). The third is $B = T \wedge N$. The first two vectors T, N define a plane P called the *osculating plane* at m. It deserves this name, because locally it is the closest to C. If we compute the distance from a point $c(s) \in C$ to P, where s is the oriented arc-length on C with origin m, then the distance $d(c(s), P)$ is at most of the order of s^3 (it means $\exists A > 0$, $\exists \epsilon$, such that $\forall |s| < \epsilon \ d(c(s), P) < A \cdot s^3$).

For all other planes, the distance will be of the order of a non-zero constant if the plane does not contain m, of the order of s if the plane contains m, but it is not tangent to C, and of the order s^2 if the plane is tangent to C, but not osculating. This discussion relies strongly on the fact that the curvature at m is not zero. We can now obtain a matrix (called the *Frenet matrix*) in the basis (T, N, B), of the vectors $\frac{dT}{ds}$, $\frac{dN}{ds}$ and $\frac{dB}{ds}$ (with columns their coordinates). As these vectors form a unit basis, the matrix is anti-symmetric. Let us write it as:

$$\begin{pmatrix} 0 & -k & 0 \\ k & 0 & \tau \\ 0 & -\tau & 0 \end{pmatrix} . \qquad (21)$$

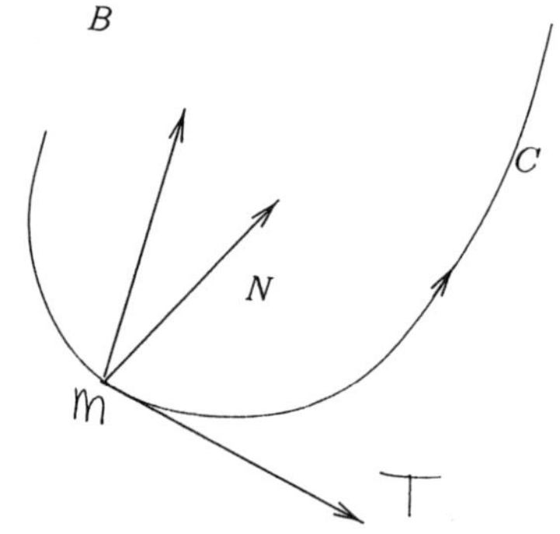

Figure 8. The Frenet frame.

The number τ is called the torsion of the curve at m. The use of k in the Frenet matrix is coherent with our previous notation $|k|$ for the curvature. The absolute value $|\tau|$ measures the speed $|\frac{dB}{ds}|$ with which the osculating plane turns. As the vector B, the torsion can be defined only at points where the curvature is different from zero.

2.3. RIBBONS ALONG A CURVE

A *ribbon* along a curve C is a section U of the unit normal bundle $\nu(C)$ of C (as in Figure 9). The name ribbon comes from the fact that U generates a thin ribbon $\{t \cdot U(m),\ 0 < t < \epsilon,\ m \in C\}$. If $\epsilon > 0$ is such that a tubular neighbouhood of radius ϵ is embedded and centred on C, we can associate to the section U the ribbon: $\{m + t \cdot U(m);\ 0 < t < \epsilon\}$. (It can be proved that, if the curve is of class $\mathcal{C}^2$, any ϵ small enough would do.) We can also see this ribbon as a union of curves $C_{U,t} = \{m + tU(m)\}$.

Using U instead of the unit normal N, we can define, now without any assumption on the curvature, a unit frame $(T, U, T \wedge U)$. Using this frame we can again construct a matrix of the derivatives $\frac{dT}{ds}$, $\frac{dU}{ds}$ and $\frac{d(T\wedge U)}{ds}$ (with columns their coordinates)

$$\begin{pmatrix} 0 & * & * \\ * & 0 & t_w \\ * & -t_w & * \end{pmatrix}, \tag{22}$$

the star denoting a number we are not interested in.

Definition 2.4 *The* **twist of the ribbon** *is defined by*

$$T_w(C, U) = \frac{1}{2\pi} \int_C t_w(s)\, ds \,. \tag{23}$$

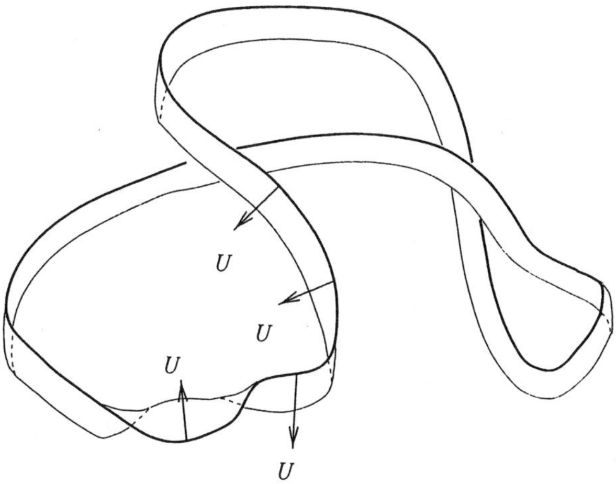

Figure 9. Ribbon associated with a section.

Notice that this twist is in general not an integer. If the curvature of the curve is everywhere different from zero, we can choose $U = B$, we get then the integral of the torsion. We get the same value if we chose $U = N$.

Parallely, we can define a linking number associated to a section of the normal bundle to C. Consider for t small enough (smaller than the radius of an embedded tubular neigbourhood) the curve $C_{U,t} = \{m + tU(m)\}$. The linking number $L(C, C_{U,t})$ does not depend on t, when it is small enough; let us call $L(C, U)$ this number.

2.4. WRITHING NUMBER

It is tempting to try to compute Gauss integral using twice the same curve C. Of course we need to remove from $C \times C$ the diagonal term.

Definition 2.5 *The* **writhing number** *of* C *is:*

$$W_r(C) = \frac{1}{4\pi} \int_{C \times C} \frac{\det\left(\frac{dx}{ds}, \frac{dy}{ds}, \frac{x-y}{|x-y|}\right)}{|x-y|^2} . \tag{24}$$

The integral on $C \times C$ has been first studied by Călugăreanu [2], [3]. Then Fuller [7] introduced the definition of "writhe" and corrected an error in Călugăreanu's work; see also [8], [12] and [15] for generalisations and other proofs. The set of tangent vectors to a curve C is a curve in the sphere S^2. Let us chose a vector $u \notin T(C)$. For t small enough, (smaller than an ϵ suitable to define an embedded tubular neighbourhood), all the curves $C_{u,t}$; $0 < t < \epsilon$ link in the same way the curve C. In particular $L(C, C_{u,t})$ does not depend on $t \in]0, \epsilon[$; we denote this value by $L(C, u)$.

Theorem 2.6 *The writhing number of C is proportional to the average of the linking number of C with the parallel curves (for t small enough) $C_{u,t}$:*

$$W_r(C) = \frac{1}{4\pi} \int_{S^2} L(C,u)\, du \,. \tag{25}$$

We shall deduce this result from another result relating the writhing number and the twist of a ribbon.

Theorem 2.7 ([7], [12]) *Let again $U = U(m)$ be a section of the unit normal bundle of the curve C. Then the difference of the linking number $L(C,U)$ and the twist $T_w(C,U)$ of the ribbon $\{m + tU(m) = \cup C_{U,t}\ ;\ 0 < t < \epsilon\}$ does not depend on the choice of the section U of the normal bundle to C. Moreover, this difference is equal to the writhing number of C:*

$$L(C, C_{U,t}) - T_w(C,U) = W_r(C) \,. \tag{26}$$

Proof (sketch). First notice that when the section of $\nu(C)$ is defined by a vector $V = V(m)$ normal to a Seifert surface S of C, the linking number $L(C,V)$ is zero. If we consider two different sections of U and V of $\nu(C)$, the difference $T_w(C,U) - T_w(C,V)$ is an integer that can be recovered considering the variation of the angle (V,U) along the curve C. Notice also that the difference $T_w(C,U) - T_w(C,V)$ measures the number of times the curve $C_{U,t}$ will intersect the Seifert surface S:

$$T_w(C,U) - T_w(C,V) = L(C, C_{U,t}) \quad (t \text{ small enough}) \,. \tag{27}$$

When a vector u is not in $T(C) \cup -T(C)$ (i.e. it is never contained in a tangent direction), one can associate to it a section U_u of the unit normal bundle of C, projecting it in each point $m \in C$ to the normal plane, and normalizing it. The twist $T_w(C,U_u)$ is the linking number $L(C,U_u)$ for small enough t. It is equal to the linking number $L(C,u)$. Averaging on all $u \notin T(C)$ we prove the result

$$\frac{1}{4\pi} \int_{S^2} L(C,u)\, du = W_r(C) \,. \tag{28}$$

QED

3. Surfaces

3.1. SMOOTH SURFACES

Let us now study the local geometry of a smooth surface M embedded in $\mathbb{R}^3$ and seen from outside (extrinsic viewpoint). A standard reference is [6]. We can recognize geometrically the osculating spheres at a point $m \in M$

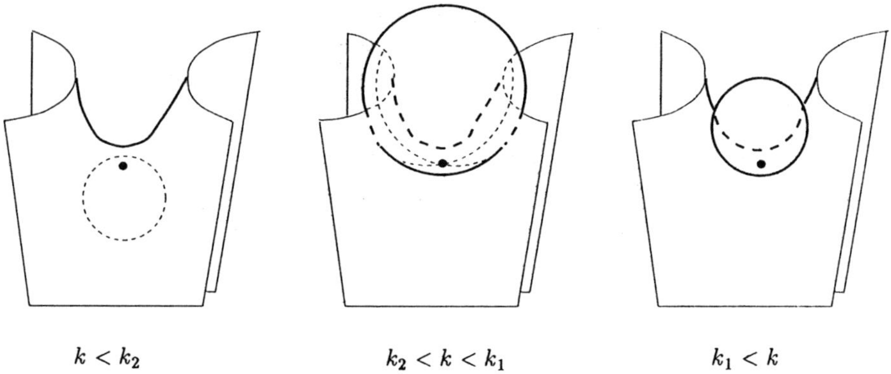

Figure 10. Center-type and saddle-type contacts between a sphere and a surface.

among the spheres tangent at m to the plane $T_m M$ (Figure 10). When the curvature $k = \frac{1}{R}$ of the sphere S_R is between k_1 and k_2, the intersection is, near the origin, the union of two arcs intersecting transversally in the origin. When $k \notin [k_1, k_2]$ the intersection of S_R and M near the origin reduces to a point, the origin. The two remaining spheres (when $k_1 \neq k_2$) are the osculating spheres.

Let us now change our viewpoint, and consider locally a unit normal vector $N(x)$ defined continuously around a point m of a surface. This defines a map γ, called the Gauss map, from the surface to the unit sphere S^2.

Definition 3.1 *The* **Gauss curvature** *at a point $x \in M$ is the Jacobian $K(x) = \det(d\gamma)(x)$ of the Gauss map.*

Proposition 3.2 *The Gaussian curvature of M at m is the product of the two principal curvatures $K(m) = k_1(m) \cdot k_2(m)$. Choose a local equation of the surface M using the tangent plane T_m and the normal vector at m:*

$$z = \frac{1}{2}[k_1 \cdot x^2 + k_2 \cdot y^2] + \text{ higher order terms }. \tag{29}$$

The eigen-directions of the diffential $d\gamma(m)$ are the eigen-directions of the quadratic part of $z = f(x, y)$.

Proof. The gradient of the function $F(x, y, z) = z - f(x, y)$ is well defined and a positive multiple of the oriented unit normal to the surface (at least in the neighbouhood of x). We have, then:

$$N(x) = \frac{\text{grad}\,(F)}{|\text{grad}\,(F)|} = \frac{1}{\sqrt{1 + \left(\frac{\partial f}{\partial x}\right)^2 + \left(\frac{\partial f}{\partial y}\right)^2)}} \begin{pmatrix} \partial f/\partial x \\ \partial f/\partial y \\ 1 \end{pmatrix} . \tag{30}$$

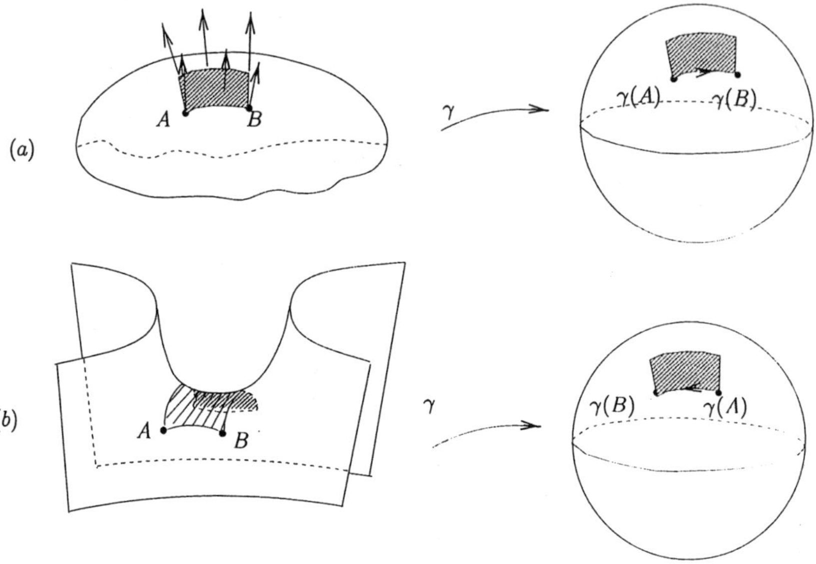

Figure 11. The Gauss map.

The matrix of $d\gamma(m)$ is then:

$$\begin{pmatrix} \frac{\partial^2 f}{\partial x^2} & \frac{\partial^2 f}{\partial x \partial y} \\ \frac{\partial^2 f}{\partial y \partial x} & \frac{\partial^2 f}{\partial y^2} \end{pmatrix} . \tag{31}$$

The latter matrix is called the *Hessian* of f. Using the coordinates x, y defined using the eigen-directions of the quadratic part of f, we get:

$$d\gamma(m) = \begin{pmatrix} k_1(m) & 0 \\ 0 & k_2(m) \end{pmatrix} . \tag{32}$$

This proves the two parts of the proposition.

QED

This Gaussian curvature has another important property: it is intrinsic, that means that $K(m)$ depends on less information than the germ of the embedding at a the point m. It depends only on the metric near m (as a infitesimal bee may see it). We shall be a little more precise in the polyhedral case. Geometrically, the function $f(x, y)$ can be seen as the projection of M (parametrised by (x,y)), on the normal vector $N(m)$. The projective space of dimension 2 ($\mathbb{P}_2$) is the set of directions of lines (or, if you prefer, lines through the origin) in $\mathbb{R}^3$. It is obtained from the sphere S^2 by identifying antipodal points. The sphere S^2 can also be seen as the set of oriented directions of lines (or oriented lines through the origin). Forgetting about orientation, we can define a projective Gauss map (Figure 11) with values in $\mathbb{P}_2$.

Proposition 3.3 *The point m is a degenerate critical point of the orthogonal projection of M on the line $L(m)$ generated by the normal $N(m)$, if and only if m is a critical point of the (usual or projective) Gauss map.*

Corollary 3.4 *The set of lines $L \in \mathbb{P}_2$ such that the projection p_L on L admits degenerated critical points is of measure zero.*

Proof. By Sard theorem these lines, critical values of the projective Gauss map, form a set of measure zero.

QED

For a direction L, denote by $|\mu(M, L)|$ the number of critical points of the orthogonal projection of M on L. The change of variable theorem implies then that there exists a neighbourhood v of m such that:

$$\int_v |K(m)|\, dm = \int_{\mathbb{P}_2} |\mu(v, L)|\, dL \; . \tag{33}$$

The result holds globally on M.

Theorem 3.5

$$\int_M |K(m)|\, dm = \int_{\mathbb{P}_2} |\mu(M, L)|\, dL \; . \tag{34}$$

The proof is similar to the one-dimensional case. Since a non-constant function on a compact set has at least a maximum and a minimum, the previous theorem implies that, when M is a closed surface, then

$$\int_M |K(m)|\, dm \geq 4\pi \; . \tag{35}$$

By using Morse theory, Chern and Lashoff obtained in 1957 the stronger result:

Theorem 3.6 *Let M be a surface of genus g embedded in $\mathbb{R}^3$, then:*

$$\int_M |K(m)|\, dm \geq 2\pi(2g + 2) \; . \tag{36}$$

We can also count "most" of the critical points with a sign. Assign to the non-degenerate critical points of the orthogonal projection of the oriented curve C on the oriented line L^+ the sign $\epsilon(m) = (-1)^{\text{index}(m)}$. As the dimension of M is 2, the sign $\epsilon(m)$ does not depend on the orientation of L. When the two unit vectors contained in L are non-degenerate values

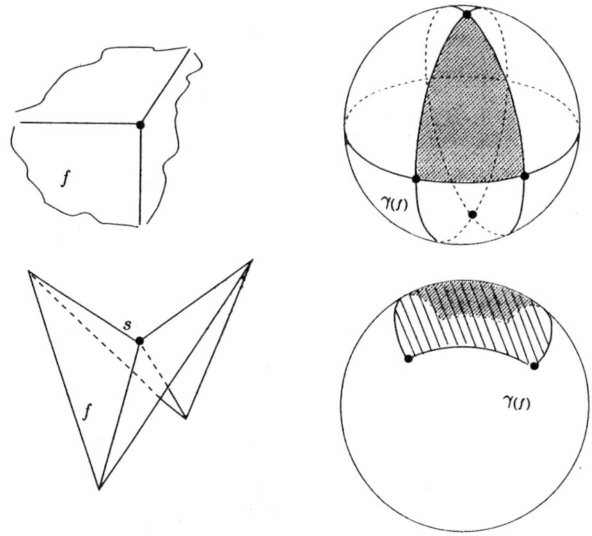

Figure 12. The Gauss image of a vertex.

of the Gauss map, we can, at each point m such that $\gamma(m) \subset L$, orient the line L using the normal $N(m)$ to define $\epsilon(m)$. Thus we get:

$$\mu(M, L) = \sum_{\gamma(m) \in L} \epsilon(m) \ . \tag{37}$$

Theorem 3.7 *If one of the integrals*

$$\int_M |K(m)| \, dm = \int_{\mathbb{P}_2} |\mu(M, L)| \, dL \tag{38}$$

is finite, then:

$$\int_M K(m) \, dm = \int_{\mathbb{P}_2} \mu(M, L) \, dL \ . \tag{39}$$

To prove this last theorem, is is enough to track the signs in the proof of the preceeding one.

3.2. POLYHEDRAL SURFACES

The "theorema egregium" (i.e. important) of Gauss is easy to prove in the polyhedral frame, using integral geometry in the spirit of the previous paragraph. For us a *polyhedral surface* will be a union of planar triangles, such that if two triangles meet, they have in common a full edge or a vertex. The union of these triangles should be topologically a surface. To prove the theorema egregium, we need to define the Gauss image of a vertex (Figure 12). Let us first define the *link* of a vertex. This is the union of the sides (opposite to a vertex) of the triangles containing the vertex. Observe that, if the vertex x_0 is 'useless', that is if x_0 is a vertex of triangles contained in the same plane P_0, then the link $\mathcal{L}(x_0)$ of the vertex is contained in the plane P_0. Therefore its intersection with any plane $P \neq P_0$ containing x_0 gives exactly two points.

Definition 3.8 *The Gauss* **image of a vertex** *is a function on* $I\!P_2$ *defined by*

$$\Gamma(P) = 1 - \frac{1}{2}[\sharp(P \cap \mathcal{L}(x_0))] . \tag{40}$$

Notice that the function $\Gamma(P)$ is identically zero when the vertex is 'useless'. We can see this function as a subset of $\mathbb{P}_2$ counted with multiplicity (and draw it as a subset of S^2).

Definition 3.9 *The total curvature of the vertex* x_0 *is:*

$$|k(x_0)| = \int_{I\!P_2} |\Gamma(P)| . \tag{41}$$

The curvature of the vertex x_0 *is:*

$$k(x_0) = \int_{I\!P_2} \Gamma(P) . \tag{42}$$

Notice that these two curvatures are intrinsic, depending only on information we can get staying inside the polyhedral surface.

Theorem 3.10 (Theorema Egregium) *The curvature at a vertex* x_0 *satisfies:*

$$k(x_0) = \int_{I\!P_2} \Gamma(P) = 2\pi - \sum \alpha_i . \tag{43}$$

Proof. The contribution of each side of the link $\mathcal{L}(x_0)$ to the integral $\int_{\mathbb{P}_2} \Gamma(P)$ is 2α, where α is the angle at x_0 of the triangle containing the edge. Summing up, we get $k(x_0) = 2\pi - \sum \alpha_i$, where the α_i are all the angles with vertex x_0.

QED

Gauss considered the smooth analog of this statement as important because it relates two definitions, one extrinsic and the other intrinsic. In the polyhedral case, the angles can be measured "inside" the surface and are therefore intrinsic.

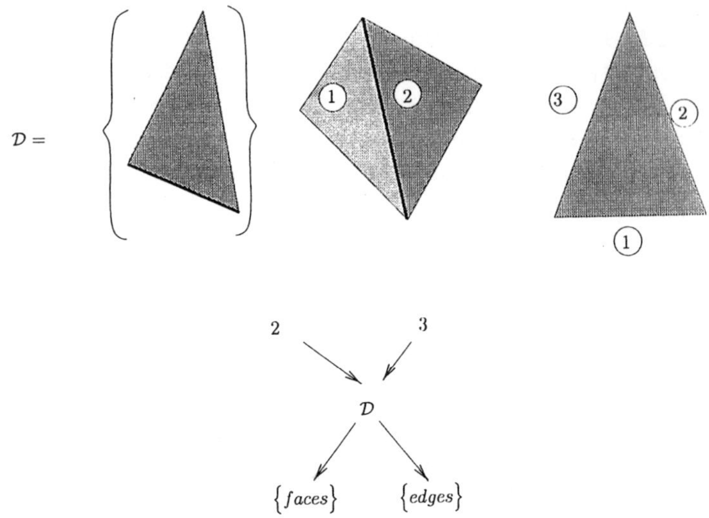

Figure 13. The pairs {edge $\subset$ face}.

Theorem 3.11 (Gauss-Bonnet theorem) *When the surface M is compact without boundary, the sum of the curvatures at all vertices is the Euler characteristic of the surface:*

$$\sum_{x \text{ vertex of } M} k(x) = 2\pi \cdot \chi(M) \ , \tag{44}$$

where the Euler characteristic $\chi(M)$ is the algebraic sum of the number of faces (f) minus the number of edges (e) plus the number of vertices (v): $\chi(M) = f - e + v$.

Proof. Let us consider the set of pairs $\mathcal{D}$ = {edge $\subset$ face} (Figure 13). Since an edge is the boundary of two faces, and a face has three boundary edges, we have $\sharp\mathcal{D} = 3f = 2e$. The sum of the curvatures over all vertices is $\sum k(x)$. Since the sum of the angles of a triangle is π, we have also:

$$\sum k(x) = \sum_{\text{vertices}} [2\pi - \sum \alpha_i] = 2\pi v - \pi f \ . \tag{45}$$

Therefore:

$$2\pi v - \pi f + 0 = 2\pi v - \pi f + 3\pi f - 2\pi e = 2\pi \cdot \chi(M) \ . \tag{46}$$

QED

Remark 3.12 The proof of the previous two theorems in the smooth case needs more knowledge. We can nevertheless notice some analogy between useful vertices of polyedral surfaces and points with non-zero Gauss curvature of a smooth surface.

Acknowledgements

The author thanks J.P. Troalen who realized the computerized pictures.

References

1. Banchoff, T. (1970) Critical points and curvature for embedded polyedral surfaces. *Am. Math. Monthly* **77**, 475–486.
2. Călugăreanu, G. (1959) L'intégrale de Gauss et l'analyse des noeuds tridimensionnels. *Rev. Math. Pures Appl.* **4**, 5–20.
3. Călugăreanu, G. (1961) Sur les classes d'isotopie des neouds tridimensionnels et leurs invariants. *Czechoslovak Math. J.* **11**(86), 588–625.
4. Călugăreanu, G. (1961) Sur les enlacements tridimensionnels des courbes fermées. *Comm. Acad. R.P. Romine* **11**, 829–832.
5. Cantarella, J. DeTurk D. & Gluck. H. (1997) Upper bounds for the writhing of knots and the helicity of vector fields. *Preprint.*
6. Do Carmo, M. (1976) *Differential Geometry of Curves and Surfaces.* Prentice Hall.
7. Fuller, F.B. (1971) The writhing number of a space curve. *Proc Nat. Acad. Sci. USA* **88**(4), 815–819.
8. Fuller, F.B. (1978) Decomposition of the linking number of a closed ribbon: a problem from molecular biology. *Proc Nat Acad. Sci. USA* **75**(8), 3557–3561.
9. Gauss, C.F. (1833) Integral formula for linking number. In *Zur Mathematischen Theorie de Electrodynamische Wirkungen. Collected Works* **5**, p. 605. Koniglichen Gesellschaft des Wissenschaften, Göttingen.
10. Milnor, J. (1969) *Topology from Differential Viewpoint.* The University Press of Virginia, 2nd edition.
11. Pohl, W.F. (1968) Some integral formulas for space curves and their generalization. *Am. J. Math.* **40**(4), 1321–1345.
12. Pohl, W.F. (1968) The self-linking number of a closed space curve. *J. Math. & Mech.* **17**, 975–985.
13. Rolfsen, D. (1976) *Knots and Links.* Publish or Perish, Wilmington, USA.
14. Rudin, W. (1970) *Real and Complex Analysis.* Mac Graw Hill.
15. White, J.H. (1969) Self-linking and the Gauss integral in higher dimensions. *Am. J. Math.* **91**, 693-728.

TOPOLOGY IN FOUR DAYS

TADASHI TOKIEDA
Département de Mathématiques, UQAM
C.P. 8888, succ. Centre-Ville
Montréal H3C 3P8, Canada

E-mail: *tokieda@math.uqam.ca*

Abstract. This is a transcription of a series of 4 lectures presented at the Newton Institute in December 2000 to an audience of fluid dynamicists and an astrophysicist. It explains how to use two of the simplest yet effective topological tools: intersection number and degree of a map.

1. Solving Problems by Deformation

Example 1.1 [Pólya] We begin with a puzzle. Can you connect AA', BB', CC' using non-intersecting curves inside the rectangle (a)?

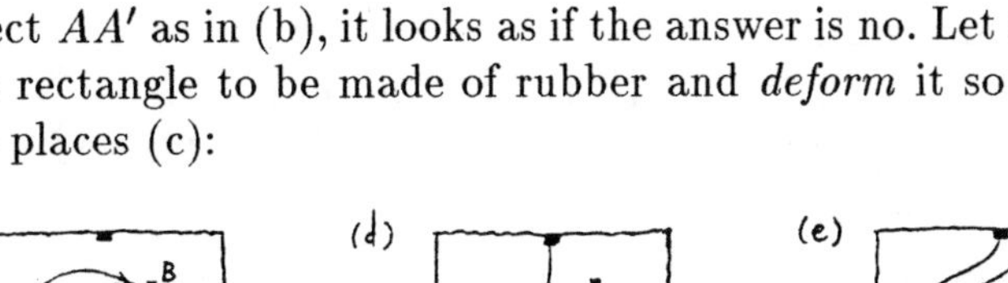

If you connect AA' as in (b), it looks as if the answer is no. Let us, however, imagine the rectangle to be made of rubber and *deform* it so that B and C exchange places (c):

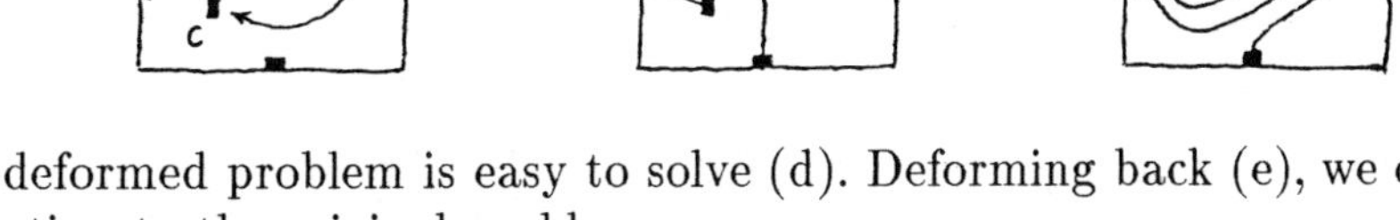

The deformed problem is easy to solve (d). Deforming back (e), we obtain a solution to the original problem.

R.L. Ricca (ed.), An Introduction to the Geometry and Topology of Fluid Flows, 35–55.

Solution sets of many hard problems (e.g. certain non-linear partial differential equations) are insensitive to deformation. One use of topology is to simplify such problems by deforming them into easier ones.

Example 1.2 For every square matrix L, $\det e^L = e^{\operatorname{tr} L}$.

There are algebraic proofs of this identity, but here is a topological argument. First, the identity holds for diagonalizable L: if $\lambda_1, \dots, \lambda_n$ denote the diagonal entries after diagonalization, both sides equal $e^{\lambda_1 + \cdots + \lambda_n}$. Now *generic* matrices are diagonalizable: a small enough perturbation of a diagonalizable matrix is still diagonalizable, and every non-diagonalizable matrix is a limit of diagonalizable ones. (Indeed, a matrix L whose characteristic polynomial χ_L has only simple roots is diagonalizable, and generic polynomials have only simple roots: when the coefficients are perturbed, multiple roots disintegrate into simple ones.) To establish the identity for possibly non-diagonalizable L, express it as a limit of diagonalizable matrices; the identity holds for each of the latter, so by continuity it holds for L as well.

Problems tend to be more tractable in the generic case. Then we can try to approach the singular cases by continuity.

Example 1.3 [Cayley-Hamilton] $\chi_L(L) = 0$.

Mimic the argument above.

Remark 1.4 In 1.2 and 1.3, the argument relies on working over $\mathbb{C}$, using as it does the possibility of splitting a polynomial into linear factors. But the resulting identities are combinatorial relations between the entries of L, so they remain correct over any commutative ring.

Motto 1.5 Deform the problem. Look at generic cases.

2. How to Build Manifolds

The smooth spaces, or manifolds, that we encounter most commonly in applications are manufactured from a few basic building blocks.

Building Blocks 2.1

n-ball $D^n = \{(x_1, \dots, x_n) \in \mathbb{R} \mid \sum x_k^2 \leq 1\}$, 

$(n-1)$-sphere $S^{n-1} = \{(x_1, \dots, x_n) \in \mathbb{R} \mid \sum x_k^2 = 1\}$, 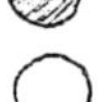

unit interval $I = [0, 1]$. 

In the contemporary mathematical dialect, it is customary to distinguish between balls (solid) and spheres (shells only).

Symbols 2.2

∂ stands for boundary: $S^{n-1} = \partial D^n$.

$\simeq$ means homeomorphic, i.e. 'topologically the same': $I \simeq D^1$.

The building blocks of 2.1 generate complicated spaces in various ways: product, quotient, surgery.

Method I: Product

Example 2.3 n-cube $I^n = \underbrace{I \times \cdots \times I}_{n} \simeq D^n$.

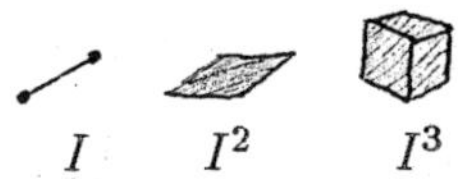

Example 2.4 n-torus $T^n = \underbrace{S^1 \times \cdots \times S^1}_{n}$.

$T^2 \simeq$ surface of a mug  $\simeq$  knotted tube.

Remark 2.5 In 2.4, the knotted tube is not 'deformable' in space to T^2. Homeomorphism is weaker, and more primitive, than deformability (cf. 3.9 *infra*).

Method II: Quotient (*alias* identification)

Example 2.6 In D^n, identify all points on ∂D^n:

$D^n/\partial D^n \simeq S^n$.

Example 2.7 In I^2, identify $(0, y) \leftrightarrow (1, y)$ and $(x, 0) \leftrightarrow (x, 1)$:

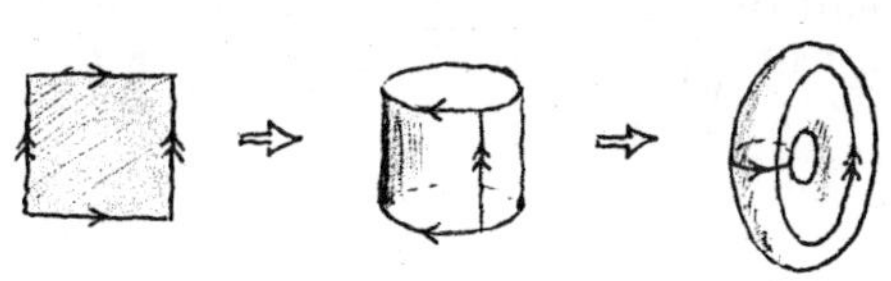

$I^2/\leftrightarrow \simeq T^2$.

Or identify $(0, y) \leftrightarrow (1, 1 - y)$:

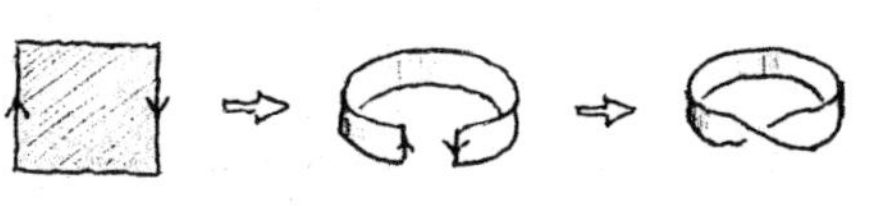

Möbius strip (non-orientable).

The result of identification $(0, y) \leftrightarrow (1, y)$ and $(x, 0) \leftrightarrow (-x, 1)$ is called the Klein bottle.

Problem 2.8 What sort of object do you get when you cut a Möbius strip along its center (= 1/2) line? along its 1/3 line?

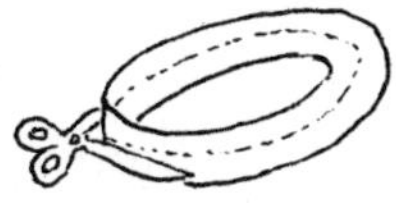

Example 2.9 In S^n, the antipodal identification $\vec{x} \leftrightarrow -\vec{x}$ results in

$\mathbb{R}P^n = S^n/\leftrightarrow$, real projective space
(orientable for n odd, non-orientable for n even). 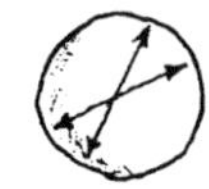

Other ways of describing $\mathbb{R}P^n$:

$\mathbb{R}P^n = (\mathbb{R}^{n+1} \setminus 0)/\propto$, where $\propto$ identifies proportional vectors, i.e.

$\mathbb{R}P^n = \{\text{lines in } \mathbb{R}^{n+1} \text{ through the origin}\}$.

In D^n, antipodally identify points of ∂D^n (contrast 2.6) to get $\mathbb{R}P^n$.

ignore north half 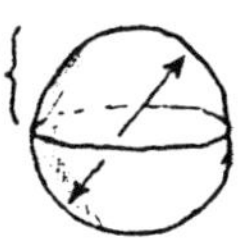 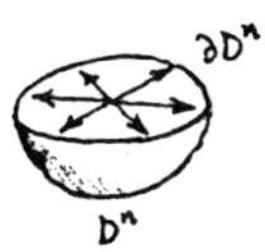

. In particular, $\mathbb{R}P^1 \simeq S^1$.

Example 2.10 Complex projective space $\mathbb{C}P^n = (\mathbb{C}^{n+1} \setminus 0)/\propto$, where $\propto$ identifies complex proportional vectors (orientable for all n).

Another way: instead of taking the quotient by identifying $\vec{z} \propto re^{i\theta}\vec{z}$, we can first restrict $\mathbb{C}^{n+1} \setminus 0$ to $S^{2n+1} = \{\vec{z} \in \mathbb{C}^{n+1} \mid |\vec{z}| = 1\}$ and identify $\vec{z} \propto e^{i\theta}\vec{z}$.

$S^{2n+1} \xrightarrow{\pi} \mathbb{C}P^n$ quotient map

$\forall p \in \mathbb{C}P^n$,
$\pi^{-1}(p) \simeq$ circle $\{e^{i\theta} \mid 0 \leq \theta < 2\pi\}$.
This is the Hopf fibration of S^{2n+1} over $\mathbb{C}P^n$ with fiber S^1.

This turns out to be the natural geometric model of a harmonic oscillator whose eigenvalues are all equal.

$\mathbb{C}P^1 \simeq S^2$, Riemann sphere.

Method III: Surgery (*alias* cut and paste)

Example 2.11 $T^2\#T^2$ (<u>connected sum</u>).

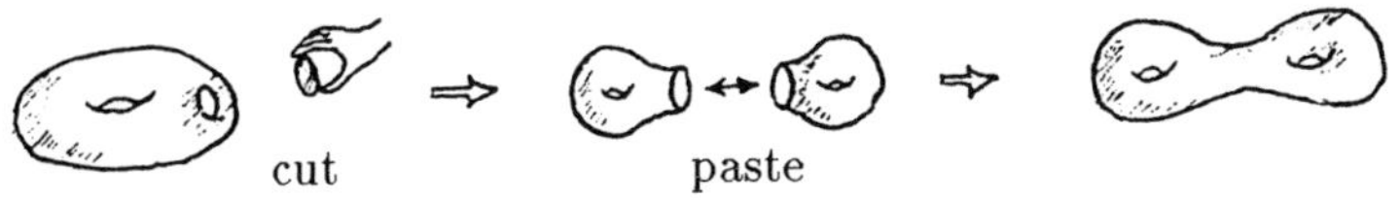

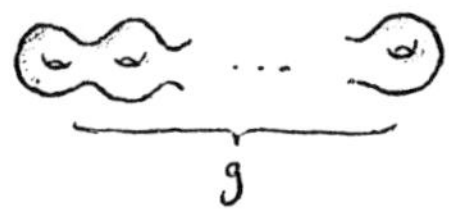

<u>Riemann surface of genus g</u>, $\Sigma_g = \underbrace{T^2\#\cdots\#T^2}_{g}$; $\Sigma_0 = S^2$.

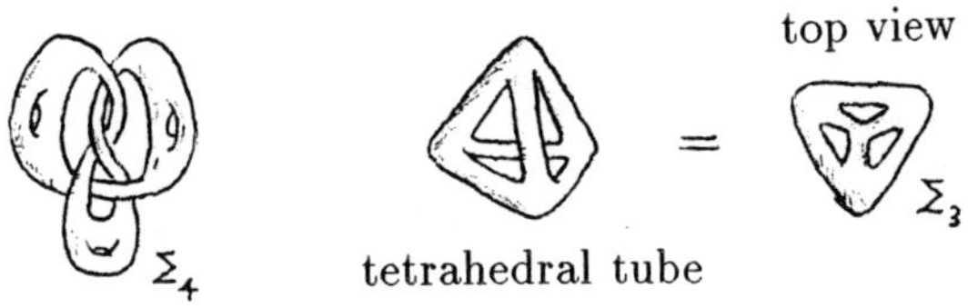

Example 2.12 $\partial($  $) = S^1 \overset{\text{paste}}{\longleftrightarrow} S^1 = \partial D^2$:
Möbius strip

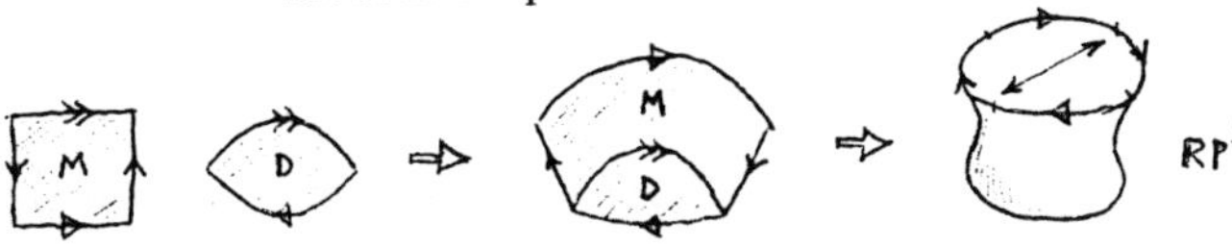

Thus, if on $\Sigma_0\backslash$ disk  we put a lid with a disk, we get $S^2 = S^2\#S^2$ (S^2 behaves like zero for #), but with a Möbius strip, we get $\mathbb{R}P^2$.

More generally, paste g Möbius strips on to $\Sigma_0 \backslash g$ disks to get N_g (non-orientable surface). This surface appears to have no traditional name.

Theorem 2.13 (Möbius) *Any closed surface $\simeq$ one of Σ_g or N_g.*

Problem 2.14 When you paste 2 Möbius strips together, you must get one of the surfaces in the classification 2.13. Which one? (In other words, calculate $\mathbb{R}P^2\#\mathbb{R}P^2$.) Develop a calculus that tells which surface you get when you drill holes in any surface and put lids on some with disks, on some with Möbius strips.

slicing a Klein bottle

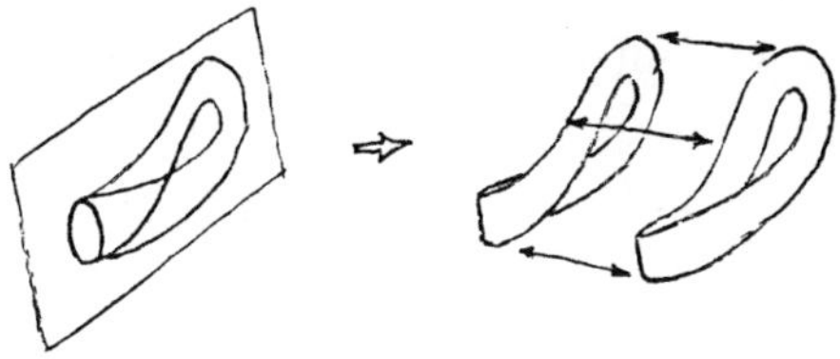

3. Homotopy

Now we give names to two kinds of deformation: one kind allows self-crossings, the other kind does not.

Definition 3.1 *Submanifolds P_0, P_1 of a manifold M are* <u>*homotopic in M*</u>, *$P_0 \sim P_1$, if they can be continuously deformed into each other within M via intermediate stages P_t allowing self-crossings.*

Aside 3.2 The 'definitions' in this article are deliberately designed to fall short of the standard of formal precision among mathematicians. It is hoped that they thereby gain in psychological digestibility.

Example 3.3

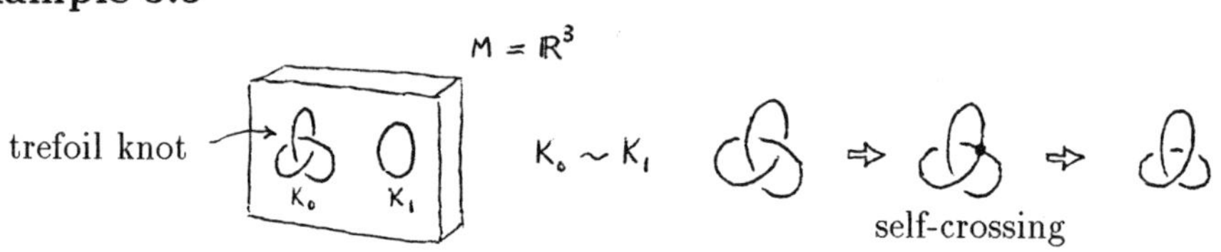

Example 3.4

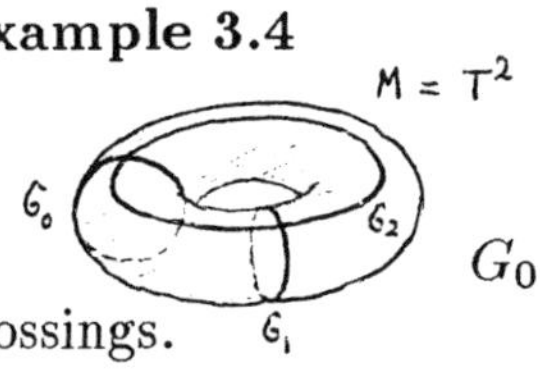

$G_0 \sim G_1$ but $G_1 \sim G_2$ in T^2 even allowing self-crossings.

Definition 3.5 *If the deformation can be achieved without self-crossings, P_0, P_1 are* **isotopic in** *M.*

Example 3.6 In 3.3, K_0 and K_1 are not isotopic in $\mathbb{R}^3$.

Remark 3.7 The concepts of homotopy and isotopy depend on the ambient space M.

In 3.3, if we replace M by $\mathbb{R}^4$, K_0 and K_1 become isotopic.

(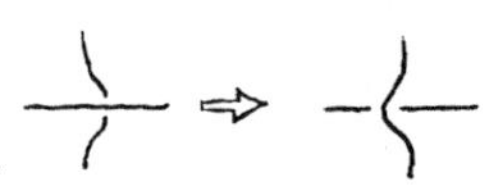 by lifting into the 4th dimension; thus knot theory is interesting only in 3-dimensional manifolds.)

In 3.4, if $M = \mathbb{R}^3$ 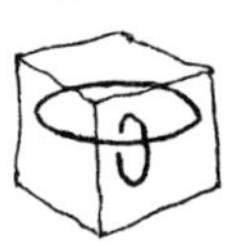, G_1 and G_2 become homotopic (even isotopic).

Problem 3.8

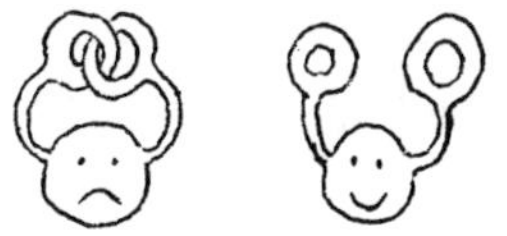

In $\mathbb{R}^3$, these surfaces (Σ_2) are homotopic. Are they isotopic?

Summary 3.9

isotopic $\Longrightarrow$ homotopic $\Longrightarrow$ homeomorphic.

The converses are false. 'Isotopic' and 'homotopic' are *extrinsic* concepts (3.7), whereas 'homeomorphic' is *intrinsic.*

4. Intersection Number

All the topological tools that we develop in these lectures are based on the idea of when geometric objects intersect and when they don't, and of how to make an intelligent count of the number of intersections when they do.

Example 4.1 How many intersections? When tallying, let us agree to take orientations into account.

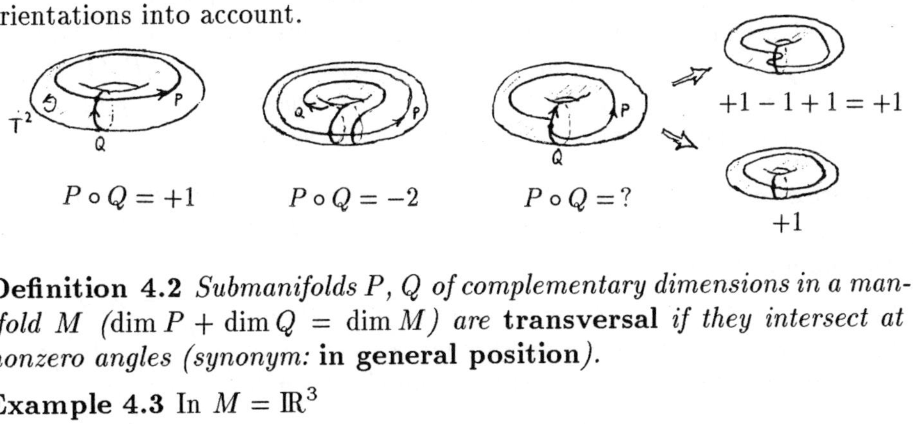

$P \circ Q = +1$ $\qquad P \circ Q = -2$ $\qquad P \circ Q = ?$

Definition 4.2 *Submanifolds P, Q of complementary dimensions in a manifold M ($\dim P + \dim Q = \dim M$) are* **transversal** *if they intersect at nonzero angles (synonym:* **in general position***).*

Example 4.3 In $M = \mathbb{R}^3$

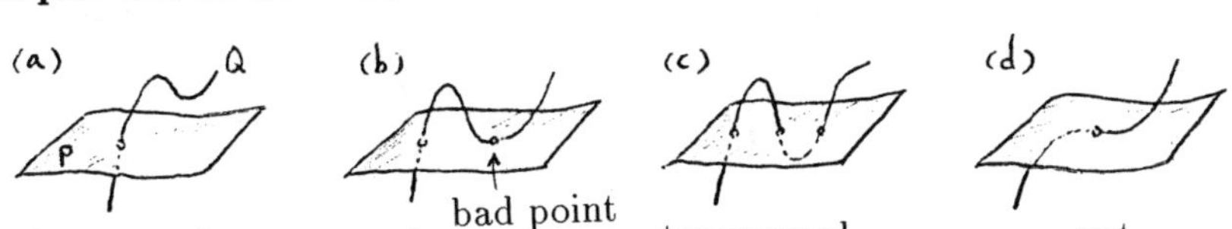

transversal $\qquad$ not $\qquad$ transversal $\qquad$ not

Principle 4.4 *Transversal intersections are* **generic***: they are robust under perturbations. In contrast, non-transversal intersections are not generic: under perturbation they become transversal ((b)→(c), (d)→(a) or (c) in 4.3) or disappear ((b)→(a) ibid.).*

Remark 4.5 There is a concept of transversality for submanifolds of general (not necessarily complementary) dimensions; we will not need it.

For reasonable P, Q in M, we expect $\dim(P \cap Q) = \dim P + \dim Q - \dim M$, so that in the scenario of 4.2, P and Q intersect in a discrete (0-dimensional) set of points.

Definition 4.6 *In an oriented manifold M, let P, Q be closed oriented transversal submanifolds. The* **intersection number** *of P and Q is*

$$P \circ Q = \sum_{x \in P \cap Q} \operatorname{sgn} x,$$

where $\operatorname{sgn} x = +1$ *or* -1 *accordingly as the joint orientation of P and Q (in this order) matches or not the orientation of M at x.*

If P, Q are not transversal, first 'shake' them into general position and then compute $P \circ Q$.

Generalization 4.7 To do this theory in the unoriented case, count mod 2 (cf. 4.14 *infra*).

Fact 4.8 $P \circ Q = (-1)^{pq} Q \circ P$, where $p = \dim P, q = \dim Q$.

Is this clear? Say the orientation of P is given by a basis (frame) $u_1, \ldots, u_p$, and that of Q by $v_1, \ldots, v_q$. The joint orientation of P, Q is given by $u_1, \ldots, u_p, v_1 \ldots, v_q$, while that of Q, P is given by $v_1, \ldots, v_q, u_1 \ldots, u_p$. To rearrange the first into the second, we need pq swaps.

Theorem 4.9 *The intersection number is homotopy-invariant: if $Q_0 \sim Q_1$, then $P \circ Q_0 = P \circ Q_1$. In particular, in 4.6, $P \circ Q$ does not depend on the way we 'shake'.*

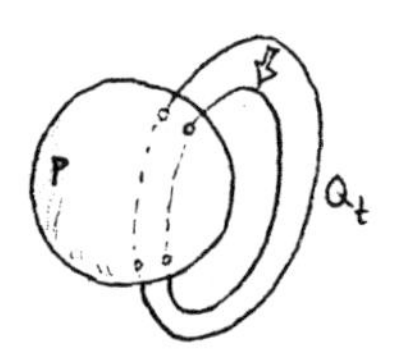

Proof. Apply 4.4. While Q_t is transversal to P, $P \circ Q_t$ does not change. When we go through a moment of non-transversality, we either create or annihilate a pair of intersections with opposite signs:

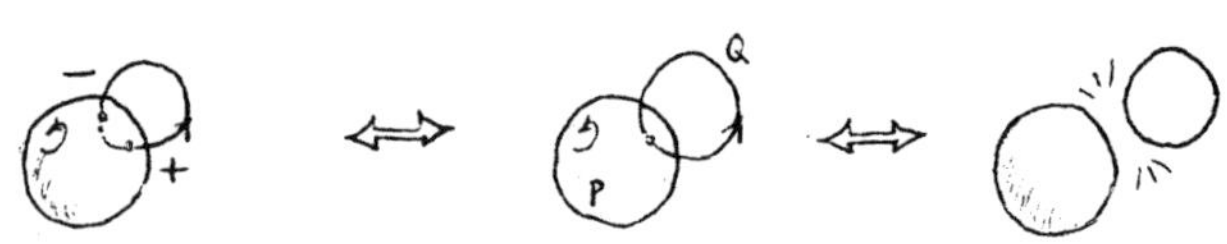

$+1 - 1$ $\qquad$ $P \circ Q = ?$ $\qquad$ 0

so $P \circ Q_t$ remains unchanged, too.

QED

Comment 4.10 4.6 requires closedness of P and Q to prevent intersections from falling off boundaries without cancellation.

transversal, should be ± 1 ouch!

Remark 4.11 It can happen that $P \circ Q = 0$ and yet P and Q are not 'unhookable' by homotopy.

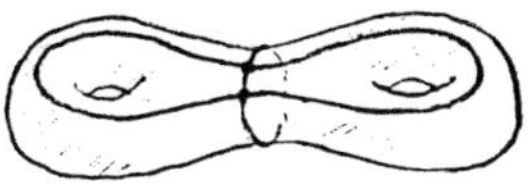

$-1 + 1 = 0$

Of course, if they are 'unhookable', then $P \circ Q = 0$.

Theorem 4.12 *In $\mathbb{R}^n$, closed submanifolds are always 'un-hookable' and so have intersection number = 0.*

Proof.

shrink away one of the submanifolds.

QED

Corollary 4.13 *Every closed hypersurface H (of dimension $n-1$) in $\mathbb{R}^n$ separates $\mathbb{R}^n$ into two parts.*

$n = 2$ gives the Jordan curve theorem. It isn't all that obvious:

Proof. If not, two nearby points x, y opposite each other across H can be joined by a curve that avoids H. Add a bridge between x and y to form a loop γ. Then $\gamma \circ H = \pm 1$, contradicting 4.12.

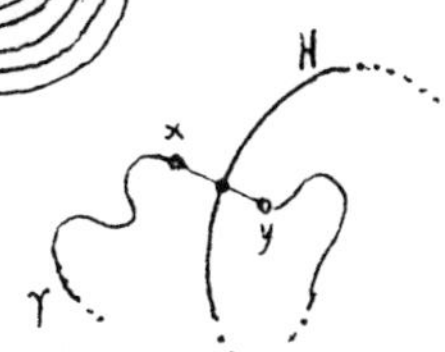

QED

Comment 4.14 Even if H is unoriented, our proof goes through with the mod 2 theory (4.7). 4.13 then implies that every closed hypersurface in $\mathbb{R}^n$ is 2-sided, hence orientable.

4.12 and 4.13 hold also on S^n $(n \geq 2)$ thanks to the shrink-away method (cf. second proof of 4.12).

Example 4.15 $\mathbb{R}P^2$, being non-orientable (2.9), cannot be realized as a surface in $\mathbb{R}^3$ without self-intersections.

It is tempting to picture manifolds as high-dimensional surfaces floating in some space. 4.15 shows that we must be careful. How much elbow room do we need to legitimize our picture?

Theorem 4.16 (Whitney) *Every closed manifold of dimension n can be realized as a submanifold of* $\mathbb{R}^{2n}$.

5. Fixed Point Theorems

Solving any species of equation (differential, integral, algebraic...) amounts to finding fixed points, via the following dictionary:

$$\text{solve } f(x) = 0 \iff \text{find fixed points of the map } x \mapsto x + f(x).$$

The fixed point problem, in turn, can be translated into the language of intersection theory.

Example 5.1 Suppose f is continuous and $-x \leq f(x) \leq 1 - x$ on I. Is $f(x) = 0$ solvable?

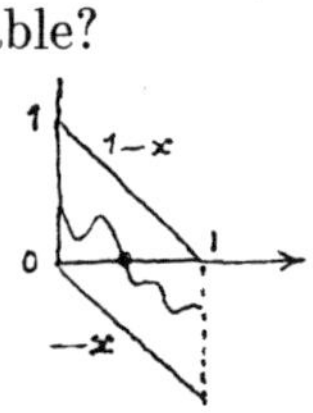

Visibly yes.

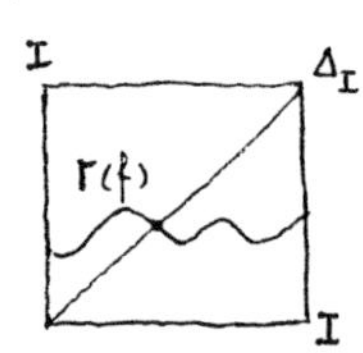

Alternatively, $x \mapsto x + f(x)$ maps I to I. And a continuous map $I \to I$ must have a fixed point where its graph $\Gamma(f)$ intersects the diagonal $\Delta_I = \{(x, x) \mid x \in I\} \subset I \times I$.

Recipe 5.2 The procedure of 5.1 is typical. To solve $f(x) = 0$:

(1) get an a priori estimate on f;

(2) use (1) to determine the set on which the map $x \mapsto x + f(x)$ acts;

(3) apply some fixed point theorem (locate $\Delta \cap \Gamma(f)$).

Example 5.3

$\Gamma(\text{rotation of } S^1 \text{ by } \alpha \equiv 0 \bmod 2\pi)$

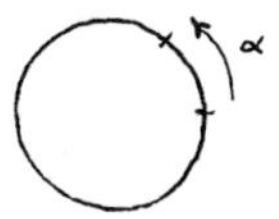

no fixed point

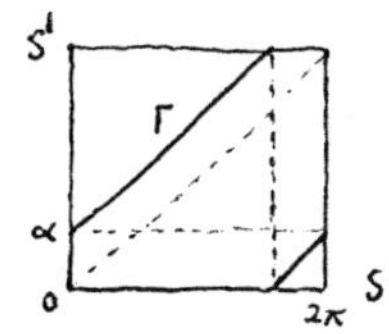

$\Gamma(z \mapsto z^n \text{ on } S^1 \subset \mathbb{C})$

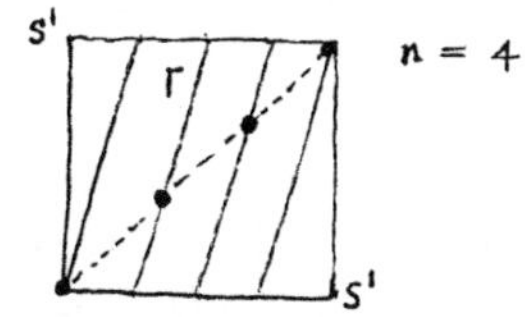

$n - 1$ fixed points

Announcement 5.4 From now on we suppose M closed and oriented. Then so is $M \times M$.

Definition 5.5 *A map $f : M \to M$ whose graph $\Gamma(f)$ is transversal to Δ_M is* **generic**.

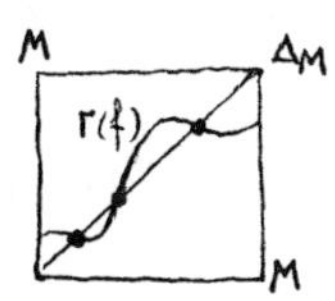

Definition 5.6 *Write a generic f in coordinates as* $\begin{pmatrix} f_1(x_1, \ldots, x_n) \\ \vdots \\ f_n(x_1, \ldots, x_n) \end{pmatrix}$.

The **Lefschetz number** *of f is*

$$L(f) = \sum_{x \text{ fixed pt of } f} \operatorname{sgn}\det \left(\frac{\partial f_i}{\partial x_j} - \delta^i_j \right)_x .$$

Example 5.7 $L\begin{pmatrix} S^1 \to S^1 \\ z \longmapsto z^n \end{pmatrix} = n - 1.$

Theorem 5.8 (Lefschetz) *If f is generic, then $L(f) = \Delta_M \circ \Gamma(f)$.*

Proof. In dimension 1.

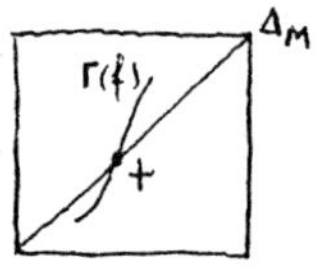

This corresponds to $\frac{df}{dx} - 1 > 0$.

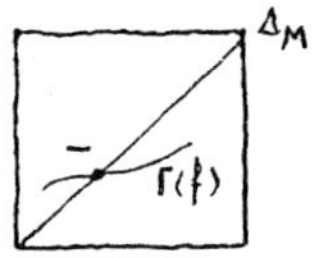

This corresponds to $\frac{df}{dx} - 1 < 0$.

QED

Corollary 5.9 *If $f : M \to M$ is deformable to a constant map $f_0 : M \to y_0$, then $L(f) = (-1)^{(\dim M)^2}$ and so f has a fixed point.*

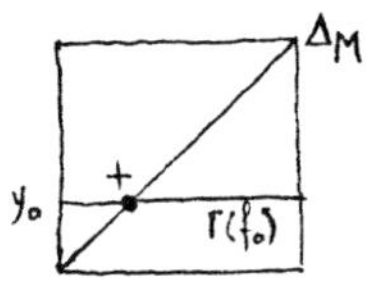

Proof. $\Gamma(f) \sim \Gamma(f_0)$, hence by 4.9, 5.8, 4.8,

$$L(f) = \Delta_M \circ \Gamma(f_0) = (-1)^{(\dim M)^2} \underbrace{\Gamma(f_0) \circ \Delta_M}_{+1} .$$

QED

Corollary 5.10 (Brouwer) *Every continuous map $f : D^n \to D^n$ has a fixed point.*

Proof. Regard D^n as the southern hemisphere of S^n and consider the north-to-south folding $g : S^n \to D^n$.

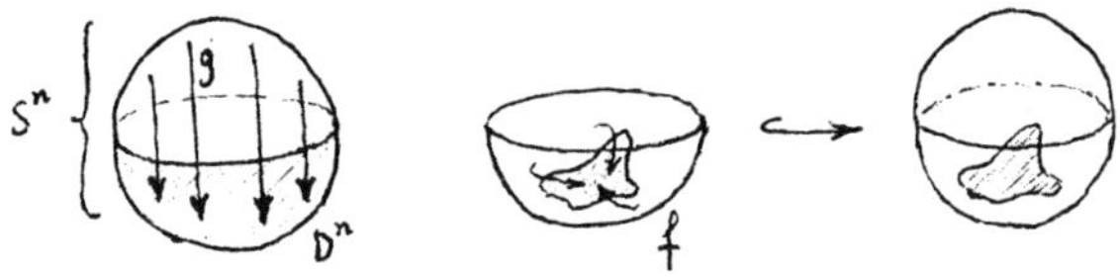

The composed map $S^n \xrightarrow{g} D^n \xrightarrow{f} D^n \xrightarrow{\text{inclusion}} S^n$ is deformable to a constant map (e.g. shrink $f(D^n)$ to the south pole). By 5.9 it has a fixed point which, located on the southern hemisphere, is also fixed by f.

QED

Warning 5.11 For us, D^n contains all its boundary points (2.1). 5.10 is false on open balls.

$\dim 1 : x \mapsto \frac{x+1}{2}$.

Example 5.12 [Perron-Frobenius] A square matrix with positive entries always admits an eigenvector whose components are all positive.

For it induces a map on the 'positive quadrant' and hence, via projection, on $\{(x_1, \ldots, x_n) \in \mathbb{R}^n \mid \text{each } x_k \geq 0, \sum x_k = 1\} \simeq D^{n-1}$.

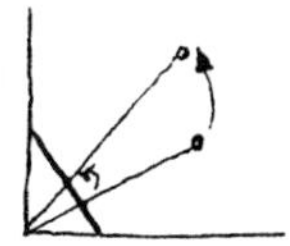

Example 5.13 [Poincaré return map]

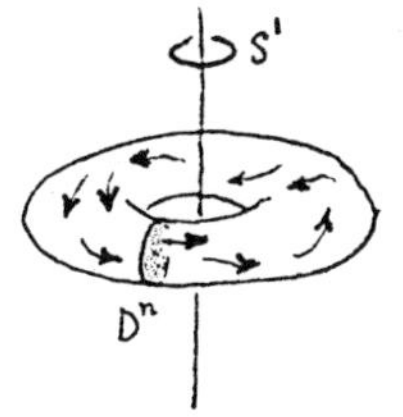

Any flow on $D^n \times S^1$ (when $n = 2$, solid torus) that has everywhere nonzero angular velocity around S^1 must contain a periodic orbit.

6. Equilibria of Vector Fields

Vector fields and their equilibria (stagnation points) are fundamental in fluid dynamics. We study them by translating them into the language of intersection theory. This section is parallel to section 5.

Reminder 6.1 The tangent bundle over a manifold M is the manifold $TM = \bigcup_{x \in M} T_x M$, where $T_x M$ = tangent space to M at x, endowed with the natural locally product topology. (Not to worry about the jargon.)

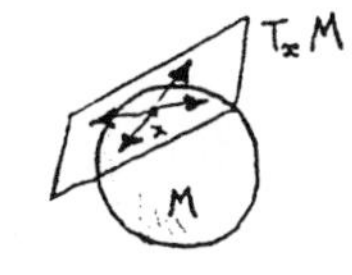

Picture 6.2

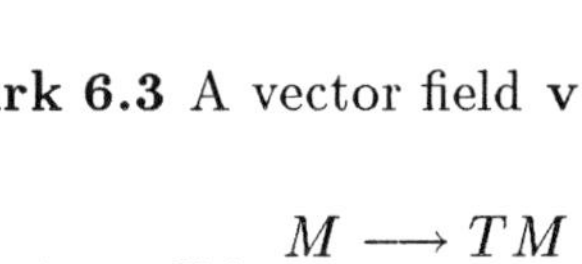

Locally a product (it is traditional to draw $T_x M$ perpendicular to M), but seldom globally. A point of TM is a pair $(x, \mathbf{v})$, where $x \in M$ and $\mathbf{v} \in T_x M$.

$$\dim TM = 2 \dim M \ .$$

Remark 6.3 A vector field $\mathbf{v}$ on M can be thought of as a map

$$\mathbf{v} : \begin{array}{l} M \longrightarrow TM \\ x \longmapsto (x, \mathbf{v}(x)). \end{array}$$

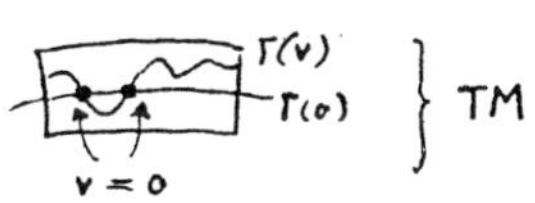

Its graph $\Gamma(\mathbf{v})$ intersects the zero section $\Gamma(0) \simeq M$ in places where $\mathbf{v} = 0$, i.e. at equilibria of $\mathbf{v}$.

Announcement 6.4 From now on we suppose M closed and oriented. Then TM is oriented, too (but open in the 'vertical' direction (cf. 6.2)).

Definition 6.5 *A vector field* $\mathbf{v}$ *whose graph* $\Gamma(\mathbf{v})$ *is transversal to* $\Gamma(0)$ *is* **generic**.

Definition 6.6 *Write a generic* $\mathbf{v}$ *in coordinates as* $\begin{pmatrix} v_1(x_1,\ldots,x_n) \\ \vdots \\ v_n(x_1,\ldots,x_n) \end{pmatrix}$.

The **index** *of* $\mathbf{v}$ *at an equilibrium* x *is*

$$\operatorname{index}_{\mathbf{v}}(x) = \operatorname{sgn}\det\left(\frac{\partial v_i}{\partial x_j}\right)_x .$$

Example 6.7

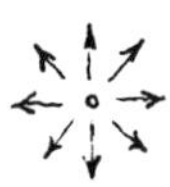 source: index=+1  sink: index= +1 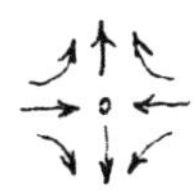 saddle: index= −1

 center: not generic: shaken into  or  both index= +1

Theorem 6.8 *If* $\mathbf{v}$ *is a generic vector field on a manifold* M*, then*

$$\sum_{x \text{ equil of } \mathbf{v}} \operatorname{index}_{\mathbf{v}}(x) = \Gamma(0) \circ \Gamma(\mathbf{v}).$$

Proof. In dimension 1.

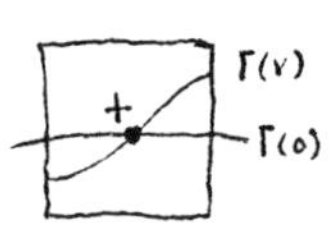

This corresponds to  , i.e. $\frac{dv}{dx} > 0$.

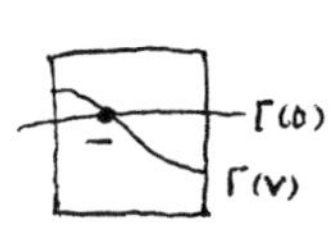

This corresponds to  , i.e. $\frac{dv}{dx} < 0$.

QED

Corollary 6.9 $\sum_{x \text{ equil of } \mathbf{v}} \operatorname{index}_{\mathbf{v}}(x)$ *does not depend on* $\mathbf{v}$*, but only on* M.

Proof. Apply 4.9: any two vector fields $\mathbf{v}$, $\mathbf{w}$ are deformable into each other via $(1-t)\mathbf{v}(x)+t\mathbf{w}(x)$, so $\Gamma(\mathbf{v}) \sim \Gamma(\mathbf{w})$.

QED

Corollary 6.10 *If* $\dim M$ = *odd, then* $\sum_{x \text{ equil of } \mathbf{v}} \text{index}_{\mathbf{v}}(x) = 0$.

Proof. By 4.8, and because $\Gamma(\mathbf{v}) \sim \Gamma(0)$,

$$(-1)^{(\dim M)^2}\Gamma(\mathbf{v}) \circ \Gamma(0) = \Gamma(0) \circ \Gamma(\mathbf{v}) = \Gamma(\mathbf{v}) \circ \Gamma(0).$$

QED

Definition 6.11 *Given an n-dimensional manifold M, triangulate it with β_0 vertices, β_1 edges, β_2 faces, ..., β_n n-dimensional tetrahedra. (In physical applications, manifolds are usually triangulable.) The* **Euler characteristic** *of M is*

$$\chi(M) = \sum_{k=0}^{n}(-1)^k\beta_k.$$

Example 6.12

$\chi(S^1) = \chi($  $) = 3-3 = 0$. In general, $\chi(S^{\text{odd}}) = 0$.

$\chi(S^2) = \chi($  $) = 4-6+4 = 2$. $\chi(S^{\text{even}}) = 2$.

$\chi(\Sigma_g) = 2-2g$, $\chi(T^n) = 0$, $\chi(\mathbb{C}P^n) = n+1$.

Theorem 6.13 (Hopf) *For every generic vector field* $\mathbf{v}$ *on a manifold M,*

$$\sum_{x \text{ equil of } \mathbf{v}} \text{index}_{\mathbf{v}}(x) = \chi(M).$$

In particular, $\chi(M)$ is independent of the choice of triangulation, and the sum of indices at equilibria is independent of $\mathbf{v}$.

Proof. In dimension 2. Triangulate M and construct a vector field as follows: source at each vertex, saddle at each mid-edge, sink at each mid-face. By 6.7, source and sink each contribute $+1$, saddle -1 to the index.

QED

Problem 6.14 By examining the gradient flow of a height function, find $\chi(\Sigma_g)$ (cf. 6.12).

Corollary 6.15 *For generic* $\mathbf{v}$, *the number of equilibria of* $\mathbf{v} \geq |\chi(M)|$; *e.g.* S^{even} *cannot carry an everywhere non-zero vector field.*

Problem 6.16 Construct an everywhere nonzero vector field on each S^{odd}.

Problem 6.17 Can S^4 carry a Minkowski metric without any singularities?

Corollary 6.18 *For odd-dimensional, closed, orientable M, $\chi(M) = 0$.*

Proof. Combine 6.10 and 6.13.

QED

7. Degree Theory

Once we discover that solutions or equilibria exist (sections 5 and 6), the next issue is to count their number.

Example 7.1

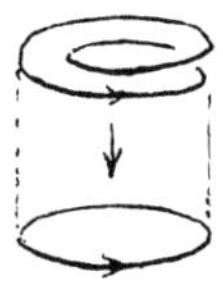

deg = +2

This map 'wraps' twice.

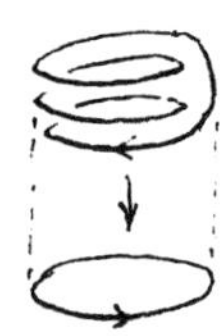

deg = −3

'Wraps against' thrice.

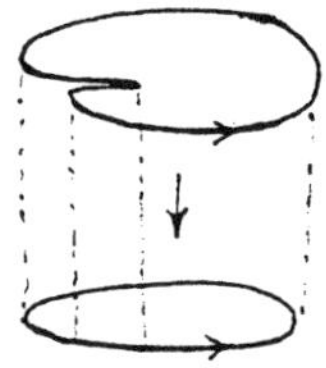

deg = +1

'Essentially' once.

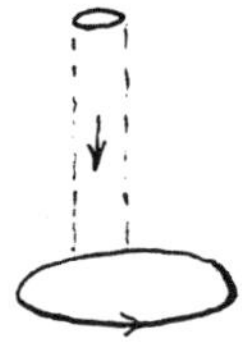

deg = 0

(deg = the number of inverse images counted with sign.)

Definition 7.2 *Let $f : M \to N$ be a map between closed oriented manifolds of equal dimension. Suppose $y_0 \in N$ is a* **regular value** *of f, i.e. at each of its inverse images* $\det(\partial f/\partial x) \neq 0$. *The* **degree** *of f is*

$$\deg f = \sum_{x \in f^{-1}(y_0)} \operatorname{sgn} \det \left(\frac{\partial f}{\partial x}\right)_x .$$

Remark 7.3 There is a theorem due to Sard guaranteeing that for every smooth f, vol{regular values of f} = vol$N$; in other words, if we pick a value in N at random, it will be regular with probability 1. So there exist plenty of regular values to choose from.

Note that according to the phrasing of 7.2, a point outside the range of f is automatically ('vacuously', as mathematicians say) regular.

Theorem 7.4 deg *does not depend on the choice of regular value* y_0, *and is homotopy-invariant.*

Proof. Denote by f_0 the constant map $M \to y_0$. $\deg f$ is just $\Gamma(f_0) \circ \Gamma(f)$. The claim follows from 4.9.

QED

Example 7.5

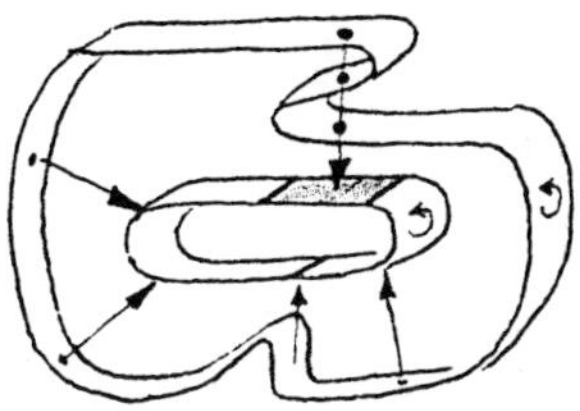

Over  3 inverse images of sgn +1, -1, +1.
Over / not regular.
(By the way, area(///) = 0 as Sard predicts.)
Elsewhere 1 inverse image of sgn +1.

Hence deg = +1.

Example 7.6

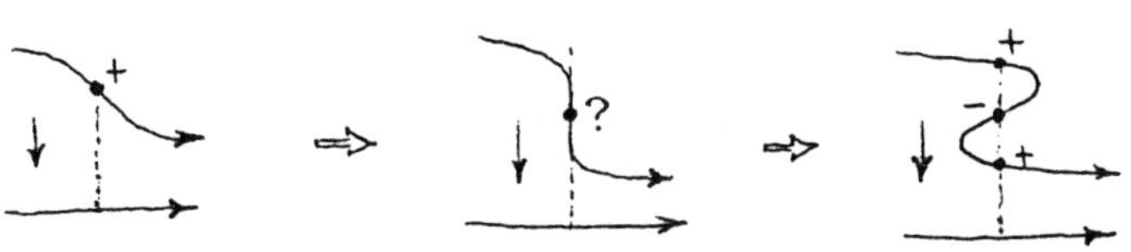

Example 7.7 The antipodal map on S^n preserves orientation if and only if $n =$ odd. So deg(antipodal)$= (-1)^{n+1}$. It follows from 7.4 and 6.16 that the antipodal map on S^n is deformable to the identity if and only if $n =$ odd.

Generalization 7.8 To do this theory in the unoriented case, count mod 2 (cf. 4.7).

Corollary 7.9 (Gauss) *Every polynomial* $f(z) = a_n z^n + \cdots + a_1 z + a_0$, $n > 0$, *has a root in* $\mathbb{C}$.

Proof. Since $\lim_{z\to\infty} f(z) = \infty$, f extends to a map from $\mathbb{C} \cup \{\infty\} \simeq \mathbb{C}P^1 \simeq S^2$ to S^2. Now deform f to $g(z) = a_n z^n$ via $tg(z) + (1-t)f(z)$. By 7.4, $\deg f = \deg g = n$ (all values are regular except 0 and ∞), so 0 must have an inverse image. QED

Remark 7.10 Thus, algebraic degree of a polynomial = degree in our sense.

Comment 7.11 In the proof of 7.9, why couldn't we deform f to any old thing $h(z)$ via $th(z) + (1-t)f(z)$? Because if the leading terms of h and f differ, this deformation produces a discontinuity at ∞.

The mod 2 theory (7.8) detects a root only when $n \equiv 0$ mod 2, i.e. $n =$ odd.

8. Gauss Map

We finish by sampling some applications of degree theory to differential geometry of surfaces.

Definition 8.1 *Let* $\mathbf{v}$ *be a vector field that is nonzero near a hypersurface* H *of* $\mathbb{R}^n$. *Write*

$$G = \frac{\mathbf{v}}{|\mathbf{v}|} : H \longrightarrow S^{n-1} \quad \textbf{(Gauss map)}.$$

The **degree** *of* $\mathbf{v}$ *on* H *is* $\deg_H \mathbf{v} = \deg G$.

Example 8.2 In $\mathbb{R}^2$

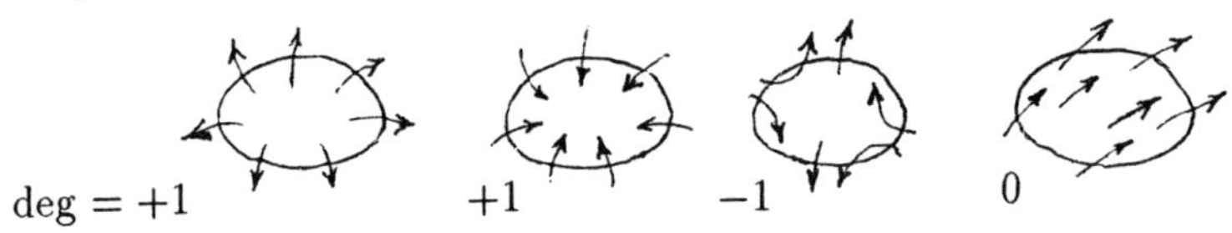

Example 8.3 In $\mathbb{R}^3$, $\mathbf{v}$ = outward unit normal on H,

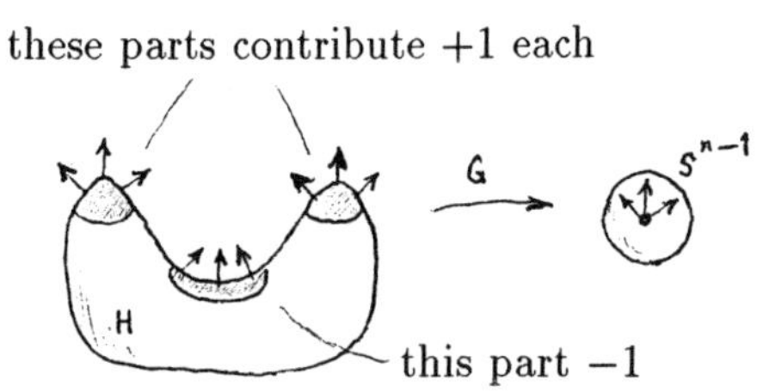

deg = +1 + 1 − 1 = +1

Theorem 8.4 (Brouwer *bis* (cf. 5.10)**)** *Every continuous map* $f : D^n \to D^n$ *has a fixed point.*

Proof. Construct a vector field $\mathbf{v}$ on D^n by drawing $\mathbf{v}(x)$ from each $x \in D^n$ to $f(x)$. No fixed point means $\mathbf{v} \neq 0$ everywhere. On ∂D^n, $\mathbf{v}$ points inward, so $\deg_{\partial D^n} \mathbf{v} \neq 0$ $(= (-1)^n$ in fact, 7.7). Now shrink ∂D^n to a tiny $(n-1)$-sphere $S \subset D^n$. If S is small enough, $\mathbf{v}$ on S is nearly parallel, so $\deg_S \mathbf{v} = 0$, contradicting the homotopy-invariance (7.4).

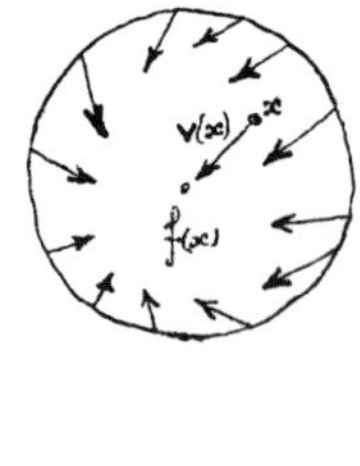

QED

Fact 8.5 The standard volume form on S^{n-1} can be written

$$\Omega = \sum_{k=1}^{n} (-1)^k x_k dx_1 \wedge \cdots \wedge \cancel{dx_k} \wedge \cdots \wedge dx_n .$$

($\mathrm{vol}(S^{n-1}) = \int_{S^{n-1}} \Omega = 2\sqrt{\pi}^{\,n}/\Gamma(n/2)$, if we must know.)

Theorem 8.6 *If* $\mathbf{v}$ *is the outward unit normal vector field on a hypersurface* H *of* $\mathbb{R}^n$, *then*

$$\deg_H \mathbf{v} = \frac{1}{\mathrm{vol}(S^{n-1})} \int_H K\, d\sigma ,$$

where K = *Gaussian curvature (product of principal curvatures) of* H, *and* $d\sigma$ = $(n-1)$*-dimensional volume element on* H *induced by the ambient euclidean metric on* $\mathbb{R}^n$.

Proof.

$$\deg_H \mathbf{v} = \frac{\deg G}{\mathrm{vol}(S^{n-1})} \int_{S^{n-1}} \Omega = \frac{1}{\mathrm{vol}(S^{n-1})} \int_H G^*\Omega = \frac{1}{\mathrm{vol}(S^{n-1})} \int_H K\, d\sigma ;$$

the second equality comes from the Jacobian formula for the change of variables ($\int_{S^{n-1}} \Omega = \int_H G^*\Omega$ if G is 1-to-1, but here G is $\deg G$-to-1).

QED

Reminder 8.7 For H parametrized by

$$\begin{pmatrix} x_1(s_1, \ldots, s_{n-1}) \\ \vdots \\ x_n(s_1, \ldots, s_{n-1}) \end{pmatrix},$$

then

$$d\sigma = \sqrt{|\det(g_{ij})|}\, ds_1 \wedge \ldots \wedge ds_{n-1}, \quad g_{ij} = \frac{\partial}{\partial s_i}\begin{pmatrix} x_1 \\ \vdots \\ x_n \end{pmatrix} \cdot \frac{\partial}{\partial s_j}\begin{pmatrix} x_1 \\ \vdots \\ x_n \end{pmatrix}.$$

Example 8.8 For a loop γ in $\mathbb{R}^2$, $\deg_\gamma \mathbf{v} = \frac{1}{2\pi}\int_\gamma K\, ds$ = rotation number of γ.

rot#= +1

Corollary 8.9 (Gauss-Bonnet)

$$\frac{1}{2\pi}\int_{\Sigma_g} K\,d\sigma = \chi(\Sigma_g)\ .$$

Proof.

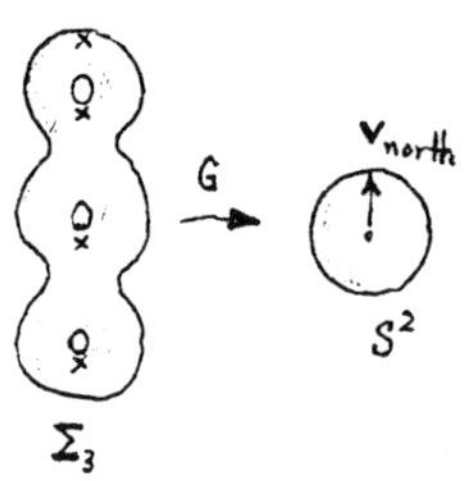

By homotopy-invariance (7.4), deforming Σ_g into the position shown does not affect $\deg G$, hence not the integral either (8.6). $\mathbf{v}_{\text{north}}$ is a regular value of G; its inverse images are the maximum and the g 'upward' saddles of Σ_g. By 8.3 they contribute $+1, -1$ each to $\deg G$. Hence by 8.6 and 6.12,

$$\frac{1}{4\pi}\int_{\Sigma_g} K\,d\sigma = 1 - g = \frac{\chi(\Sigma_g)}{2}\ .$$

QED

Remark 8.10 We derived 8.9 for $\Sigma_g \subset \mathbb{R}^3$, but it can be shown to be correct even in cases where the metric on Σ_g is not induced by an ambient euclidean metric.

Corollary 8.11 Σ_g *of genus* $g > 1$ *(respectively* $g = 1$, $g = 0$*) cannot carry a metric of everywhere positive (respectively non-zero, negative) curvature.*

9. Suggested Reading

A beautiful book at a similar level as these lectures is [4]. [3] is a systematic textbook. [5] is semi-popular and entertaining. For further topics in topology, [7] is recommended; [2] is heftier and covers more. For fluid dynamicists, a natural sequel to these 4 days would be infinite-dimensional topological methods in partial differential equations; they are nicely explained in [8]. The practical calculation of topological invariants requires algebraic machines; an efficient little book is [6]. And every graduate student in geometry and topology seems to have read [1].

Acknowledgements

Thanks are due to K. Moffatt, H. Aref, D. Mackay, R. Ricca, and E. Spiegel for their kind hospitality in Cambridge and generous feedback.

References

1. Bott, R. & Tu, L. (1982) *Differential Forms in Algebraic Topology.* Springer.
2. Dubrovin, B.A., Fomenko, A.T. & Novikov, S.P. (1984–1990) *Modern Geometry: Methods and Applications.* Volumes I–III, Springer.
3. Guillemin, V. & Pollack, A. (1974) *Differential Topology.* Prentice-Hall.
4. Milnor, J. (1965) *Topology from the Differentiable Viewpoint.* University Press Virginia.
5. Prasolov, V.V. (1994) *Intuitive Topology.* Amererican Mathematical Society.
6. Sato, H. (1999) *Algebraic Topology: An Intuitive Approach.* American Mathematical Society.
7. Schwarz, A.S. (1994) *Topology for Physicists.* Springer.
8. Smoller, J. (1994) *Shock Waves and Reaction-Diffusion equations.* Springer.

ELEMENTS OF CLASSICAL KNOT THEORY

CLAUDE WEBER
Section de mathématiques
2-4 rue du Lièvre, Case postale 240
CH-1211 Genève 24, Switzerland

E-mail: *Claude.Weber@math.unige.ch*

Abstract. The aim of this article is to present an elementary introduction to classical knot theory. The word classical means two things. First, it means the study of knots in the usual 3D space R^3 or S^3. It also designates knot theory before 1984. In section 1 we describe the basic facts: curves in 3D space, isotopies, knots, links and knot types. We then proceed to knot diagrams and braids. Finally we introduce the useful notion of tangle due to John Conway. In section 2 we present some important problems of knot theory: classification, chirality, search for and computation of invariants. In section 3 we give a brief description of some knot families: alternating knots, two-bridge knots, torus knots. Within each family, the classification problem is solved. In section 4 we indicate two ways to introduce some structure in knot types: via ideal knots and via the knot complement.

1. Basics

1.1. CURVES IN SPACE

A **curve** in space is the trajectory of a moving particle. We shall require usually that the curve is **closed**, i.e. that the trajectory is periodic, say of period 1. In mathematical terms, $c(s)$ designates the position of the particle at time s and one has the equality $c(s) = c(s+1)$ for all s. In some instances, the map $c : R \to R^3$ is called a parametrization of the curve, the curve itself being the image of c. Such a curve is called **simple** if the trajectory does not cut itself (besides the fact that it is periodic). In other words one requires that if $c(s) = c(s')$, then $s' = s + m$ with m integer. Henceforth we shall deal with simple closed curves in space. We request the map c to be at least

R.L. Ricca (ed.), An Introduction to the Geometry and Topology of Fluid Flows, 57–75.

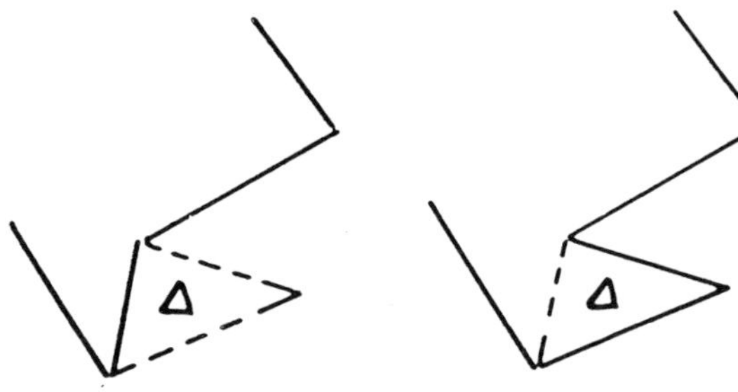

Figure 1. The slide operation.

continuous. If one wishes, one can add more conditions on c: differentiable, several times differentiable, derivative never equal to zero, curvature never equal to zero, etc.. An important special case is provided by **polygonal curves**, i.e. curves which are the union of a finite number of line segments.

1.2. EQUIVALENCES OF SIMPLE CLOSED CURVES

Imagine that the curves are made of rubber. We define (informally) two simple closed curves as being **rubber-equivalent** if we can transform one into the other by moves which are allowed by the rubber. Scale changes are also allowed. Of course, this is not a mathematical definition. Here are two ways to proceed mathematically.

1. Slides of polygonal curves. By definition two polygonal curves are **slide equivalent** if one can pass from one to the other by a finite number of slides. A **slide** is the following move (Figure 1). Let C be a polygonal curve. Let Δ be a triangle in space such that the intersection of Δ with C is equal to one side of Δ. Replace this side of Δ by the other two sides and do not change C anywhere else . Literally, one slides C through the triangle. The inverse operation is also a slide by definition.
2. Isotopies. Again we resort to physics. Suppose that one has a fluid filling the whole of R^3 and that the equations of motion can be solved for all t between $t = 0$ and $t = 1$. Fix a $t \in [0,1]$. Let Φ_t be the map of R^3 to itself, that sends a point x, represented by a tiny drop of fluid at time $t = 0$, to the position occupied by the drop at time t. The flow map Φ_t is continuous. It has an inverse which is also continuous. In topology, such a map is called a **homeomorphism** of R^3.
   The map Φ_0 is the identity. The family of homeomorphisms Φ_t for all $t \in [0,1]$ can be thought of as a path joining Φ_1 to the identity Φ_0. One calls it an **isotopy** between Φ_0 and Φ_1. One can prove that this implies that Φ_1 keeps the orientation of R^3 fixed. Conversely, for any

homeomorphism of R^3 which keeps the orientation fixed, there is an isotopy between Φ and the identity.
Suppose now that the fluid is inviscid, incompressible and homogeneous. Suppose that there is a vortex line C_0 at time $t = 0$ which is a simple closed curve. Then $C_1 = \Phi_1(C_0)$ is a vortex line at time $t = 1$ (Helmholtz second law of vortex motion). One can extract from this the following mathematical definition. Two simple closed curves C_0 and C_1 are **isotopic** if there exists a homeomorphism Φ of R^3, keeping the orientation of R^3 fixed, and such that $C_1 = \Phi(C_0)$.

One can prove (this is non-trivial) that two polygonal curves are slide equivalent if and only if they are isotopic. One moral of this is the following. Slide equivalence is supposed to mimic rubber equivalence, which has a strong intuitive bearing. In practice, however, for technical reasons we often need to use the isotopy relation.

1.3. KNOTS, LINKS AND TYPES

Let us now introduce the usual definitions. A simple closed curve in R^3 is called a **knot**. Hence, there are two expressions for the same object, but the context is different. If a topologist uses the word 'knot', his next step will be to move it by an isotopy. On the other hand, if a geometer uses the words 'simple closed curve', the curve will remain fixed as it is (as in differential geometry).

The isotopy class of a knot is called its **knot type**. For instance, the frozen-in-the-fluid theorem mentioned above means that the knot type of a closed vortex line is constant in time (but of course, the knot itself is not constant in time). This topological conservation law was discovered by Kelvin and Tait and was the starting point of Tait's work on knots, that began in 1867.

A knot is said to be **tame** if it is isotopic to a polygonal knot. Non-tame knots exist. They are called **wild** (Figure 2). The set of tame knot types is countable, but there are uncountably many wild knot types. There is little hope to classify wild knots, but they can be fun. One usually considers only tame knots and accordingly one discards the adjective tame.

A finite union of disjoint simple closed curves is called a **link**. The number of these curves is the number of **components** of the link. Most of the concepts presented here are valid also for links. One talks of **link type**, **tame links**, etc..

Remark 1.1 Smooth versus polygonal knots. Let us define a knot to be **smooth** if it has a non-vanishing tangent at every point, which moves continuously as the point moves on the curve. It is equivalent to ask that the parametrization has a non-vanishing derivative everywhere, and that

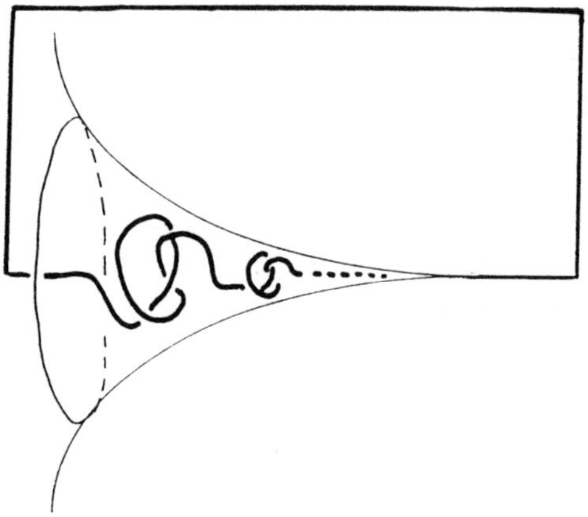

Figure 2. A wild knot which has a non-vanishing tangent everywhere, but with the tangent discontinuous at one point.

the derivative is continuous. The process which consists of 'smoothing the corners' of a polygonal curve shows that a polygonal curve is isotopic to a smooth one. The converse is also true: a smooth curve is isotopic to a polygonal one. The proof requires however some care. One must definitely use an appropriate argument which relies on the fact that the derivative is continuous. Indeed, it is possible to produce a wild knot which has a non-vanishing tangent everywhere, the tangent being continuous except at one point, as shown by the example of Figure 2.

As a consequence, the theory of tame knot types can be developed with polygonal or with smooth knots as well. We have chosen polygonal curves because they are closer to combinatorics (think of the definition of a slide). Besides, computers generate polygonal curves!

1.4. HOW DOES ONE REPRESENT A KNOT?

- By exhibiting a parametrization $c(t) = (c_1(t), c_2(t), c_3(t))$, where the c_i are periodic functions $R \to R$. Usually this is not very convenient. However some interesting (and tricky) questions can arise: the c_i being given, determine which knot does the parametrization represent. See, for instance, the Lissajoux knots.
- If the parametrization is twice differentiable, the curvature and torsion functions (of arc length) determine the knot up to isometry.
- In case of a polygonal curve, curvature is concentrated at vertices, and torsion at the middle of the edges. Arc-length is given by the length of the edges.

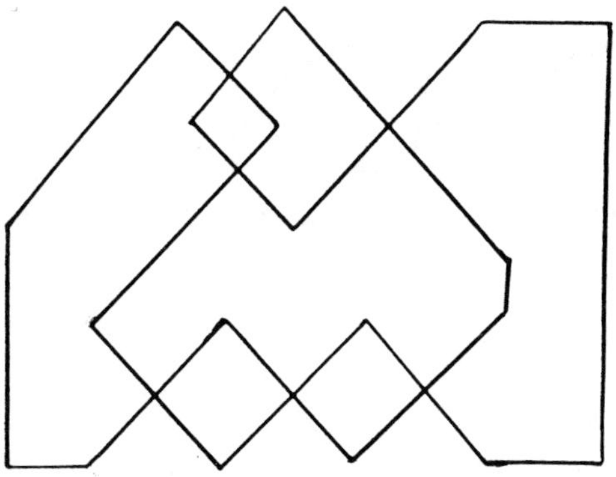

Figure 3. A polygonal nodal curve.

- For a polygonal knot, one can simply give the finite list of the coordinates of the vertices. The vertices are ordered travelling along the curve, once we choose a starting point and an orientation on the curve.
- One can proceed like architects, giving two different projections of the knot. This can be very powerful. For instance, J. Dubochet's laboratory at the University of Lausanne has obtained beautiful representations of DNA in space by this method. The feat is to obtain two simultaneous pictures via electronic microscopy.
- The information contained in two projections is redundant. Hence, one can use only one projection and give the missing height function. This is particularly easy if the knot is polygonal and if one requires that the image of the knot by the projection is a **nodal curve** (Figure 3). By this one means that the only self-intersections of the image are transversal double points. It is then only necessary to give the height at the vertices.

1.5. HOW DOES ONE REPRESENT A KNOT TYPE?

1.5.1. *Knot diagrams*

Suppose we are given a tame knot type. We choose a polygonal knot representative for it. We project it in such a way that the image is a nodal curve. Since we wish to represent the knot up to isotopy, we do not need to know the actual height of the vertices. We only need to know at each crossing point which line segment lies above the other in space. The picture thus obtained is called a **knot diagram**. One defines **link diagrams** in the same way. The diagram on the left-hand-side of Figure 4 represents the unknot, the one at the centre the right-handed trefoil knot, the one on the right-hand-side the four-crossing knot.

Figure 4. Three diagrams of knots: the unknot (left-hand-side), the right-handed trefoil knot (centre) and the four-crossing knot (right-hand-side).

Definition 1.2 *The (minimal)* **crossing number** $c(K)$ *of a given knot type* K *is the minimum of crossings among all diagrams representing* K.

Of course there is an infinite number of diagrams representing K. It is in general quite difficult to determine $c(K)$. See, however, the case of alternating knots (§ 3.1).

1.5.2. *Closed braids*
Let $b_i : [0,1] \to R^2$ for $i = 1, ..., n$ be a finite number of continuous maps such that:

1. $b_i(t) \neq b_j(t)$ if $i \neq j$ for all $t \in [0,1]$.
2. $b_i(0) = (i, 0)$
3. There exists a permutation σ of $\{1, ..., n\}$ such that $b_i(1) = (\sigma(i), 0)$.

Let $B_i : [0,1] \to R^3$ be defined by $B_i(t) = (b_i(t), t)$. The union of the images of the B_i for $i = 1, ..., n$ is by definition a **braid** on n strings. Polygonal braids are defined in the obvious way. One can interpret the

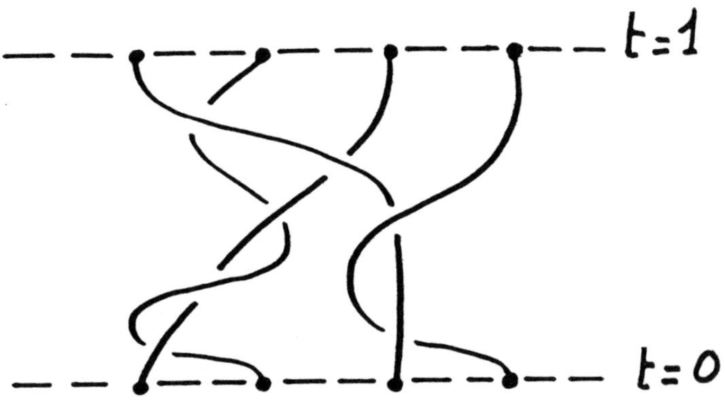

Figure 5. Diagram of a braid: the vertical axis represents time.

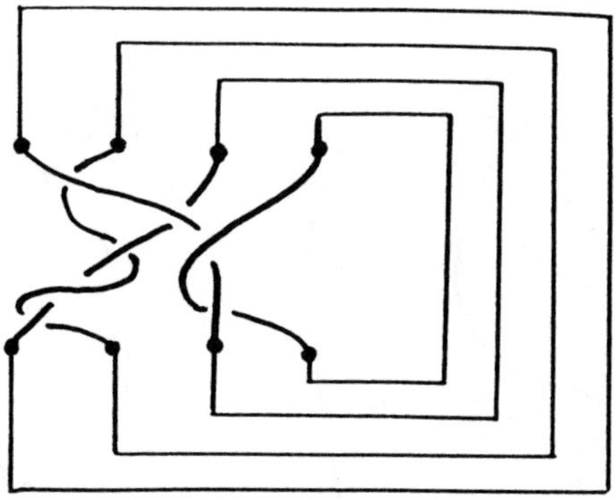

Figure 6. Closure of a braid.

b_i's as trajectories of points moving in the plane, with no collisions. The B_i's are the trajectories in space-time (2 space dimensions plus time; see Figure 5 which shows a braid projected onto a plane in which the vertical axis represents time).

One can close a braid to obtain a **closed braid**, as in Figure 6, which shows the closure of the braid of Figure 5. The type of the knot or link thus obtained is well defined. One can also prove that any link type can be represented by closed braids. This can be done in infinitely many different ways.

Definition 1.3 *The* **braid index** *of a link type L is the minimum number of strings among all closed braid representations of L.*

The determination of the braid index is quite difficult.

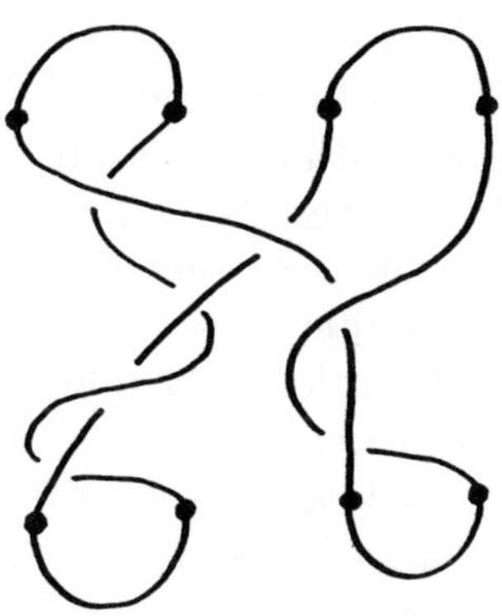

Figure 7. A plat.

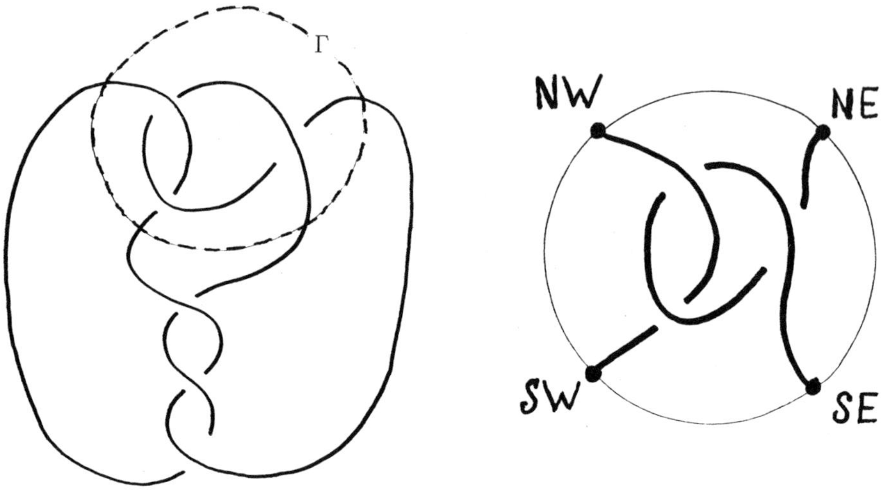

Figure 8. The intersection of Γ with D (on the left) determines the tangle shown on the right.

1.5.3. *Plats*

If a braid has an even number of strings, one can close it in a different way as shown in Figure 7 called a **plat**. It is not difficult to prove that every link type L can be represented by plats.

Definition 1.4 *The* **bridge number** *of a knot type K is one half of the minimum number of strings among all plat representations of K.*

This number is equal to the total curvature of K (divided by 2π) as defined by J. Milnor. Recall that the total curvature of K is the infimum of the total curvature of all knots representing the knot type K. Again, it is difficult to determine the bridge number.

1.5.4. *Tangles*

Tangles are both an important concept and a useful tool, due to J. Conway. Let D be a link diagram drawn in a plane Π. Let Γ be a simple closed curve in Π intersecting D transversaly in 4 points (none of them being a crossing point of D): see the left-hand-side diagram of Figure 8. The region of Π limited by Γ is homeomorphic to a disc. This disc, together with its intersection with D, is by definition a **tangle**. One should be a bit more precise: the model disc considered here is a disc in the (x, y)-plane. The four intersection points on its boundary are the cardinal points NE, NW, SW and SE (see the right-hand-side diagram of Figure 8).

Two tangles are considered as being equivalent only if one can go from one to the other by a homeomorphism of model discs which respects the

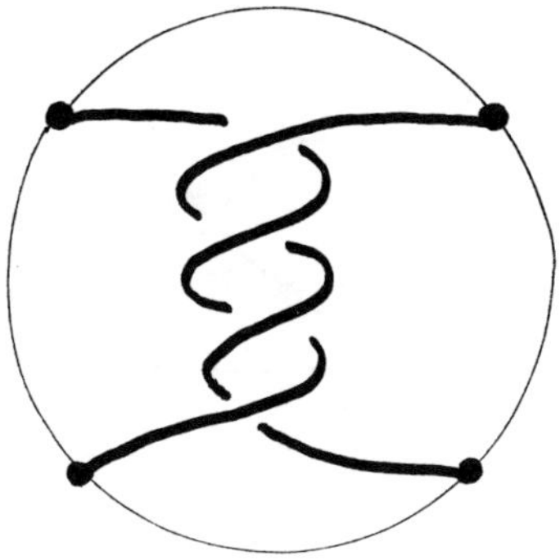

Figure 9. A vertical twist with $m = 4$.

cardinal points and their position (i.e. NE goes to NE, and so on). The following examples of tangles will be of use in the next pages. Let m be an integer. The **vertical m-twist** is the tangle shown in Figure 8. The helix is, by definition, right-handed if m is positive, left-handed if m is negative. Figure 9 shows a right-handed helix with $m = 4$. The **horizontal m-twist** is the object obtained from the vertical one by a rotation of 90 degrees.

Here are some uses of tangles.

1. One can close a tangle in two ways, without introducing new crossings to obtain a link diagram (Figure 10). Hence tangles are another way of presenting knot (or link) types.
2. Given a link diagram, one can draw a little circle around each crossing point. Then one can replace the crossing point inside each circle by tangles of our choice. By coding the original diagram and the tangles used, we get a combinatorial way to generate new diagrams from old

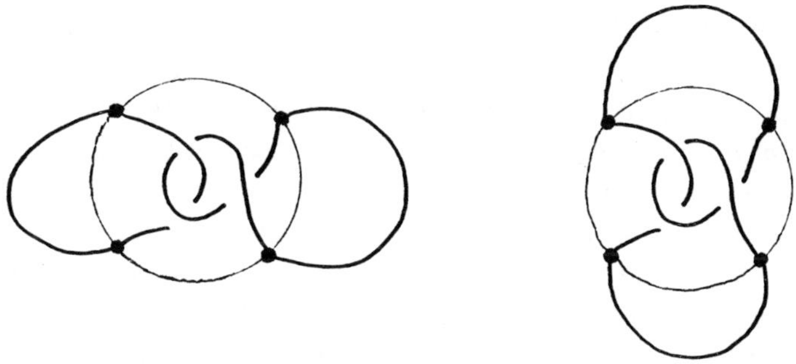

Figure 10. This example shows how the closure of tangles produces links.

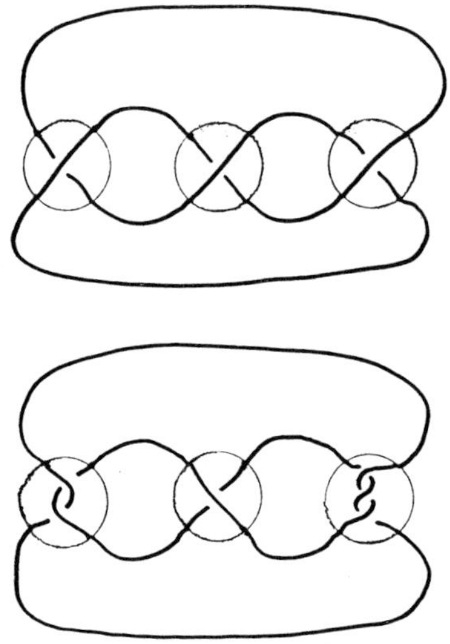

Figure 11. Bretzel link (bottom diagram) obtained from the top diagram.

ones and to encode them. An example is given in Figure 11: one starts from the top diagram and then replaces each crossing point by some vertical twist, as defined above, obtaining the so-called Bretzel link (bottom diagram).

3. A **tangle move** consists of replacing a given tangle in a link diagram with a prescribed one. Tangle moves are present in knot classification (cf. flype moves, § 3.1). They are also the key tool used to model the action of certain enzymes acting on DNA (work of De W. Sumners).

2. Some Problems in Knot Theory

2.1. KNOT CLASSIFICATION

The problem of the classification of knots up to isotopy is adressed in three steps. It has some similarity with the setting-up of a herbarium in botany, with the proviso that the number of species is infinite.

1st step: the enumeration problem. It consists of trying to make a list of all knot types, by exhibiting at least one representative of each isotopy class. As this list is infinite, one attempts to display it by the increasing degree of complexity, this ususally represented by the crossing number $c(K)$.

2nd step: the duplication removal problem. Here, one wants to have exactly one representative for each isotopy class, or one wants to put together the knots of the first list which represent the same type. A sub-problem is the separation problem: given two knots, one wants to know if they represent the same knot type or not. The main tool used here are the **knot** (or **link) invariants**.

Definition 2.1 *A* **knot invariant** *is a mathematical object (often of an algebraic nature) associated with each knot, in such a way that the object attributed to two isotopic knots is the same.*

The crossing number is a knot invariant by nature (the definition forces it to be an invariant). The same goes for the braid index and for the bridge number. Somehow one pays the price of cheating with a definition of an invariant which is hard to compute. For a discussion of polynomial invariants we refer the reader to Kauffman's article in this volume.

3rd step: the recognition problem. Given a knot, one wants to know to which knot in our list it is isotopic. In botany, this problem is familiar to all amateurs. Here again, invariants prove useful.

None of these problems is solved today.

2.2. CHIRALITY

Suppose that we have a knot in R^3 representing some knot type K. Its image by a reflection through a plane is naturally called the **mirror image** of the knot. As the reflexion is a homeomorphism which reverses the orientation of R^3, it makes sense to ask the following question: is the given knot isotopic to its mirror image? If yes, one says that the knot type is **achiral**. If no, it is **chiral**. This problem can be seen as part of the classification problem. It is known since the beginning of the 20th century that the trefoil is chiral. There are indeed two trefoils: a left-handed one and a right-handed one. The adjectives left and right are purely conventional. To get a diagram for the left-handed trefoil, change all over-passes to under-passes in the second diagram of Figure 4. P.G. Tait proved that the four-crossing knot is achiral. Informally, one can say that most knots are chiral.

2.3. NUMERICAL INVARIANTS

The problem is: to obtain for each numerical invariant (such as the crossing number, the braid index or the bridge number) a procedure for its computation. Another classical numerical invariant one would like to compute is the unknotting number.

Definition 2.2 *Given a knot diagram D, it is always possible to change some over-crossings to under-crossings (or vice versa) in order that the new diagram represents the unknot* (can you prove it?). *Following Tait, define the* **beknottedness** *of the diagram D as the least number of changes necessary to obtain the unknot. Now, the* **unknotting number** *of a knot*

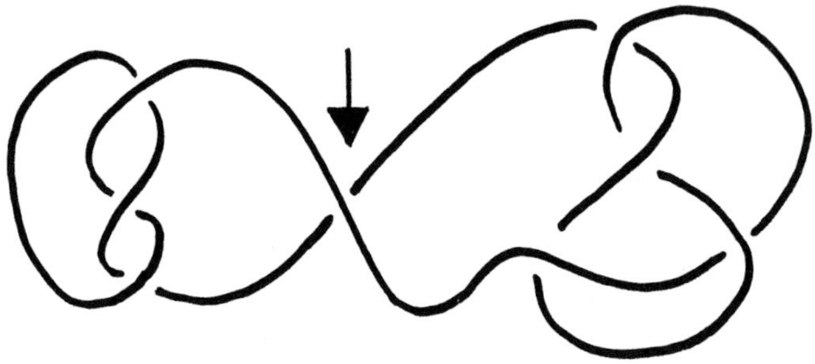

Figure 12. A knot diagram with a nugatory crossing.

type K is, by definition, the minimum of the beknottedness among all diagrams representing K.

The **stick number** of a knot type K is defined as the smallest number of line segments among all polygonal knots representing K. What is known about this invariant is close to nothing. These problems (among many others) can be seen as challenges. They serve as tests for the new tools and invariants discovered by topologists.

3. Some Knot Families

Knot families play an important rôle. They provide 'large' samples of knots. They are usually rather well understood. For instance, in several cases, the classification problem is solved within the family. Hence, they can serve as testing material. Here are some well-known families.

3.1. ALTERNATING KNOTS

Definition 3.1 *A knot diagram D is* **alternating** *if, when we proceed along the nodal curve, we pass alternately over, under, over, and so on, at each crossing.*

The diagrams of Figure 4 are alternating, while the one of Figure 7 is not.

Definition 3.2 *A knot type K is alternating if there exists a diagram representing the knot K that is alternating.*

If it doesn't, a knot type is said to be **non-alternating**. But beware: it is easy to see that any knot type can be represented by some non-alternating diagram. Hence, the problem of deciding whether a given knot type is alternating or not can be difficult. Some knot invariants can help. Up to 7 crossings, all knot types are alternating. One can prove that torus knots with $u \geq 3$ are non-alternating (see § 3.3 for a definition of torus knots).

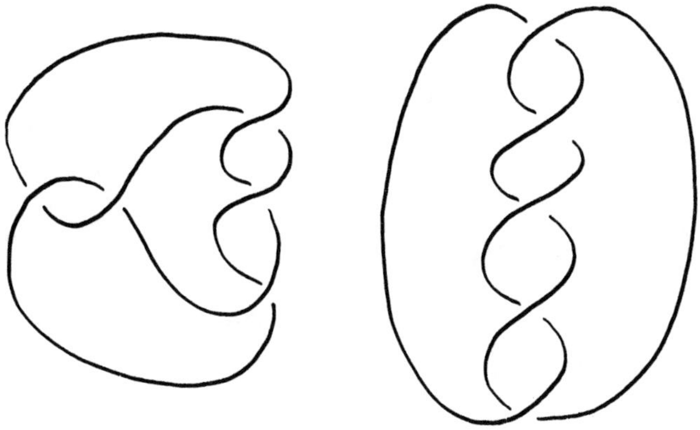

Figure 13. Two alternating diagrams without nugatory crossings.

Remark 3.3 Alternating knots have been discovered and studied by the physicist P.G. Tait in the last third of 19th century. His work relied on several 'principles', that have only been proved since 1984! As a consequence, one understands today alternating diagrams quite well and several classification questions can be answered for them.

Here are some examples. Following Tait, a crossing in a diagram is termed **nugatory** if the diagram can by decomposed into two parts joined by this crossing as in Figure 12. Clearly one can get rid of such a crossing by an isotopy, thus obtaining a diagram for the same knot type with one crossing less.

Theorem 3.4 *Let K be an alternating knot type. Then all alternating diagrams representing K without nugatory crossings have the same number of crossings and this number is equal to $c(K)$.*

The proof of the theorem makes use of the Jones polynomial in its representation given by L.H. Kauffman. The statement of the theorem is due to Tait, but the proof proposed by him was incomplete. Before 1984, the statement was known as one of "Tait conjectures". This result can be employed to answer practical questions, without apparent use of knot invariants (they are hidden in the proof of the theorem). Here is an example. Let us look at the three diagrams shown in Figure 4. They are alternating, without nugatory crossings. The theorem implies that $c(K)$ can be computed from these diagrams (indeed $c(K)$ can easily be computed for all alternating knots!). Since $c(K)$ is different for each diagram, these diagrams represent different knot types. Hence the trefoil is knotted and non-isotopic to the four-crossing knot.

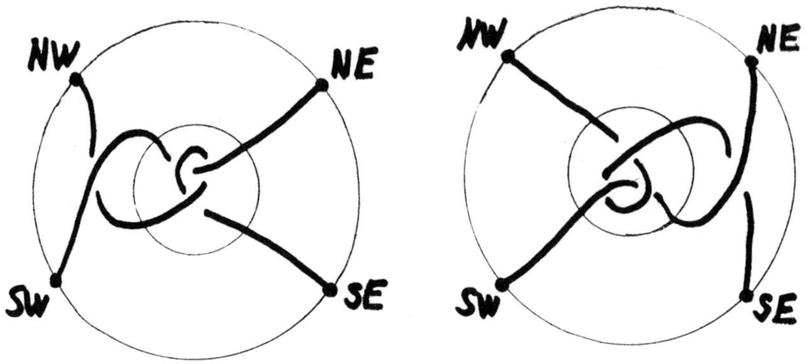

Figure 14. Illustration of a flype move.

Let us now look at the diagrams of Figure 13. It shows two alternating diagrams without nugatory crossings. The preceeding argument cannot be used to distinguish their corresponding knot type, because the two diagrams have the same number of crossings. Following W. Menasco and M. Thistlethwaite (who proved one of the "Tait conjectures"), however, one can argue that if the diagrams were representing the same knot type, then one could transform one into the other by a finite number of **flype** moves (which Tait called **distorsions**), shown in Figure 14. One can check that by performing flype moves on the diagrams of Figure 13 one cannot transform one diagram into the other, hence the two diagrams represent distinct knot types.

3.2. TWO BRIDGE KNOTS AND LINKS

It is easy to see that a link with bridge number equal to 1 is indeed a knot: just untwist the braid while proceeding along it. So the first interesting case deals with two bridge knots and links. One can prove that it is always possible to find representative diagrams which look like the diagrams of Figure 15. In these diagrams, notice the rôle played by the tangles we called vertical twists. The first diagram is alternating; one can check that this is equivalent to say that the signs of the twists alternate when one proceeds along the braid. From these diagrams, the classification of two bridge knot and link types is completely known. Let's see how.

Let $(m_1, m_2, ..., m_k)$ be the twists of the braid, with m_1 positive. Write $a_j = |m_j|$. Now consider the rational number obtained by the continued fraction

$$\frac{p}{q} = \cfrac{1}{a_1 + \cfrac{1}{a_2 + \dots}} .$$

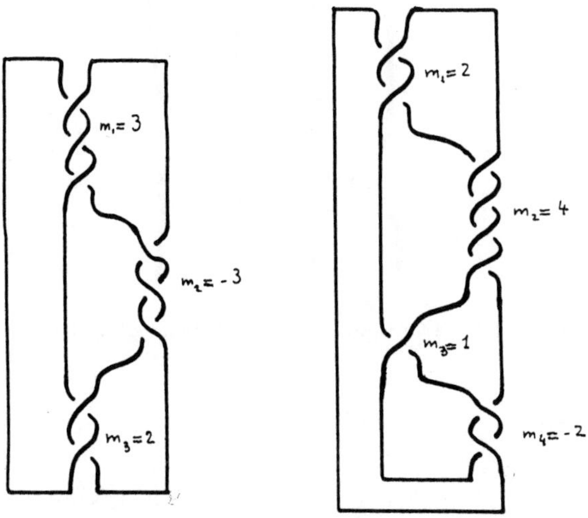

Figure 15. Two bridge knots: the first diagram (on the left) is alternating, whereas the other is non-alternating. Notice the rôle played by the vertical twists.

The interest for considering this rational number lies in the next theorem.

Theorem 3.5 *1. Two rational numbers defined as above represent the same knot or link type if and only if: i)* $p = p'$ *and* $q \equiv q' \pmod p$ *or ii)* $p = p'$ *and* $qq' \equiv 1 \pmod p$*. 2. One has a two component link if and only if* p *is even. Otherwise one has a knot.*

The chirality question can be solved for these knots. Up to 7 crossings, all knot types are two bridge ones.

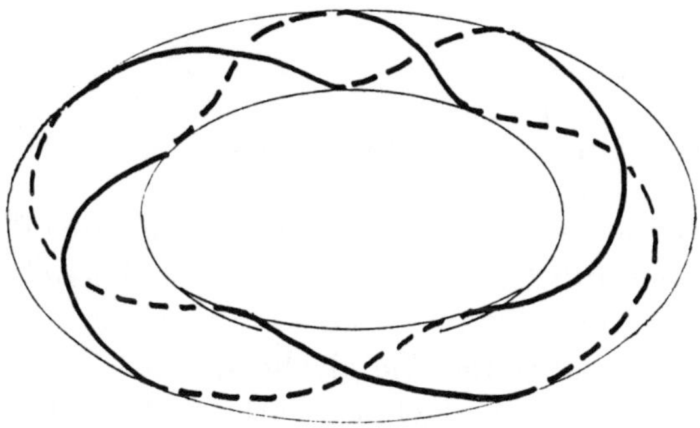

Figure 16. A torus knot with $u = 2$ and $v = 5$.

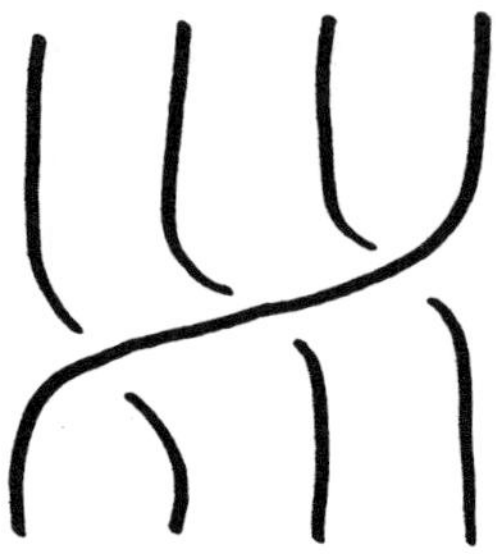

Figure 17. The right-handed full twist of u strings.

3.3. TORUS KNOTS

A knot which sits (i.e. which is embedded) in an unknotted torus in R^3 is called a **torus knot**. Figure 16 shows a torus knot with $u = 2$ and $v = 5$ in the classification below. A non trivial torus knot can be represented up to isotopy by a closed braid in the following way. Let $u \geq 2$ be an integer. Figure 17 shows the braid called right-handed full twist on u strings. Now let v be an integer. Consider the braid $\beta(u, v)$ on u strings which is the product of v right-handed full twists if v is positive (Figure 18) and of $|v|$ left-handed full twists if v is negative.

Let $\hat{\beta}(u, v)$ be the closed braid associated to $\beta(u, v)$.

Theorem 3.6 *$\hat{\beta}(u, v)$ represents a knot if and only if v is prime to u. A non-trivial torus knot is isotopic to one and only one closed braid $\hat{\beta}(u, v)$ with $u \geq 2$ and $|v| > u$.*

Remark 3.7 The mirror image changes v to $-v$. Hence all non-trivial torus knot types are chiral.

4. Internal Structures for Knot Types

The classification results obtained by knot theorists are often boring. They look like the phone directory: a list of names and numbers, which gives little insight about the individuals. This is to be contrasted, for instance, with the classification of simple Lie algebras, which rests on the study of the structure carried by such algebras: Cartan sub-algebras, root systems, etc.. It is tempting to try the same approach for knot types, by bringing in some 'internal structure'. Here are two ways.

Figure 18. The closed braid $\hat{\beta}(3,4)$.

4.1. IDEAL KNOTS

This approach tries to find a canonical representative within each knot type. Remember how the search for a canonical representative is fruitful in Nielsen-Thurston theory for isotopy classes of homeomorphisms of surfaces. Parenthesis: Nielsen-Thurston theory can be used with success to study an important sub-class of knots, called fibered knots. Most of today's attempts are based on the following scheme. Let E be the space of all knots representing a given knot type K and let $\phi : E \to R$ be a function of interest. For instance, ϕ may be an energy function of some sort. Another possibility is the "thickness" of a knot C, defined as follows. Let r be the largest radius of an embedded tube centered on C and let s be the length of C. The **thickness** of C is, by definition, equal to $\frac{r}{s}$. Then one looks for the appropriate extrema of ϕ: minima for an energy or maxima for the thickness. One hopes that there exists one and only one such extremum (up to some symmetries). The corresponding knot will be canonical, with respect to ϕ. It should enjoy interesting features. All this is in need of proofs. In any case the study of local extrema is a problem of great interest, clearly related to relaxation questions within a given topological type. For more details, see the book by A. Stasiak.

4.2. THE KNOT COMPLEMENT

In this subsection, it is better to think of knots embedded in the 3-sphere S^3, which is nothing else than the space R^3 compactified by adding one point at infinity. Let K be a (polygonal) knot in S^3. The **knot complement** is by definition (as the name indicates) the complement of the knot in the

3-sphere, i.e. $(S^3 - K)$. Now, let K and K' be two isotopic knots in S^3, and let $\Phi : S^3 \to S^3$ be a homeomorphism of S^3, keeping the orientation fixed, and such that $\Phi(K) = K'$. The restriction $\Phi|(S^3 - K) \to (S^3 - K')$ is also a homeomorphism. Hence: two isotopic knots have homeomorphic complements (note that here we definitely need the more complicated definition of isotopy given via the homeomorphisms Φ_t, rather than the intuitive one via slides). A long standing conjecture asserted that the converse is also true. It has been proved in 1989 by Gordon-Luecke (no new invariants needed). A similar statement for links is known to be false. However, what follows is also valid for links.

Typical of classical knot theory is the rôle played by surfaces embedded in a knot complement. Of particular interest are Seifert surfaces, which are orientable surfaces, whose boundary is equal to a given knot. The work of several topologists since 1960 (Haken, Waldhausen, Johannson, Jaco-Shalen, Thurston, ...) has succeeded in providing each sufficiently large 3-dimensional manifold with a rich structure. Knot and link complements are such manifolds. As such, they can be canonically decomposed into pieces, and each piece can be equipped with a geometry in a very precise sense. For knot complements they are essentially of two kinds, the more spectacular one being hyperbolic geometry. See the appropriate references in the next section, for striking examples of use of these geometries.

5. A Bibliography With Comments

Basic facts on classical knot theory can be found in the books by Rolfsen and Burde-Zieschang. Surfaces related to knots are abundantly displayed in the first one, while several classification results quoted in this article are proved in the second one. A good presentation of knot theory since 1984 is provided by Lickorish's book, which gives also a special treatment of alternating knots. For ideal knots, see the book with the same title by A. Stasiak. De W. Sumners's work on tangle moves modelling the action of some enzymes is well explained in his article quoted below. The article by Hoste-Thistlethwaite-Weeks is a present day review of classification problems, that contains also a good description of the use of hyperbolic geometry. Thurston theory is presented in many places; see for instance the article by Scott or the article by Thurston himself. Waldhausen-Johannson-Jaco-Shalen decomposition theorem is explained in the article by Jaco-Shalen.

Acknowledgements

The author would like to acknowledge financial support from the Fonds National Suisse de la Recherche Scientifique and the kind hospitality of the Isaac Newton Institute for Mathematical Sciences, Cambridge (UK).

References

1. Burde, G. & Zieschang, H. (1985) *Knots.* Studies in Mathematics, de Gruyter .
2. Hoste, J., Thistlethwaite, M. & Weeks, J. (1998) The first 1,701,935 Knots. *Math. Intell.* **20**, 33–48.
3. Jaco, W. & Shalen, P. (1979) Seifert fibered spaces in 3-mainfolds. In *Geometric Topology*, pp. 91–99. Academic Press.
4. Lickorish, R. (1997) *An Introduction to Knot Theory.* Graduate Texts in Mathematics **175**, Springer.
5. Rolfsen, D. (1976) *Knots and Links.* Publish or Perish.
6. Scott, P. (1983) The geometries of 3-manifolds. *Bull. London. Math. Soc.* **15**, 401–487.
7. Stasiak, A. (1999) *Ideal Knots.* World Scientific.
8. Sumners, De W. (1990) Untangling DNA. *Math. Intell.* **12**, 71–80.
9. Thurston, W. (1979) Hyperbolic Geometry and 3-manifolds. In *Low-Dimensional Topology*, pp. 9–25. Lecture Notes **48**, London Math. Society.

AN INTRODUCTION TO KNOT THEORY

LOUIS H. KAUFFMAN
Dept. Mathematics, Statistics & Computer Science (m/c 249)
University of Illinois at Chicago
851 South Morgan St., Chicago, IL 60607-7045, USA
E-mail: *kauffman@uic.edu*

Abstract. This paper concentrates on the construction of invariants of knots, such as the Jones polynomials and the Vassiliev invariants, and the relationships of these invariants to other mathematics (such as Lie algebras).

1. Introduction

This essay constitutes an introduction to the theory of knots as it has been influenced by developments concurrent with the discovery of the Jones polynomial in 1984 and the subsequent explosion of research that followed this signal event in the mathematics of the twentieth century. The theme of a relationship of knots with algebra and physics begins already with the Jones polynomial (Section 4) and the bracket model for the Jones polynomial (Section 5). Sections 6 and 7 provide an introduction to Vassiliev invariants and the remarkable relationship between Lie algebras and knot theory.

2. Knot Tying and the Reidemeister Moves

The key result that makes it possible to begin a combinatorial theory of knots is the Theorem of Reidemeister [25] that states that *two diagrams represent equivalent loops if and only if one diagram can be obtained from the other by a finite sequence of special deformations called the Reidemeister moves.* The Reidemeister moves are illustrated in Figure 1. The upshot of Reidemeister's theorem is that the topological problems about knots can all be formulated in terms of knot diagrams.

R.L. Ricca (ed.), An Introduction to the Geometry and Topology of Fluid Flows, 77–104.

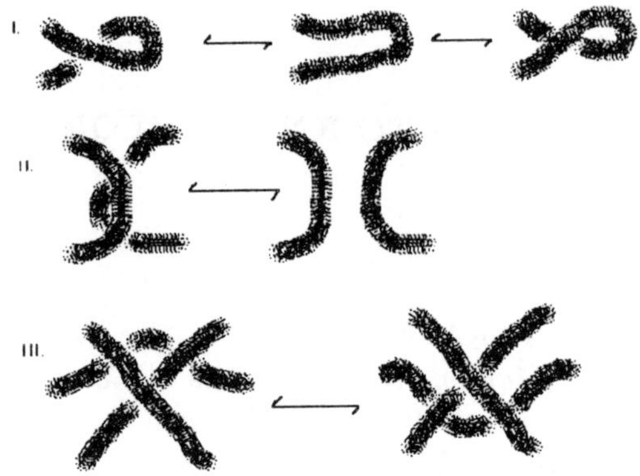

Figure 1. Reidemeister moves.

The moves shown in Figure 1 are intended to indicate changes that are made in a larger diagram. These changes modify the diagram only locally as shown in the Figure. Move zero is as important as the other Reidemeister moves, but since it does not change any essential diagrammatic relationships it is left in the background of the discussion.

3. Invariants of Knots and Links. A First Pass

We want to be able to calculate numbers (or bits of algebra such as polynomials) from given link diagrams in such a way that these numbers do not change when the diagrams are changed by Reidemeister moves. Numbers or polynomials of this kind are called *invariants* of the knot or link represented by the diagram. If we produce such invariants, then we are finding topological information about the knot or link. The easiest example of such an invariant is the *linking number* of two curves, which measures how many times one curve winds around another (for a definition see the article of R. Langevin in this volume). In order to calculate the linking number we *orient* the curves. This means that each curve is equipped with a directional arrow, and we keep track of the direction of the arrow when the curve is deformed by the Reidemeister moves.

It can be shown that the linking number is invariant under the Reidemeister moves. That is, if we take a given diagram D (representing the curves A and B) and change it to a new diagram E by applying one of the Reidemeister moves, then the linking number calculation for D will be the same as the calculation for E. The calculation is unaffected by the first Reidemeister move because self-crossings of a single curve do not figure in the calculation of the linking number. The second Reidemeister move either creates or destroys two crossings of opposite sign, and the third move

Figure 2. The Whitehead link.

rearranges a configuration of crossing without changing their signs.

With these observations we can in fact *prove* that two singly linked rings are indeed linked! There is no possible sequence of Reidemeister moves from these rings to two separated rings because the linking number of separated rings is equal to zero, not to plus or minus one.

One of the most fascinating aspects of the linking number is its limitations as an invariant. Figure 2 shows the Whitehead link, a link of two components with linking number equal to zero. The Whitehead link is indeed linked, but it requires methods more powerful than the linking number to demonstrate this fact.

3.1. THREE-COLORING A KNOT

There is a remarkable proof that the trefoil knot is knotted. This proof goes as follows: color the three arcs of the trefoil diagram with three distinct colors. Lets say these colors are red, blue and purple. Note that in the standard trefoil diagram three distinct colors occur at each crossing. Now adopt the following coloring rule:

Coloring Rule 3.1 *Either three colors or exactly one color occur at any crossing in the colored diagram.*

Call a diagram *colored* if its arcs are colored and they satisfy this rule. Note that the standard unknot diagram is colored by simply assigning one color to its circle. A coloring does not necessarily have three colors on a given diagram. Call a diagram *3-colored* if it is colored and three colors actually appear on the diagram.

Theorem 3.2 *Every diagram that is obtained from the standard trefoil diagram by Reidemeister moves can be 3-colored. Hence the trefoil diagram is inequivalent to the unknot.*

Exercise 3.3 Prove Theorem 3.2 for the trefoil knot, keeping in mind that each time a Reidemeister move is performed, it is possible to extend the

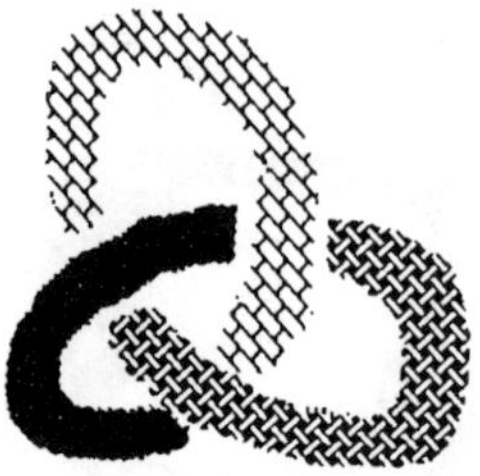

Figure 3. The three-colored trefoil.

coloring from the original diagram to the diagram that is obtained from the move. These extensions of colorings involve only local changes in the colorings of the original diagrams.

Note that in general, although a color may disappear from a region of the knot diagram during the process, this color must appear elsewhere in the diagram or else it is not possible for the two arcs in the move to have different colors (since there is a path along the knot from one local arc to the other). Thus 3-coloration is preserved under Reidemeister moves, whether they make the diagram simpler or more complicated. As a result, every diagram for the trefoil knot can be colored with three colors according to our rules. This proves that the trefoil is knotted, since an unknotted trefoil would have a simple circle among its diagrams, and the simple circle can be colored with only one color.

3.2. THE QUANDLE AND THE DETERMINANT OF A KNOT

There is a wide generalization of this coloring argument. We shall replace the colors by arbitrary labels for the arcs in the diagram and replace the coloring rule by a method for combining these labels. It turns out that a good way to articulate such a rule of combination is to make the label on one of the undercrossing arcs at a crossing a product (in the sense of this new mode of combination) of the labels of the other two arcs. In fact, we shall assume that this product operation depends upon the orientation of the arcs as shown in Figure 4.

In Figure 4 we show how a label a on an undercrossing arc combines with a label b on an overcrossing arc to form $c = a * b$ or $c = a\#b$ depending upon whether the overcrossing arc is oriented to the left or to the right for an observer facing the overcrossing line and standing on the arc labelled a. This operation depends upon the orientation of the line labelled b so that $a * b$ corresponds to b pointing to the right for an observer approaching the crossing along a, and $a\#b$ corresponds to b pointing to the left for the same observer.

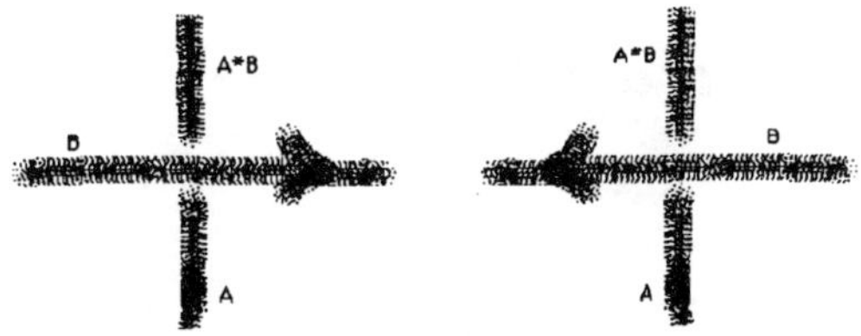

Figure 4. The quandle operation.

The binary operations $*$ and $\#$ are not necessarily associative. For example, our original color assignments of R (red), B (blue) and P (purple) for the trefoil knot correspond to products $R * R = R$, $B * B = B$, $P * P = P$, $R * B = P$, $B * P = R$, $P * R = B$. Then $R * (B * P) = R * R = R$ while $(R * B) * P = P * P = P$.

We shall insist that these operations satisfy a number of identities so that the labeling is compatible with the Reidemeister moves. An algebraic system satisfying these rules is called a *quandle* [14] (see Figure 5; see also [6, 28]).

1. $a * a = a$ and $a\#a = a$, for any label a.
2. $(a * b)\#b = a$ and $(a\#b) * b = a$, for any labels a and b.
3. $(a * b) * c = (a * c) * (b * c)$ and $(a\#b)\#c = (a\#c)\#(b\#c)$, for any labels a, b, c.

These rules correspond, respectively to the Reidemeister moves 1, 2 and 3. Labelings that obey these rules can be handled just like the 3-coloring that we have already studied. In particular a given labeling of a knot diagram means that it is possible to label (satisfying the rules given above for the labels) any diagram that is related to it by a sequence of Reidemeister moves. However, not all the labels will necessarily appear on every related diagram, and for a given coloring scheme and a given knot, certain special restrictions can arise.

To illustrate this, consider the color rule for numbers: $a * b = a\#b = 2b - a$. This satisfies the axioms. Figure 6 shows how, on the trefoil, such a coloring must obey the equations $a * b = c$, $c * a = b$, $b * c = a$. Hence $2b - a = c$, $2a - c = b$, $2c - b = a$. For example, if $a = 0$ and $b = 1$, then $c = 2b - a = 2$ and $a = 2c - b = 4 - 1 = 3$. We need $3 = 0$. Hence this system of equations will be satisfied for appropriate labelings in $Z/3Z$, the integers modulo three. In general we can consider color values in Z/nZ where n is any positive integer modulus.

The modular number system $Z/3Z$ reproduces exactly the 3-coloring of the trefoil, and we see that the number 3 emerges as a characteristic of the equations associated with the knot. In fact, 3 *is the value of a determinant*

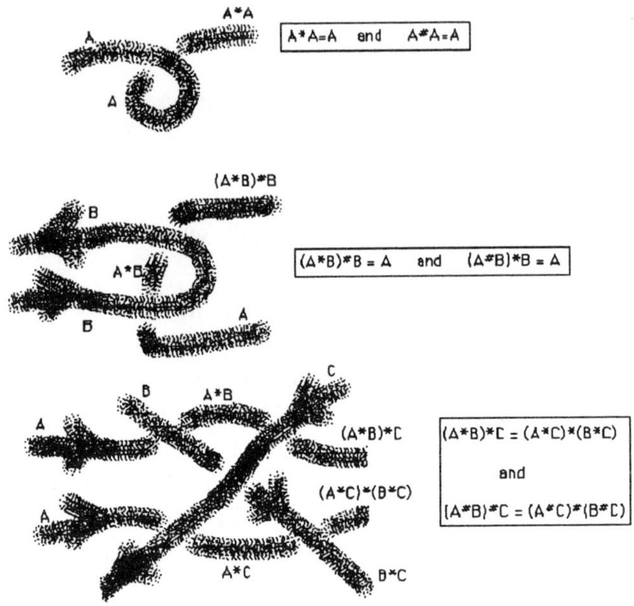

Figure 5. Quandle identities.

that is associated with these equations, and its absolute value is an invariant of the knot (for more about this construction, see [20], Part 1, Chapter 13).

In [21] we define the *coloring number* of a knot or link K to be the least number of colors (greater than 1) needed to color it in the $2b - a$ fashion for any diagram of K. It is a nice exercise to verify that the coloring number of the figure eight knot is indeed 4. In general the coloring number of a knot or link is not easy to determine. This is an example of a topological invariant that has subtle combinatorial properties. Other knots and links that we have mentioned in this section can be shown to be knotted and linked by the modular method. The reader should try it for the Borommean rings

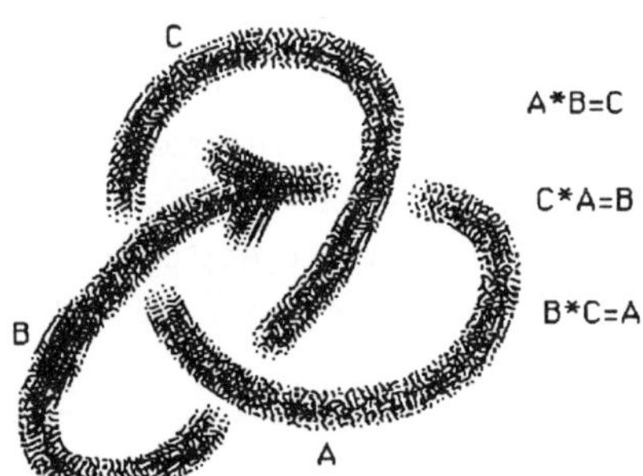

Figure 6. Equations for the trefoil knot.

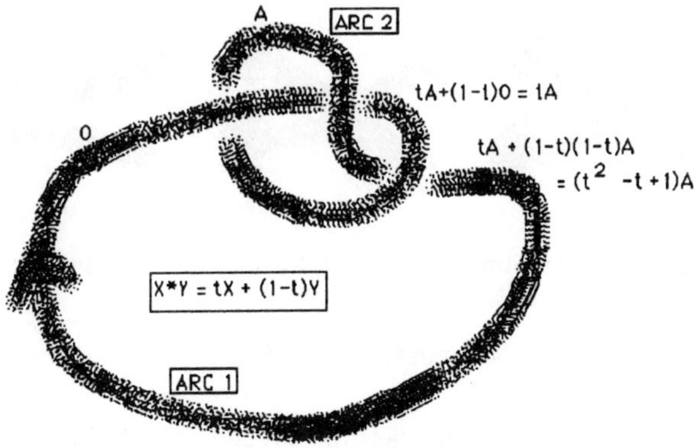

Figure 7. Alexander polynomial of the trefoil knot.

and the Whitehead link.

3.3. THE ALEXANDER POLYNOMIAL

The modular labeling method has a marvelous generalization to the *Alexander polynomial* [2] of the knot. This comes about through generalized coloring rules $a * b = ta + (1 - t)b$ and $a\#b = t^{-1}a + (1 - t^{-1})b$, where t is an indeterminate. It is a nice exercise to verify that these rules satisfy the axioms for the quandle. This algebraic structure is called the *Alexander Module*.

The case $t = -1$ gives the rule $2b - a$ that we have already considered. By coloring diagrams with arbitrary t, we obtain a polynomial that generalizes the modulus. This polynomial is the Alexander polynomial. Alexander [2] described it differently in his original paper, and there is a remarkable history to the development of this invariant (for more information see, for example, [5, 7, 16, 18]). The circularity inherent in the knot diagram results in relations that must be satisfied by the module action. In Figure 7 we see directly by labeling the diagram that if arc 1 is labeled 0 and arc 2 is labeled a, then $(t + (1 - t)^2)a = 0$. In fact, $t + (1 - t)^2 = t^2 - t + 1$ is the Alexander polynomial of the trefoil knot. The Alexander polynomial is an algebraic modulus for the knot.

4. The Jones Polynomial

Our next topic describes an invariant of knots and links of quite a different character than the modulus or the Alexander polynomial of the knot. It is a "polynomial" invariant of knots and links discovered by Vaughan Jones in 1984 [9]–[12]. Jones's invariant, usually denoted $V_K(t)$, is a polynomial

Figure 8. Denoting different states for the Jones polynomial.

in the variable $t^{1/2}$ and its inverse $t^{-1/2}$. One says that $V_K(t)$ is a Laurent polynomial in $t^{1/2}$.

The Alexander polynomial was discovered in the 1920's and until 1984 no one had found another polynomial invariant of knots and links that was not a simple generalization of the Alexander polynomial. Vaughan Jones discovered a new polynomial invariant of knots and links that had some very remarkable properties. The Alexander polynomial cannot detect the difference between any knot and its mirror image. What made the Jones polynomial such an exciting discovery for knot theorists was the fact that it could detect the difference between many knots and their mirror images. Later other properties began to emerge. It became a key tool in proving properties of alternating links (and generalizations) that had been conjectured since the last century [17, 23, 22].

Here are a set of axioms for the Jones polynomial. The polynomial was not discovered in the form of these axioms. The axioms are in a format analogous to the framework that John H. Conway [5, 15, 16], discovered for the Alexander polynomial. I start with these axioms because they give a quick access to the polynomial and to sample computations.

Axioms 4.1 (for the Jones Polynomial)

1. *If two oriented links K and K' are ambient isotopic, then $V_K(t) = V_{K'}(t)$.*
2. *If U is an unknotted loop, then $V_U(t) = 1$.*
3. *If K_+, K_-, and K_0 are three links with diagrams that differ only as shown in the neighborhood of a single crossing site for $K+$ and $K-$ (see Figure 8), then $t^{-1}V_{K_+}(t) - tV_{K_-}(t) = (t^{1/2} - t^{-1/2})V_{K_0}(t)$.*

The axioms for $V_K(t)$ are a consequence of Joness original definition of his invariant. He was led to this invariant by a trail that began with the study of von Neumann algebras [9] (a branch of algebra directly related to quantum theory and to statistical mechanics) and ended in braids, knots and links. The Jones polynomial has a distinctly different flavor from the Conway-Alexander polynomial even though it can be axiomatised in a very similar way. In fact, this similarity of axiomatics points to a common generalization (the Homfly(Pt) polynomial) [8, 24] and to another generalization

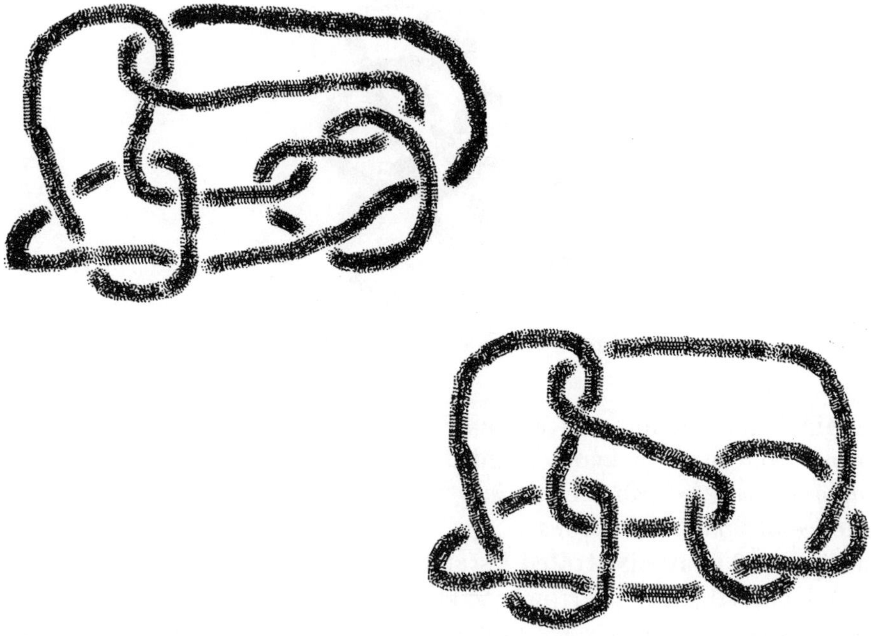

Figure 9. The Conway knot (left) and the Kinoshita-Terasaka knot (right) are inequivalent pairs of knots that have the same Jones polynomial.

(the Kauffman polynomial) [19, 20], and then to further generalizations in the connection with statistical mechanics [13, 1].

To this date no one has found a knotted loop that the Jones polynomial does not declare to be knotted. Thus one can make the

Conjecture 4.2 *If a single component loop K is knotted, then $V_K(t)$ is not equal to one.*

While it is possible that the Jones polynomial is able to detect the property of being knotted, it is not a complete classifier for knots. There are inequivalent pairs of knots that have the same Jones polynomial. Such a pair is shown in Figure 9. These two knots, the Kinoshita-Terasaka knot and the Conway knot, both have the same Jones polynomial but are different topologically. Incidentally these two knots are examples whose knottedness cannot be detected by the Alexander polynomial.

Lets get to work and use the axioms to compute the Jones polynomial for the trefoil knot. To this end, there is a useful device called the *skein tree.* A skein tree is obtained from a given knot or link diagram by recording the knots and links obtained from this diagram by smoothing or switching crossings. Each node of the tree is a knot or link. The nodes farthest from

Figure 10. A standard unknot.

the original knot or link are unknotted or unlinked. Such a tree can be produced from a given knot or link by using the fact that *any knot or link diagram can be transformed into an unknotted (unlinked) diagram by a sequence of crossing switches.*

In Figure 10 I have illustrated a standard unknot diagram. This diagram is drawn by starting at the arrowhead in the Figure and tracing the diagram in such a way that one always draws an over crossing before drawing an undercrossing. This is the easiest possible knot diagram to draw since one never has to make any corrections - just pass under when you want to cross an an already created line in the diagram. Standard unknot diagrams are always unknotted.

Using the fact that standard unknot diagrams are available, we can use the difference between a given diagram K and a standard unknot with the same plane projection to give a procedure for switching crossings to unknot the diagram K (see Figure 11). This switching procedure can be used to produce a skein tree for calculating the Jones polynomial of K.

As is illustrated in Figure 12 for the trefoil knot, the tree reduces the

Figure 11. An unlinked pair of circles with no twists (U0) obtained by smoothing the crossing of unknots with single positive (U+) and negative (U-) twists in them.

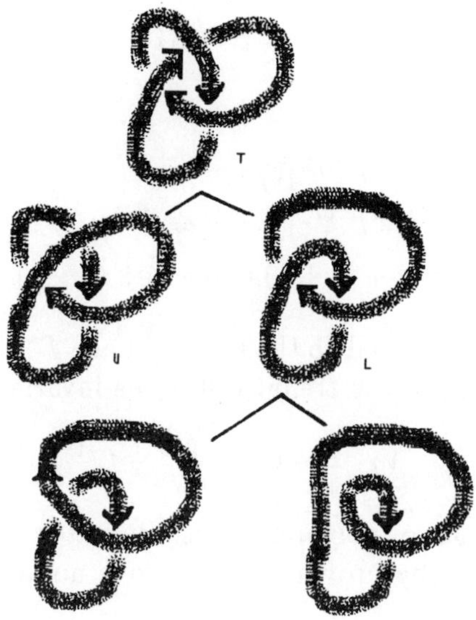

Figure 12. A skein tree for the computation of the Jones polynomial of the trefoil knot.

calculation of the Jones polynomial of the trefoil diagram to the calculation of certain unknots and unlinks. In order to see how to calculate an unlink it is useful to observe the behaviour of the axioms in this case:

$$t^{-1}V_{U_+} - tV_{U_-} = (t^{1/2} - t^{-1/2})V_{U_0} ,$$

where U_+ and U_- denote unknots with single positive and negative twists in them and U_0 is an unlinked pair of circles with no twists, as shown in Figure 11.

Therefore

$$(t^{-1} - t)1 = (t^{1/2} - t^{-1/2})V_{U_0} .$$

Hence

$$d = V_{U_0} = (t^{-1} - t)/(t^{1/2} - t^{-1/2}) = -(t^{1/2} + t^{-1/2}) .$$

Thus we see that an extra unknotted component of the link multiplies the invariant by $d = -(t^{1/2} + t^{-1/2})$. Here T denotes the trefoil knot, U denotes the unknot and L denotes the link of two unknotted circles as shown in Figure 12. With this fact in place, we find that

$$t^{-1}V_T - tV_U = (t^{1/2} - t^{-1/2})V_L ,$$

and

$$t^{-1}V_L - td = (t^{1/2} - t^{-1/2})V_U .$$

Thus

$$V_L = t(td + (t^{1/2} - t^{-1/2})) = -t^{5/2} - t^{1/2} .$$

Hence

$$\begin{aligned} V_T &= t(t + (t^{1/2} - t^{-1/2})V_L) \\ &= t(t + (t^{1/2} - t^{-1/2})(-t^{5/2} - t^{1/2})) \\ &= t(t - t^3 - t + t^2 + 1) = t(-t^3 + t^2 + 1) = -t^4 + t^3 + t . \end{aligned}$$

The same calculation applied to the mirror image T^* (obtained by reversing all the crossings of T) of the trefoil yields the invariant

$$V_{T^*} = -t^{-4} + t^{-3} + t^{-1} .$$

This shows how the Jones polynomial discriminates between the trefoil and its mirror image, thereby proving that there is no ambient isotopy from T to T^*.

This method of calculating the Jones polynomial from its axioms does not tell us why the invariant works. It is possible to analyze this method of calculation and show that it does not depend upon the choices that one makes in the process and that it gives topological information about the knot or link in question. There is a different way to proceed that leads to a very nice formula for the Jones polynomial as a sum over "states" of the diagram. In this formulation, the polynomial is well defined from the beginning, and we can see the topological invariance arise in the course of adjusting certain parameters of a well-defined function.

5. The Bracket State Sum

In the last section we gave axioms for the Jones polynomial and showed how to compute it by skein calculations from these axioms. In this section we shall show one way to prove that the Jones polynomial is well-defined by these axioms, and that it is an invariant of ambient isotopy of links in three dimensional space. In order to accomplish this aim, we shall give a different definition of the polynomial as a certain summation over combinatorial configurations associated with the given link diagram. This summation will be called a state *summation model* for the Jones polynomial.

In fact, we shall first construct a state summation called the *bracket polynomial* [17], and then explain how to modify the bracket polynomial to obtain the Jones polynomial. The bracket polynomial has a rather natural development, and is defined for unoriented link diagrams. We work with diagrams for unoriented knots and links. To each crossing in the diagram assign two *local states* with labels A and B (see Figure 13: in the A-state

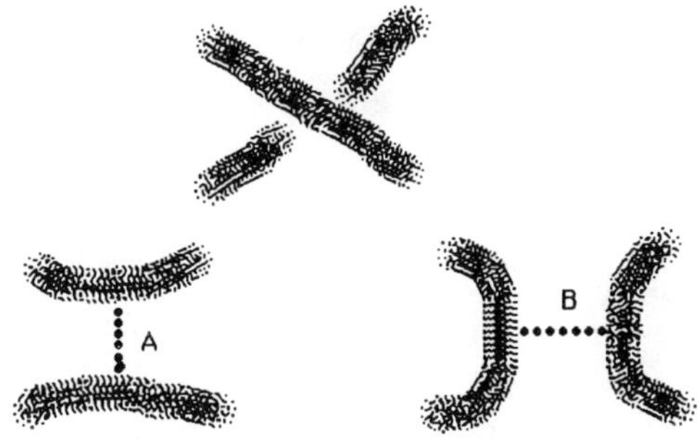

Figure 13. Labeling local states for the bracket state sum.

the regions swept out by a counterclockwise turn of the overcrossing line are joined; in the B-state the regions swept out by a clockwise turn of the overcrossing line are joined).

A state S of a diagram K consists in a choice of local state for each crossing of K. Thus a diagram with N crossings will have 2^N states. Two states S and S' of the trefoil diagram are indicated in Figure 14. States are evaluated in two ways. These ways are denoted by $< K|S >$ and by $||S||$. The *norm* of the state S, $||S||$, is defined to be one less than the number of closed curves in the plane described by S. In the example in Figure 14, we have $||S|| = 1$ and $||S'|| = 0$. The evaluation $< K|S >$ is defined to be the product of all the state labels (A and B) in the state. Thus in Figure 14 we have

$$< K|S >= A^3 \quad \text{and} \quad < K|S' >= A^2 B \ .$$

Taking variables A, B and d, we define the state summation associated to a given diagram K by the formula

$$< K >= \sum_S < K|S > d^{||S||} \ .$$

In other words, for each state we take the product of the labels for that state multiplied by d raised to the number of loops in the state. $< K >$ is

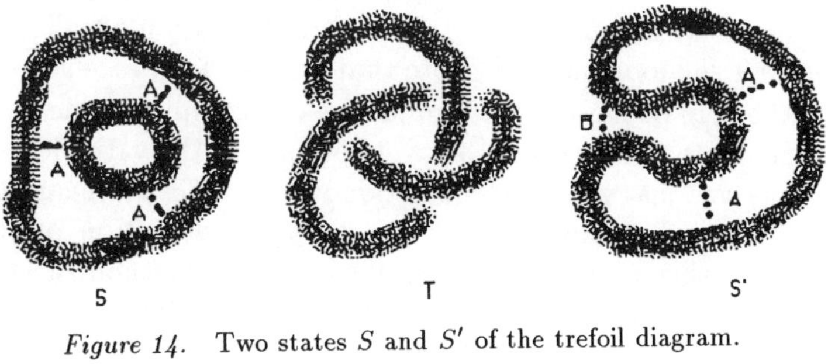

Figure 14. Two states S and S' of the trefoil diagram.

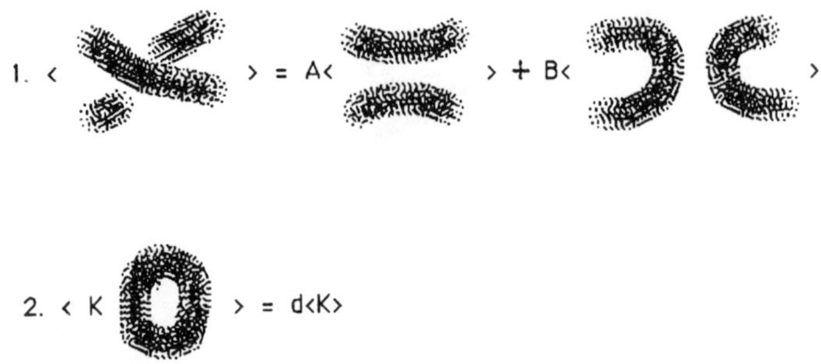

Figure 15. Bracket equations: here the small diagrams indicate parts of larger diagrams that are otherwise identical.

the summation of this state evaluation over all the states in the diagram for K. (The notation Σ_S means "sum over all instances of S".) We will show that the state summation $< K >$ is invariant under the second and third Reidemeister moves if we take $B = 1/A$ and $d = -(A^2 + A^{-2})$. A normalization then enables us to obtain invariance under all three Reidemeister moves, and hence topological information about knots and links (see [17] for more information about the bracket and its relationship with the Jones polynomial). There is a great deal of topological information in the calculations that ensue from the bracket polynomial. In particular, one can distinguish many knots from their mirror images, and it is possible that the bracket calculation can detect whether a given diagram is actually knotted.

The first constructions related to the bracket polynomial are quite elementary. There are two basic formulas that are reminiscent of the exchange relations we have already seen for the Jones polynomial. These formulas are as shown in Figure 15. Formula 1 just says that the state summation breaks up into two sums with respect to a given crossing in the diagram. In one sum, we have made a smooothing of type A at the crossing, while in the other sum we have made a smoothing of type B. The factors of A and B indicated in the formula are the contributions to the product of vertex weights from this crossing. All the rest of the two partial sums can be interpreted as bracket evaluations of the smoothed diagrams. Formula 2 just states that an extra simple closed curve in a diagram multiplies its bracket evaluation by the loop value d. Note that a single loop receives the value 1.

With the help of these two formulas, we can compute some basic bracket evaluations. Note that we have not yet specialized the variables A,B and d. We shall analyze just what specialization will produce an invariant of knots and links. The advantage to having set up the definition of the bracket polynomial in this way is exactly that we have a method of labeling link diagrams with algebra, and it is possible to then adjust the evaluation so

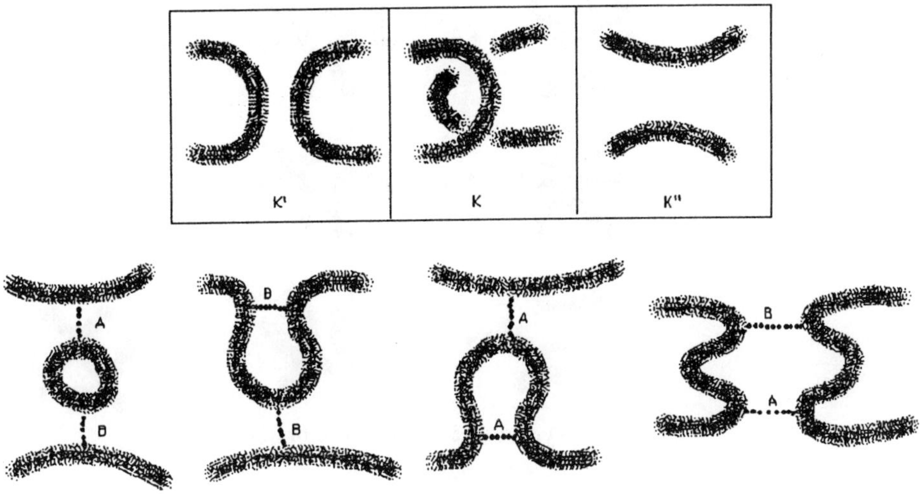

Figure 16. Type 2 changes.

that it is invariant under Reidemeister moves. To this end, the next Lemma tells us how the general bracket behaves under a Reidemeister move of type 2. Essential diagrams for this Lemma are in Figure 16.

Lemma 5.1 *Let K be a given link diagram, and let K' denote a diagram that is obtained from K by performing a type 2 Reidemeister move in the simplifying direction (eliminating two crossings from K). Let K'' be the diagram obtained from K by replacing the site of the type 2 move by two arcs in the opposite pattern to the form of the simplified site in K'. (The diagrams in Figure 16 illustrate this construction.) Then*

$$< K > = AB < K' > + (ABd + A^2 + B^2) < K'' > \quad .$$

Proof. Consider the four local state configurations that are obtained from the diagram K on the left hand side of the equation, as illustrated in Figure 16. The formula follows from the fact that one of these states has coefficient AB and the other three have the same underlying diagram and respective coefficients ABd (after converting the loop to a value d), A^2 and B^2. This completes the proof of the Lemma.

QED

With the help of this Lemma it is now obvious that if we choose $B = 1/A$ and $d = A^2 + B^2$, then $< K >$ is invariant under the second Reidemeister move. Once this choice is made, the resulting specialized bracket is invariant under the third Reidemeister move, as illustrated in Figure 17.

Finally, we can investigate bracket behaviour under the first Reidemeister move.

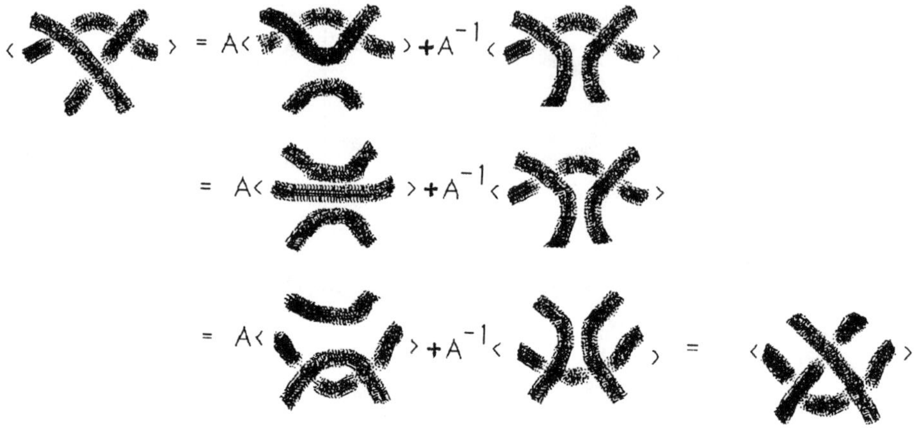

Figure 17. Type 3 invariance of the bracket.

Lemma 5.2 *Let $< K >$ denote the bracket state sum with $B = A^{-1}$ and $d = -(A^2 + A^{-2}$.) Then $< K >$ is invariant under the Reidemeister moves 2 and 3 and on move 1 behaves as shown below.*

1. $< K(+) >= (-A^3) < K >$,

2. $< K(-) >= (-A^{-3}) < K >$.

Here $K(+)$ denotes a diagram with a simplifying move of type 1 available where the crossing that is to be removed has type $+1$. K is the diagram obtained from $K(+)$ by doing the type 1 move. Similarly, $K(-)$ denotes a diagram with a simplifying move of type 1 available where the crossing that is to be removed has type -1. *Proof.* See Figure 18 for the behaviour under type 1 move. We have already verified the other statements in this Lemma.

QED

5.1. FRAMING PHILOSOPHY: WRITHE

Is it unfortunate that the bracket is not invariant under the first Reidemeister move? No, it is fortunate! First of all, the matter is easy to fix by a little adjustment: let K be an oriented knot or link, and define the *writhe* of K, denoted $w(K)$, to be the sum of the signs of all the crossings in K. Thus the writhe of the right-handed trefoil knot is three. The writhe has the following behaviour under Reidemeister moves:

1. $w(K)$ is invariant under the second and third Reidemeister moves.
2. $w(K)$ changes by plus or minus one under the first Reidemeister move:

$$w(K(+)) = w(K) + 1\,, \quad \text{and} \quad w(K(-)) = w(K) - 1\,.$$

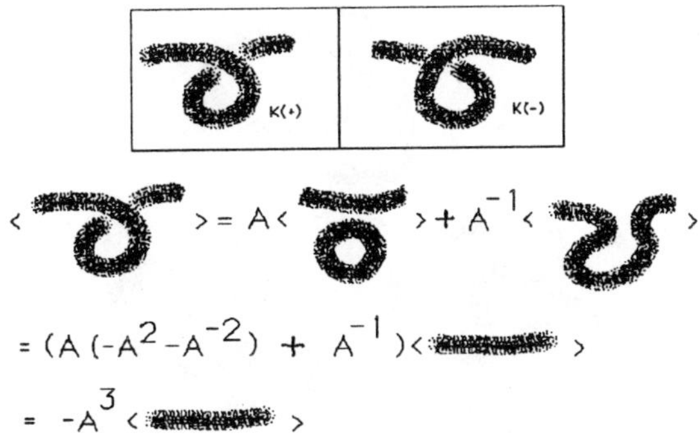

Figure 18. Bracket under type 1 move.

Here we use the notation of the previous Lemma as shown in Figure 18. Thus the writhe behaves in a parallel way to the bracket on the type 1 move, and we can combine writhe and bracket to make a new calculation that is invariant under all three Reidemeister moves. We call the fully invariant calculation the *f-polynomial* and define it by the equation

$$f_K(A) = (-A^3)^{-w(K)} < K > (A) \ .$$

Up to this normalization, the bracket gives a model for the original Jones polynomial. The precise relationship is that $V_K(t) = f_K(t^{-1/4})$ where $w(K)$ is the sum of the crossing signs of the oriented link K, and $< K >$ is the bracket polynomial obtained by ignoring the orientation of K.

5.2. CALCULATING THE BRACKET

In Figure 19 we show a tree for calculating the bracket polynomial of the trefoil knot T. It follows at once from the behaviour of the bracket on curls that the contributions of the three (farthest from the trefoil itself) branches of this tree add to give the bracket polynomial of the trefoil:

$$< T >= A^2(-A^3) + AA^{-1}(-A^{-3}) + A^{-1}(-A^{-3})^2 = -A^5 - A^{-3} + A^{-7} \ .$$

Hence

$$f_T = (-A^3)^{-w(T)} < T >= A^{-4} + A^{-12} - A^{-16} \ .$$

Note that we managed with only three branches in the tree for this calculation rather than the full expansion into the eight states. A savings like this is always possible, because we know how the bracket behaves on curls. The resulting expansion gives a sum of monomials and is useful for thinking about the properties of the invariant.

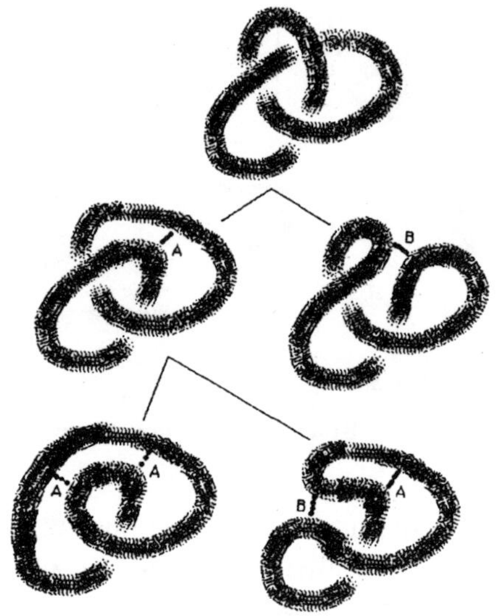

Figure 19. Tree for the bracket of the trefoil.

5.3. MIRROR MIRROR

The knot K^* obtained by reversing all the crossings of K is called the mirror image of K. K^* is the mirror image of the knot that would ensue if the plane on which the knot is drawn were a mirror. It is easy to see that $< K^* > (A) = < K > (A^{-1})$ and that $f_{K^*}(A) = f_K(A^{-1})$. Thus, if K is ambient isotopic to K^* (all three Reidemeister moves allowed), then

$$f_K(A) = f_{K^*}(A) = f_K(A^{-1}) .$$

Returning to the evaluation of the f-invariant for the trefoil, note that $f_T(A^{-1})$ is not equal to $f_T(A)$. Therefore, the trefoil knot T and its mirror image T^* are topologically distinct. The proof that we have given is the simplest proof known to this author. Note that we have given a complete proof of this fact, starting with the Reidemeister moves, constructing and applying the bracket invariant.

Now lets verify that the bracket does indeed give a model for the Jones polynomial. To see this, consider $f_K(A) = (-A^3)^{-w(K)} < K > (A)$. Since the writhe, $w(K)$, is obtained by summing signs over all the crossings of K, we can interpret the factor $(-A^3)^{-w(K)}$ as the product of contributions of $(-A^3)$ or $(-A^3)^{-1}$, one from each crossing, and depending upon the sign of the crossing. Thus we can write an oriented state expansion formula for f_K as shown below where K_+ and K_- denote links with corresponding sites with oriented crossings, K_0 is the result of smoothing the crossing in an oriented fashion and $K_\&$ is the result of smoothing the crossing against the orientation.

$$f_{K_+} = (-A^3)^{-1} A f_{K_0} + (-A^3)^{-1} A^{-1} f_{K_\&} .$$

Hence,

$$f_{K_+} = -A^{-2} f_{K_0} - A^{-4} f_{K_\&} ,$$

and similarly, for a negative crossing

$$f_{K_-} = -A^2 f_{K_0} - A^4 f_{K_\&} .$$

Letting $V_K(t) = f_K(t^{-1/4})$, we have

$$V_{K_+} = -t^{1/2} V_{K_0} - t V_{K_\&} ,$$

and

$$V_{K_-} = -t^{-1/2} V_{K_0} - t^{-1} V_{K_\&} .$$

Therefore,

$$t^{-1} V_{K_+} - t V_{K_-} = (t^{1/2} - t^{-1/2}) V_{K_0} .$$

Exercise 5.3 Verify that $V_K(t)$ is the Jones polynomial. (Hint: check that it has the right behaviour on unknotted loops.)

6. Vassiliev Invariants

We have seen how it is fundamental to take the difference of an invariant at a positive crossing and at a negative crossing, leaving the rest of the diagram alone. The earliest instance of this is the Conway polynomial [5], $\nabla_K(z)$ with its exchange identity

$$\nabla_{K_+} - \nabla_{K_-} = z \nabla_{K_0} .$$

Vassiliev [27] gave new meaning to this sort of identity by thinking of the structure of the entire space of all mappings of a circle into three-dimensional space. This space of mappings includes mappings with singularities where two points on a curve touch. He interpreted the equation

$$Z_{K_+} - Z_{K_-} = Z_{K_\#} ,$$

as describing the difference of values across a singular embedding $K_\#$ where $K_\#$ has a transverse singularity in the knot space as illustrated in Figure 20. (In a transverse singularity the curve touches itself along two different directions.)

Figure 20. State notation for the difference equation.

The Vassiliev formula serves to define the value of the invariant on a singular embedding in terms of the the values on two knots on either side on this embedding. This Vassiliev formula serves to describe a method of extending a given invariant of knots to a corresponding invariant of embedded graphs with controlled singularities of this transverse type. The key concept is the idea of an invariant of finite type.

Definition 6.1 *We shall say that Z_G is an* **invariant of finite type** *i if Z_G vanishes for all graphs with greater than i nodes.*

This concept was extracted from Vassiliev's work by Birman and Lin [4]. A (*rigid vertex*) invariant of knotted graphs is a *Vassiliev invariant of finite type i* if it satisfies the identity $Z_{K_+} - Z_{K_-} = Z_{K_\#}$ and it is of finite type i. In *rigid vertex isotopy* the cyclic order at the vertex is preserved, so that the vertex behaves like a rigid disk with flexible strings attached to it at specific points.

Vassiliev invariants form an extraordinary class of knot invariants. It is an open problem whether the Vassiliev invariants are sufficient to distinguish knots that are topologically distinct. We begin an analysis of the combinatorial conditions on graph evaluations that could support such invariants. The key observation is the

Lemma 6.2 *If Z_G is a Vassiliev invariant of finite type i, then Z_G is independent of the embedding of the graph G when G has i vertices.*

Proof. Suppose that G is an embedded graph with i nodes. If we switch a crossing in G to form G' then the exchange relation for the Vassiliev invariant says that $Z_G - Z_{G'} = Z_{G''}$ where G'' has one more node than G or G'. Then G'' has $i + 1$ nodes and hence $Z_{G''} = 0$. Therefore $Z_G = Z_{G'}$. This shows that we can switch crossings in any embedding of G without changing the value of Z_G. It follows from this that Z_G is independent of the embedding and depends only on the graph G. This completes the proof of the Lemma.

QED

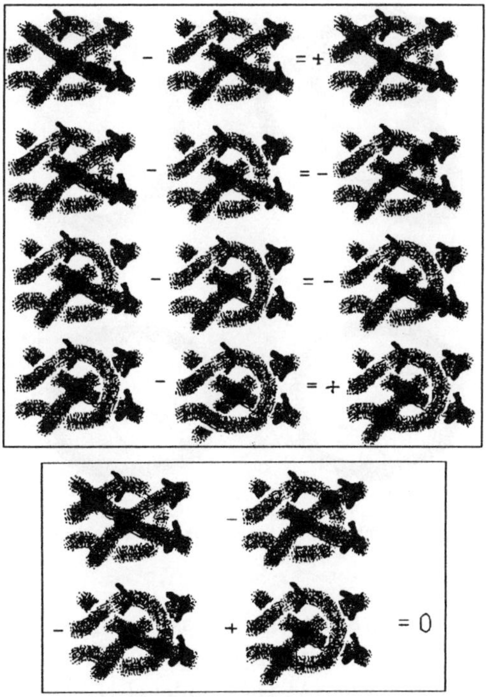

Figure 21. Embedded four term relation.

For a Vassiliev invariant of type i, there is important information in the values it takes on graphs with exactly i nodes. These evaluations do not depend upon the embedding type of the graph. However, not just any such graphical evaluation will extend to give a topological invariant of knots and graphs. There are necessary conditions. Vassiliev found a version of these conditions through his analysis of the knot space and Ted Stanford [26], a student of Joan Birman, discovered the beautiful topological meaning of these conditions in relation to the switching identity. Stanford's argument goes as follows: consider a singular crossing that has an arc from the diagram passing underneath it, as shown in Figure 21. Four crossing switches will take that arc above the singular crossing and return the diagram to a position that is topologically equivalent to its original position. Each crossing switch gives an equation. There are four equations. Add them up and you get an identity among the values of the invariant on four diagrams. Call this the *four-term relation.* This identity is illustrated in the second box in Figure 21.

Now recall from the Lemma we proved above that for a Vassiliev invariant of type i, the graphs with i nodes have values that are independent of their embeddings in three dimensional space. This means that at the

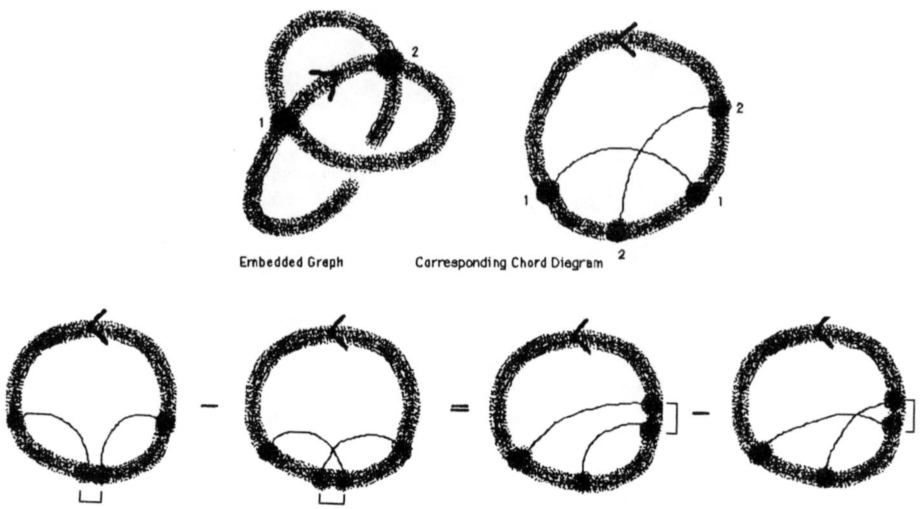

Figure 22. Abstract four term relation via chord diagrams.

top level — the i noded graphs for a Vassiliev invariant of type i will be called the *top level* — the four term relations will be relations among the evaluations of abstract graphs. At the top level the four term relations will be purely combinatorial conditions related to the topology.

How shall we think of abstract 4-valent graphs corresponding to singular embeddings of a knot? An abstract knot is just a circle. An abstract singular knot is a circle with pairs of points marked that become the singular points in the embedding. Indicate these paired points by arcs between them. Call the resulting structure a *chord diagram* (see the example at the beginning of Figure 22). In the language of the chord diagrams the 4-term relation at the top level becomes the equation shown in Figure 22. This can be seen by translating the relation in Figure 21 into the language of chord diagrams. In Figure 22 we have indicated parts of the chord diagram that are neighbors by showing an outer bracket connecting them. Those sites that are neighbors can have no other chords between them. Otherwise there can be many chords in these diagrams that are not indicated, just so long as the diagrams in the equation for the four-term relation differ only as shown in the Figure.

If you can write down a top level evaluation of chord diagrams that satisfies the 4-term relation, then you have the raw data for a Vassiliev invariant. Such an evaluation of chord diagrams is called a *weight system* for a Vassiliev invariant. By the Theorems of Kontsevich and Bar-Natan [3], this raw data guarantees the existence of at least one invariant that satisfies the top level evaluation.

The world is rife with Vassiliev invariants. Birman & Lin [4] showed directly that the Jones polynomial and its generalizations give rise to Vassiliev invariants. In the case of the Jones polynomial here is an easy proof of their result:

Theorem 6.3 *Let $V_G(t)$ denote the Jones polynomial extended to rigid vertex 4-valent graphs by the formula $V_{K_+} - V_{K_-} = V_{K_\#}$. Let $v_i(G)$ denote the coefficient of x^i in the expansion of $V_G(e^x)$. Then $v_i(G)$ is a Vassiliev invariant of type i.*

Proof. Use the identities from the end of Section 5.

1. $V_{K_+} = -t^{1/2}V_{K_0} - tV_{K_\&}$.
2. $V_{K_-} = -t^{-1/2}V_{K_0} - t^{-1}V_{K_\&}$.

Substitute $t = e^x$. It follows at once that $V_{K_\#} = V_{K_+} - V_{K_-}$ is divisible by x. Hence V_G is divisible by x^i when G has i nodes. This implies that the coefficients $v_i(G) = 0$ if G has more than i nodes. Hence the coefficient invariants $v_i(G)$ are of finite type, proving the Theorem.

QED

With the help of theorems of this type it is possible to study Vassiliev invariants by studying the structure of known invariants of knots and links. In particular it is possible to justify the structure of many weight systems in terms of known invariants.

7. Vassiliev Invariants and Lie Algebras

The subject of Lie algebras is an algebraic study with a remarkable connection with the topology of knots and links.

7.1. DEFINITION OF A LIE ALGEBRA

Recall that an $n \times n$ matrix A is an array of numbers A_{ij} (real or complex) $A = (A_{ij})$, where i and j range in value from 1 to n. One defines the product of two matrices by the formula $(AB)_{ij} = \Sigma_k A_{ik}B_{kj}$ where k runs from 1 to n in this summation. For our purposes it is essential to have a diagrammatic representation for matrix multiplication. This representation is illustrated in Figure 23. Each matrix is represented by a labeled box with one arrow that enters the box and one arrow that leaves the box. The entering arrow corresponds to the left index i in A_{ij}, while the right arrow corresponds to the right index j. In multiplying two matrices A and B together to form AB *we tie the outgoing arrow of A to the ingoing arrow of B*. By convention, an arrow that has no free ends connotes the summation over all possible choices of index for that arrow.

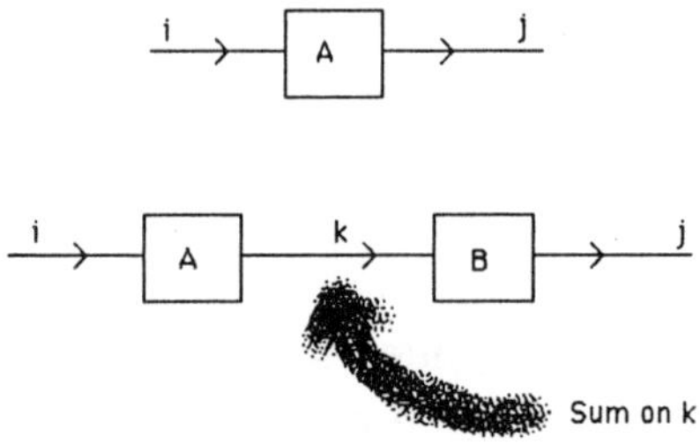

Figure 23. Diagrammatic representation for matrix multiplication.

Exercise 7.1 Show by a diagrammatic proof that $\mathrm{tr}(AB) = \mathrm{tr}(BA)$, where $\mathrm{tr}(A)$ denotes the trace of A, i.e. the sum of the diagonal entries A_{ii}, where i ranges from 1 to n.

Many facts about matrices become quite transparent in this notation. We can ask when a matrix A can be written in the form

$$A = e^B = E + (1/1!)B + (1/2!)B^2 + (1/3!)B^3 + \dots ,$$

for some other matrix B. Since $e^B = \lim(E + B/m)^m$ where the limit is taken as m approaches infinity, we can regard $(E + B/m)$, for m large, as an "infinitesimal" version of the matrix A, and one refers to B as an *infinitesimal generator* for A. It is interesting and mathematically significant to compare the algebraic properties of A and B. The key property for this comparison is the determinant equation

$$\mathrm{Det}(e^B) = e^{\mathrm{tr}(B)} .$$

(One way to prove this identity is to use the Jordan canonical form for the matrix and the fact that similar matrices have the same trace and determinant.) For example, if $\mathrm{Det}(e^B) = 1$ then we need that $tr(B) = 0$. This means that the matrices of determinant one are the exponentials of matrices with trace equal to zero.

Let $sl(n)$ denote the set of $n \times n$ matrices with trace equal to zero. The set $sl(n)$ is not closed under matrix multiplication, but it is closed under the *Lie bracket* (or *commutator*) operation

$$[B, C] = BC - CB .$$

If $\mathrm{tr}(B) = \mathrm{tr}(C) = 0$, then

$$\mathrm{tr}[B, C] = \mathrm{tr}(BC - CB) = \mathrm{tr}(BC) - \mathrm{tr}(CB) = \mathrm{tr}(BC) - tr(BC) = 0 ,$$

since $\mathrm{tr}(BC) = \mathrm{tr}(CB)$ for any matrices B and C.

Thus, if B and C belong to $sl(n)$, then $[B,C]$ also belongs to $sl(n)$. This closure under the bracket operation leads directly to the notion of a Lie algebra:

Definition 7.2 *A* **Lie algebra** *is a vector space L over a field F that is closed under a binary operation, called the* **Lie bracket** *and denoted by $[B,C]$ for B and C in L. The bracket is assumed to satisfy the following axioms:*

1. $[X,Y] = -[Y,X]$, *for all X and Y in L.*
2. $[aX + bY, Z] = a[X,Z] + b[Y,Z]$, *for all a and b in F and X, Y, Z in L.*
3. $[X,[Y,Z]] + [Z,[X,Y]] + [Y,[Z,X]] = 0$.

This last identity is called the *Jacobi identity.* It is easy to verify that the bracket operation $[B,C] = BC - CB$ on the vector space of all $n \times n$ matrices over F (e.g. $F = R$) satisfies the axioms given above. Thus, we have so far seen that $sl(n)$ is a Lie algebra that is naturally associated with the group of matrices $Sl(n)$ of determinant one. In fact, $sl(n)$ generates $Sl(n)$ by exponentiation.

7.2. RELATION WITH VASSILIEV INVARIANTS

In our context the Lie algebras turn out to be related to the formation of weight systems for Vassiliev invariants. One way to see this is to just take the case of matrix Lie algebras with commutator brackets and interpret diagrammatically the formula that states that the Lie algebra is closed under the bracket operation. This formula states that there is a basis $\{T_1, T_2, \ldots, T_m\}$ for the Lie algebra as a vector space over F such that each T_a is an $n \times n$ matrix and such that

$$T_aT_b - T_bT_a = \sum_{c=1,\ldots,n} f^c_{ab}T_c \;,$$

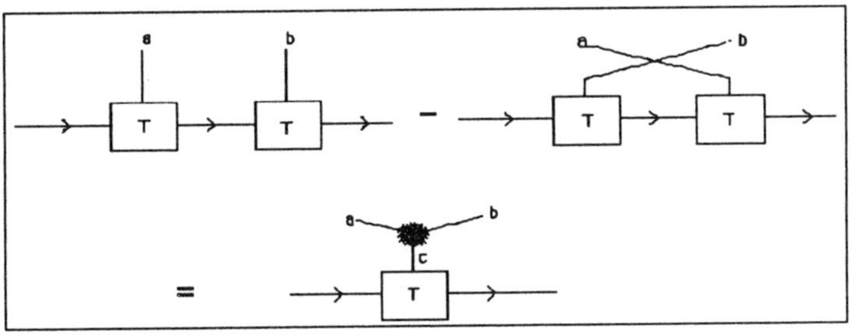

Figure 24. Diagrammatic form of the commutator equation.

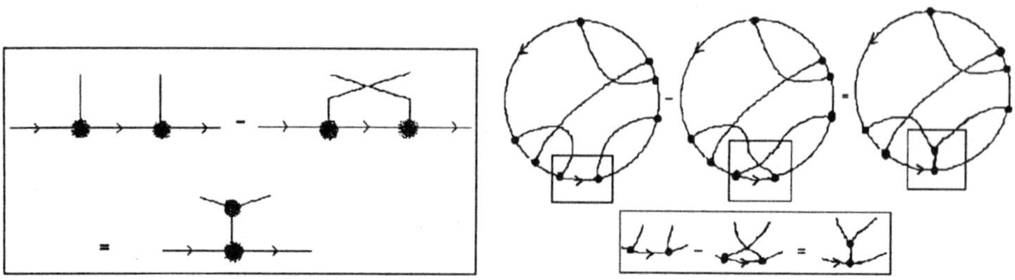

Figure 25. Diagrammatic modification of the commutator equation.

where f_{ab}^{c} is a set of constants in F depending on the three indices a, b, c (running from 1 to n). The right-hand-side of this equation connotes a summation over all values of the index $c = 1, \ldots, n$. The left hand side is the commutator of T_a and T_b for any given choice of a and b.

In Figure 24 we have diagrammed this equation using the conventions for diagrammatic matrix multiplication explained in this section. The structure constants f_{ab}^{c} are represented by a graphical vertex with three lines attached to it, one for a, one for b and one for c. For the purpose of discussion, we shall assume that f_{ab}^{c} is dependent only on the cyclic order of abc. It is convenient to regard the graphical vertex as representing a "tensor" that has this cyclic invariance since this means that we can slide the diagram for the structure constant tensor around in the plane so long as we keep the cyclic order of its legs unchanged. Such bases can be obtained in many cases of matrix Lie algebras, and the results that we outline can be generalized in any case.

Now view Figure 25. You will see, first, a formal version of the commutator relation of Figure 24, except that the labels and indices have been

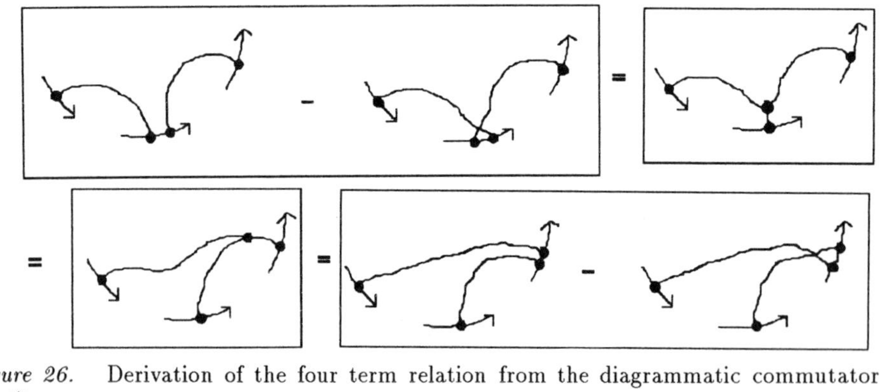

Figure 26. Derivation of the four term relation from the diagrammatic commutator identity.

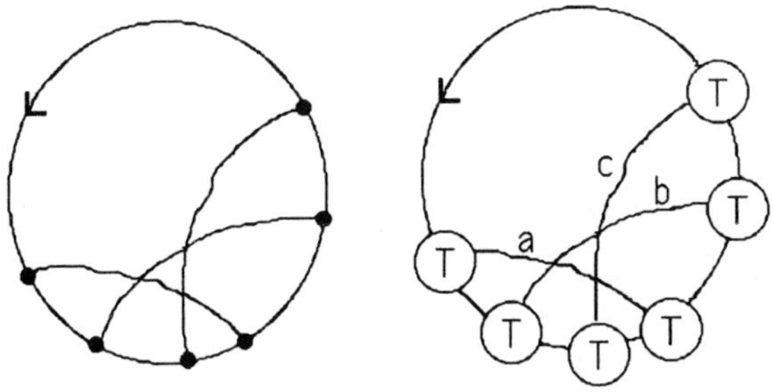

Figure 27. Chord diagram and sum of traces as represented by the second diagram.

removed and the boxes for matrix elements have been replaced by graphical vertices. Imagine that the terms in this formal version of the commutator relation are parts of chord diagrams as illustrated with examples in this figure. In other words, recall the method of chord diagrams from the last section and imagine that along with the chords there are also trivalent graphical vertices among the chords, and that these vertices are related to commutators as shown in the Figure. Finally, view Figure 26 and you will see a formal derivation of the four term relation for chord diagrams from the diagrammatic commutator identity. This means that the four-term relation, that we derived from topological considerations in the last section, is intimately related to the basic structure of a Lie algebra. This is the essence of the relationship of Vassiliev invariants with Lie algebras and their generalizations.

Concretely, the relationship we have just described means that it is possible to construct weight systems for Vassiliev invariants by using matrix Lie algebras. To see how this works view Figure 27. Here we have indicated a chord diagram D and a corresponding diagram involving matrices T_a from a Lie algebra basis. The second diagram represents the sum of traces

$$wt(D) = \sum_{a,b,c} \mathrm{tr}(T_a T_b T_c T_a T_b T_c) \ ,$$

where we are summing over all values for the indices a, b and c. This second diagram represents the weight, $wt(D)$, that is assigned to the first diagram. It follows from our considerations in Figure 27 that this weight system satisfies the four term relation and hence, by the Theorem of Kontsevich [3], is the top row evaluation for a Vassiliev invariant.

Acknowledgements

Research supported in part by NSF Grant DMS 9802859.

References

1. Akutsu, Y. & Wadati, M. (1987) Knot invariants and critical statistical systems. *J. Phys. Soc. Japan* **56**, 839–842.
2. Alexander, J.W. (1923) Topological invariants of knots and links. *Trans. Amer. Math. Soc.* **20**, 275–306.
3. Bar-Natan, D. (1995) On the Vassiliev knot invariants. *Topology* **34**(2), 423–472.
4. Birman, J. & Lin, X.S. (1993) Knot polynomials and Vassiliev's invariants. Invent. Math. **111**(2), 225-270.
5. Conway, J.H (1970) An enumeration of knots and links and some of their algebraic properties. In *Computational Problems in Abstract Algebra* (ed. A. Leech), pp. 329–358. Pergamon Press, New York.
6. Fenn, R.A. & Rourke, C.P. (1992) Racks and links in codimension two. *J. Knot Theory and Its Ramifications* **1**(4), 343–406.
7. Fox, R.H. (1962) A quick trip through knot theory. In *Topology of 3-Manifolds* (ed. M.K. Fort Jr.), pp. 120–167. Prentice Hall.
8. Freyd, P., Yetter, D., Hoste, J., Lickorish, W.B.R., Millett, K.C. & Ocneanu, A. (1985) A new polynomial invariant of knots and links. *Bull. Amer. Math. Soc.* **12**, 239–246.
9. Jones, V.F.R. (1983) Index for subfactors. *Invent. Math.* **72**, 1–25.
10. Jones, V.F.R. (1985) A polynomial invariant for links via von Neumann algebras. *Bull. Amer. Math. Soc.* **129**, 103–112.
11. Jones, V.F.R. (1986) A new knot polynomial and von Neumann algebras. *Notices of Am. Math. Soc.* **33**, 219–225.
12. Jones, V.F.R. (1987) Hecke algebra representations of braid groups and link polynomials. *Ann. of Math.* **126**, 335–338.
13. Jones, V.F.R. (1989) On knot invariants related to some statistical mechanics models. *Pacific J. Math.* **137**(2), 311–334.
14. Joyce, D. A classifying invariant of knots, the knot quandle. *J. Pure Appl. Alg.* **23**, 37–65.
15. Kauffman, L.H. (1980) The Conway polynomial. *Topology* **20**, 101–108.
16. Kauffman, L.H. (1983) *Formal Knot Theory.* Lecture Notes **115**. Princeton University Press.
17. Kauffman, L.H. (1987) State Models and the Jones Polynomial. *Topology* **26**, 395–407.
18. Kauffman, L.H. (1987) *On Knots.* Annals of Mathematics Studies Number **115**, Princeton University Press.
19. Kauffman, L.H. (1990) An invariant of regular isotopy. *Trans. Amer. Math. Soc.* **318**(2), 417–471.
20. Kauffman, L.H. 1993 *Knots and Physics.* World Scientific.
21. Kauffman, L.H. & Harary, F. (1999) Knots and Graphs I. *Advances in Applied Math.* **22**, 312–337.
22. Menasco, W. & Thistlethwaite, M. (1993) The classification of alternating links. *Annals of Math.* **138**, 113–171.
23. Murasugi, K. (1987) The Jones polynomial and classical conjectures in knot theory. *Topology* **26**, 187–194.
24. Przytycki J.H. & Traczyk, P. (1987) Invariants of links of Conway type. *Kobe J. Math.* **4**, 115–139.
25. Reidemeister, K. (1948) *Knotentheorie.* Chelsea Pub. Co. N.Y. . [(1932) Copyright Julius Springer, Berlin.]
26. Stanford, T. (1992) Finite-type invariants of knots, links and graphs. *Preprint.*
27. Vassiliev, V. (1990) Cohomology of knot spaces. In *Theory of Singularities and Its Applications* (ed. V.I. Arnold), pp. 23–69. American Mathematical Society.
28. Winker, S. (1984) *Quandles, Knot Invariants and the N-fold Branched Cover.* Ph.D. Thesis, University of Illinois at Chicago.

FLUID MECHANICS AND MATHEMATICAL STRUCTURES

PHILIP BOYLAND
Department of Mathematics
University of Florida
Gainesville, FL 32611-8105, USA

E-mail: *boyland@math.ufl.edu*

To Ed Spiegel on his 70th birthday.

Abstract. This paper provides an informal survey of the various mathematical structures that appear in the most basic models of fluid motion.

1. Introduction

Fluid mechanics is the source of many of the ideas and concepts that are central to modern Mathematics. The vestiges of these origins remain in the names of mathematical structures: flows, currents, circulation, frames, ..., and in the names of the distinguished mathematician-scientists which pervade the terminology of both fields: Euler, Bernoulli, Stokes, Cauchy, Lagrange, ... Mathematicians have abstracted and vastly generalized basic fluid mechanical concepts and have created a deep and powerful body of knowledge that is unfortunately now mostly inaccessible to fluid mechanicians, while mathematicians themselves have lost all but a passing knowledge of the physical origins of many of their basic notions.

This paper provides an informal survey of the various mathematical structures that appear in the most basic models of fluid motion. As befits this volume it is strictly pedagogical. There are no original results, and most of what we say here has been known for at least a century in one guise or another. In style it attempts to reproduce the informal tutorials that took place during the Newton Institute program in Fall, 2000. Mathematical objects are described informally with an emphasis on their meaning in Fluid Mechanics. Many formulas are stated without proof. The paper is meant to be a friendly introduction to basic concepts and ideas and will

R.L. Ricca (ed.), An Introduction to the Geometry and Topology of Fluid Flows, 105–134.

hopefully provide an intuitive foundation for a more careful study of the literature. Readers are strongly encouraged to consult the references given in each section.

In describing the fluid model we give primacy to the actual evolution of the fluid, to the fluid flow, as opposed to the velocity fields. This agrees with our direct experience of seeing fluids move as well as being very natural in the mathematical progression we describe. While conceptually valuable, this point of view is computationally very impractical. The basic equations of Fluid Mechanics are framed in terms of the velocity fields. Solutions of the equations are difficult to obtain, but formulas for the fluid motions are virtually nonexistent.

One of the goals of the paper is to introduce students of fluid mechanics to the "world view" of modern Mathematics. A common view among mathematicians is that Mathematics is the study of sets with structures and of the transformations between them. From this point of view, the basic model of Fluid Mechanics is the set consisting of the fluid body and the various additional structures that allow one to discuss such properties as continuity, volume, velocity and deformation. The flow itself is a transformation of the fluid body to itself and a basic object of study is the transport of structures under the flow.

It is considered a mathematical virtue to use just those structures which are needed in a given situation and no more. The goal is not to be mindlessly abstract but rather to discover and illustrate what is essential and fundamental to the task at hand. This has the added advantage that conclusions have the widest applicability. Thus we develop the fluid model step by step, introducing structures to fit certain needs in the modelling process. When possible coordinate free terminology and notation is used. We invoke the usual argument in its favor: anything fundamental shouldn't depend on the choice of coordinates and being forced to treat computational objects as global entities provides a new and sometimes valuable perspective on familiar operations. We do, however, try to connect new objects with familiar formula, and we try to be clear when formulas are only valid in usual Euclidean space.

We have attempted to keep the paper accessible to a wide audience, but some fluid mechanical terminology is used without explanation. General readers may consult any standard text; the mathematically inclined may prefer [22], [7], [16] or [5].

2. Basic Kinematics and Mathematical Structures

We begin by describing the most fundamental mathematical structures used in modelling the simplest type of fluid behavior. Unless otherwise stated the fluid is three-dimensional; two-dimensional flow can be treated similarly.

2.1. THE FLUID REGION, FLUID MAPS AND CARDINALITY

The first step is to model the fluid region itself. In Fluid Mechanics it is usually said that the fluid is a continuum. In mathematical language this says that, at least locally, the fluid looks like usual two or three-dimensional space. We allow the possibility that on large scales the fluid is not flat, but rather can curve back on itself like the surface of the (almost) spherical earth. Objects with a local structure like the plane or three space, but perhaps nontrivial global behavior are called 2- or 3-dimensional manifolds. A fluid particle is a mathematical point in the manifold. The generic fluid region or fluid body is denoted B.

The modelling of the fluid as a manifold has many implications and some obvious problems. It is clearly false on very small scales as it completely ignores the molecular nature of fluids. In addition, a mathematical point obviously has no physical meaning. Nonetheless, the theories and technologies based on continuum models have been wildly successful and we can proceed with confidence using a manifold model.

As a fluid flows particles are transported. In terms of the model the fluid motion is defined as the collective motion of the particles. It is assumed that the individual particles do not split into pieces, and each has a well defined future and remains distinguished for all times. This means that the evolution is described by the transformation that takes the initial position of a particle as input and gives the position after a time T as the output. Thus for each time T the evolution of the fluid after time T is described by a function (or map), f, from the initial region occupied by the fluid, B_0, to the region occupied after time T, B_T. The usual notation for this is $f : B_0 \to B_T$, indicating that f is a function with domain B_0 and range B_T. The function f is called the *time-T fluid map*. The typical point in a fluid region is labelled $\mathbf{p}$, $\mathbf{q}$, etc. Note that these label the geometric position in the region and not the fluid particles, and so they do not move with the flow.

During their evolution it is further assumed that fluid particles do not overlap, coalesce or collide. Thus if two points are distinct, then their positions after time T are different, or $\mathbf{p} \neq \mathbf{q}$ implies $f(\mathbf{p}) \neq f(\mathbf{q})$, and so f is a one-to-one (or injective) function. Since by definition, B_T is the fluid region after time T, we have that $f : B_0 \to B_T$ is an onto (or surjective) function, which is just to say that for any $\widehat{\mathbf{p}}$ in B_T there is a $\mathbf{p}$ in B_0 with $f(\mathbf{p}) = \widehat{\mathbf{p}}$, i.e. the particle at $\widehat{\mathbf{p}}$ came from somewhere.

Summarizing, the most basic assumptions on a fluid map is that it is one to one and onto, i.e. it is a bijection, or what is sometimes termed a one-to-one correspondence. This means that f preserves the "number" of fluid points. This idea is formalized as the most fundamental property of a set, its *cardinality.* By definition, two sets have the same cardinality exactly when there is a one-to-one correspondence between them. While this may not seem like much information about the fluid map it is worth noting that there are sets with different cardinality.

Exercise 2.1 Show that there is a one-to-one correspondence between the set of points in the interval, $0 \leq x \leq 1$, and the set of points in the unit square in the plane, $0 \leq x \leq 1$, $0 \leq y \leq 1$, but that these sets have a different cardinality from the set of integers $0, \pm 1, \pm 2, \ldots$.

A bijection f always has an inverse map, usually denoted f^{-1}, which by definition satisfies $f \circ f^{-1} = f^{-1} \circ f = \mathrm{id}$. While the inverse of a fluid map always exists, there is no assumption in general that it is a physically reasonable fluid map, though in certain cases it can be.

Elementary set theory is covered in the first sections of most textbooks in Algebra or Topology, [13] is a standard text, and [26] is recreational but informative.

2.2. STRUCTURES, TRANSPORT, AND CATEGORIES

At this point our fluid is described by the most basic of mathematical objects, sets and a transformation. We pause now to introduce some general notions about mathematical structures and how they are transported by transformations.

Mathematical structures take many forms. The most basic examples are an *order structure* which declares which elements are larger than others, a *topology* which is a distinguished collection of subsets, and an *algebraic* structure which is a binary operation on the set which satisfies certain rules. Other common structures are made out of various kinds of functions into or out of the set. For example, a velocity field is a map out of the set, it assigns a vector to each point in the set. A loop is a map from the circle into a set. The collection of all tangent vectors to a space are turned into the tangent bundle, which in turn is the basis for other structures like tensors, Riemannian metrics and differential forms. The general intuition is that a structure is something that "lives on top of " the set, representing additional information and properties.

If a transformation between underlying sets is sufficiently well behaved it induces a map on the various structures, transforming a structure on one set to the same kind of structure on the other set. For example, in ideal magnetohydrodynamics the fluid flow transports the magnetic vector field at time zero to the magnetic vector field after a time T. The general mathematical rubric is that transport in the direction of the map is called *push forward* while *pull back* is an induced map that goes in the direction opposite to the given map. If the map we are studying is a fluid map, then the push forward transports the structure in forward time while pull back refers to transport from the future backward, usually back to the initial state of the system. The standard notation puts a star on the given map to indicate the induced map on structures, with a subscript indicating a push forward and a superscript indicating a pull back. Thus if f is a fluid map and α is a structure (say a magnetic field) then $f_*(\alpha)$ is this vector field pushed forward by the fluid evolution and $f^*(\alpha)$ indicates its pull back.

Since a fluid map f is invertible, pulling back is the same as pushing forward by the inverse f^{-1}, and we will freely pass between the two. Depending on the structures it is usually most natural to either pull back or else push forward. As a general guideline when a structure is defined by a map into the set it is pushed forward. For example, a loop in the fluid region is a map $\Gamma : S^1 \to B_0$, and this naturally pushes forward to the same kind of structure on B_T since the composition $f \circ \Gamma$ is a map $S^1 \to B_T$ called $f_*\Gamma$. On the other hand, a structure defined by maps out of the set, say a scalar field α on B_T, is most naturally pulled back because $\alpha \circ f = f^*\alpha$ is a map out of B_0. Another basic pull back is that of a set. If N is a subset of the fluid region B_T, then its pull back is a subset of B_0 defined by $f^*(N) = \{\mathbf{p} \text{ in } B_0 : f(\mathbf{p}) \text{ is in } B_T\}$; this set is also often denoted $f^{-1}(N)$.

The notion of a mathematical category and its morphisms formalizes many of the notions of sets, structures and transformations. The most basic category is the collection of sets and other categories consist of sets with some specific additional structure such as topological spaces, vector spaces, smooth manifolds, or flows on manifolds. The *morphisms* associated with a category are the maps between sets that preserve the structure, i.e. the induced map on structures is defined and a transported structure is compatible with the pre-existing one. For example, linear transformation are the morphisms of vector spaces and continuous functions are the morphisms of topological spaces.

An *isomorphism* in a category is a morphism that is a bijection and its inverse is also a morphism. Most of the commonly studied categories have their own traditional name for their isomorphism such as bijection in the category of sets, homeomorphism in the category of topological spaces, and isometry in the case of geometries. Within a given category isomorphic objects are considered indistinguishable, and indeed a rather formal definition of a mathematical area of study is that it studies those properties that are preserved by the morphisms in its particular category.

The text [15] is a standard reference.

2.3. CONTINUITY AND TOPOLOGY

Exercise 2.1 clearly indicates that a bijection can scramble a fluid domain very badly, prompting the requirement that fluid motions preserve the continuum nature of the fluid, not tearing or distorting it too wildly. This requirement is the essence of Topology and its morphisms, the continuous functions, and is expressed in the assumption that fluid particles which are sufficiently close initially should still be close after evolving for time T. The technicalities arise from defining "sufficiently close" and "still be close".

A topology as a structure on a set is simply a collection of distinguished subsets called the *open sets* (there are conditions which are not important here). Open sets are the allowable neighborhoods of the point and *neighborhood* is used synonymously with open set. A property is true for points sufficiently close to $\mathbf{p}$ if it is true for all open sets containing $\mathbf{p}$. In the category of sets with topologies (topological spaces) the morphisms or structure preserving maps are the continuous functions. A function $g : X \to Y$ between two topological spaces is called *continuous* if the pull back of an open set is open. Thus for each open set U in Y it is required that $f^*(U)$ is an open set in X, or by acting on all open sets, it is required that the pull back of the topology on Y is compatible with the topology on X.

Sequences provide a simple way to connect this abstraction with the more physical condition that nearby fluid particles should stay near each other. In terms of a given topology, the sequence $(\mathbf{p}_n)$ converges to the point $\mathbf{p}_0$, written $\mathbf{p}_n \to \mathbf{p}_0$, when any neighborhood V that contains $\mathbf{p}_0$ also contains the tail of the sequence, i.e. there is an N so that $n > N$ implies that $\mathbf{p}_n \in V$. If f is the time T fluid map, the flow preserving the continuum structure clearly requires that $\mathbf{p}_n \to \mathbf{p}_0$ implies $f(\mathbf{p}_n) \to f(\mathbf{p}_0)$. This says that any neighborhood U of $f(\mathbf{p}_0)$ must contain the tail of the sequence $f(\mathbf{p}_n)$. But transporting this data back to the fluid at time zero we see that the convergence after time T requires that $\mathbf{p}_n \in f^*(U)$ for $n > N$. But since $\mathbf{p}_n \to \mathbf{p}_0$, this would be achieved by requiring that $f^*(U)$ is itself a neighborhood, which is the case if f is continuous in the sense defined above.

Exercise 2.2 In the standard topology on $\mathbb{R}^n$ a set U is open when for every $\mathbf{p} \in U$ there is an r, so that the ball $B_r(\mathbf{p}) = \{\mathbf{q} : |\mathbf{q} - \mathbf{p}| < r\}$ is wholly contained in the set U. Show that in this case the definition of continuity of a function using the pull back of open sets is the same as the usual one: f is continuous at $\mathbf{p}$ if for all ϵ there exists a δ so that

$$|\mathbf{q} - \mathbf{p}| < \delta \text{ implies } |f(\mathbf{q}) - f(\mathbf{p})| < \epsilon \ .$$

In the fluid model particles which are sufficiently close after time T should come from particles which were close at time zero. Thus we assume that the inverse of a fluid map f^{-1} is also continuous. Such a continuous bijection with a continuous inverse is called a *homeomorphism.* Homeomorphisms are the isomorphisms in the category of topological spaces. Under a homeomorphism the pull back and the push forward of each open set is an open set, and so the pull back of the topology on Y is exactly the topology on X, and vice versa. Thus the homeomorphism has altered only labels and has altered nothing related to the topology. Thus it makes sense to define Topology as the study of those properties that are invariant under homeomorphisms. These properties include the connectedness, the number of holes, and the dimension of the space. So in the model these topological properties of the fluid region do not change as the fluid evolves.

In requiring fluid maps to be homeomorphisms we are excluding some common fluid behaviors. When we sip our coffee, the fluid tears, and we see the droplets of rain coalescing into the continuum of the creek. It is also worth noting that what is commonly called "The Continuity Equation" in Fluid Mechanics is not a statement about the kind of continuity discussed in this section, but rather a statement about the preservation of the mass of the continuum during its evolution (see sections 2.5 and 4.6).

The books [20] and [8] are standard Topology texts.

2.4. FLUID MOTIONS AND MATHEMATICAL FLOWS

At this point a fluid map describes the evolution of the fluid up to fixed, but arbitrary time. The complete evolution is described by a family of maps, ϕ_t, one for each time t. The particle at position $\mathbf{p}$ at time 0 is at the position $\phi_t(\mathbf{p})$ after time t, and so what was denoted f above is now denoted ϕ_T. If we fix $\mathbf{p}$ and vary t, the positions $\phi_t(\mathbf{p})$ sweep out the *trajectory, flow line or path* of the particle. For simplicity of exposition we assume that the fluid motion can be defined for all forward and backwards time and so ϕ_t is defined for all $t \in \mathbb{R}$. In terms of fluid maps, *slip boundary conditions* say that the fluid region is preserved by the fluid motion, $B_t = B$ for all t. The *no slip* conditions require that on the boundary of B, $\phi_t = \text{id}$ for all t. Henceforth we always assume that our fluid maps satisfy at least slip conditions.

The family of fluid maps ϕ_t is best described as simple map on a bigger space including both the fluid domain and time, $\phi : B \times \mathbb{R} \to B$, with $\phi(\mathbf{p}, t) = \phi_t(\mathbf{p})$. In the previous subsection we required that ϕ_t is a homeomorphism for each t. It is clear that we must also require continuity as the time is varied so that fluid particles don't suddenly jump during their evolution. This is accomplished by requiring that the full function ϕ is continuous, i.e. continuous in both variables.

Remark 2.3 It is traditional in Fluid Mechanics to distinguish the Eulerian and the Lagrangian perspective. In the first, the labels are pinned down in the fluid region, in mathematical language they are local coordinates in the manifold modelling the fluid. In the Lagrangian perspective, the fluid particle is labelled and that label remains on the particle as it moves. In addition, the Eulerian perspective is a field theory, focusing on the velocity vector field while the Lagrangian focuses on the motions or flow lines of particles. The point of view here is mixed: we always label using a fixed, pinned down coordinate system in the fluid region and never use Lagrangian or advected coordinates, but the primary focus is on the motion of particles. It is also traditional in Fluid Mechanics to use $\mathbf{x}$ for the Eulerian coordinate, and to denote a trajectory as $\mathbf{x}(t)$, indicating how the coordinate varies as a function of time. This use of the same symbol to indicate both a coordinate and a function of time makes many mathematicians uncomfortable; it represents an unacceptable confusion of different classes of mathematical objects. Thus we denote the time evolution by a distinct function, ϕ, which encapsulates all the time and space evolution. To avoid confusion with the training of fluid mechanicians, while maintaining the notation $\mathbf{x} = (x_1, x_2, x_3)$ for the local coordinates, $\mathbf{p}$ and $\mathbf{q}$ are used here to denote the typical point in the fluid region (labelled by the fixed coordinate system) rather than the more common mathematical x or $\mathbf{x}$.

There is a useful distinction among fluid motions based on how the future of particles depends on the starting time. Pick a fixed, but arbitrary location in the fluid region and monitor the future of the particle that is there at time 0 and also that of the particle that is there at some later time. If these futures are always the same the flow is *steady*, otherwise it is *unsteady*.

This distinction has a nice formulation in terms of the algebra of the time parameter, i.e. of the real line. Begin the evolution of the fluid particle at $\mathbf{p}$ and flow for time t, now at this point begin again and flow for time s, so we are at the point $\phi_s(\phi_t(\mathbf{p}))$. On the other hand, we could start at $\mathbf{p}$, flow for time t and then continue without restarting time for another time interval s, ending up at the point $\phi_{s+t}(\mathbf{p})$. The flow is steady exactly when these points are the same for all s, t and points $\mathbf{p}$, or in terms of the fluid maps, $\phi_s \circ \phi_t = \phi_{s+t}$. This group law coupled with the fact that ϕ_0 is the identity map says that a steady flow is an *action* of the real numbers on the region B. This implies, in particular, that the inverse maps be computed by reversing time, $\phi_t^{-1} = \phi_{-t}$

Within most of mathematics the word *flow* refers to this kind of action of the reals on a space. In particular, mathematical flows always correspond to steady fluid flows. For this reason the general, perhaps unsteady, evolution of a fluid is called here a *fluid motion* reserving the word flow for steady flows.

Two simple examples of one-dimensional flows are the linear flow, $\phi_t(p) = p + rt$, and the exponential flow, $\phi_t(p) = pe^{rt}$, where r is a real number. In these cases the group law of the action corresponds to the standard distributive and exponential laws, respectively. Flows correspond to solutions of time-independent differential equations, and it is remarkable that any such solution yields a flow with nice algebraic properties.

The simplest unsteady fluid motions are periodic. A fluid motion (or fluid map ϕ_t) is called *periodic* if starting at an initial time t at the point $\mathbf{p}$ yields the same future as starting at the same point at a time P later, where P is called the *period.* This implies that $\phi_{t+P} = \phi_t \circ \phi_P$, and letting $t = (n-1)P$ for an integer n, that $\phi_{nP} = \phi_P \circ \phi_P \circ \ldots \circ \phi_P$ (n times). Thus if we define the *Poincáre* map as $g = \phi_P$, its iterates (repeated self compositions) satisfy $g^n = \phi_{nP}$, and so they describe the evolution of the fluid. The Poincáre map is often called the stroboscope map, as it corresponds to viewing the fluid under a light that turns on once each period.

If we have a fluid motion ϕ_t on a region B and a homeomorphism h from B to another region $\hat{B}$, we can push forward the flow to one on $\hat{B}$ by the rule $h_*(\phi_t) = h \circ \phi_t \circ h^{-1}$. The motion for time T on $\hat{B}$ is obtained by first coming back to B via h^{-1}, then flowing from that point via ϕ_T, and then going back to B via h. If the homeomorphism is from B to itself, it represents a change of coordinates, and if $h_*\phi_t = \psi_t$ one says that two flows ϕ_t and ψ_t are the same fluid motion up to change of coordinates by h.

2.5. INCOMPRESSIBILITY, MASS CONSERVATION AND MEASURES

If there are no mechanisms for inflow or outflow as a fluid subregion evolves its mass stays the same. Similarly, if there are no mechanisms to compress or expand regions of the fluid, then the volume of fluid subregions is preserved under the evolution. Both of these properties can be included in the model by postulating measures that are invariant under the fluid motion.

A measure is a rule which assigns a non-negative number to various subsets of the main set (again, there are additional technicalities that are not crucial here). In a fluid model, the measure of a subset could represent the mass of the fluid in the subset, or perhaps the volume of the subset. Thus if $m(U)$ is the mass of the fluid contained in the region U, then mass conservation of the flow ϕ_t is expressed by $m(\phi_t(U)) = m(U)$ for all measurable sets U and all times t. Similarly, if $\nu(U)$ is the usual three-dimensional volume (Lebesgue measure) of a set U in $\mathbb{R}^3$, then a fluid flow is *incompressible* if $\nu(\phi_t(U)) = \nu(U)$ for all measurable sets U and all times t.

Measures are most commonly used for integration, and one writes $\int_U \alpha \, d\nu$ for the usual integral of the real valued function α over the region U, and thus $\nu(U) = \int_U d\nu$. A *density* ρ is a real-valued function which represents the "measure per unit volume", or more precisely, ρ is a density for the measure m, exactly when $m(U) = \int_U \rho\, d\nu$. One then writes $dm = \rho\, d\nu$.

If $g : X \to Y$ is a function, we can use it to transport measures. Given a measure m on X, its *push forward* to Y is $g_* m$. It is defined by the rule that the measure $g_* m$ assigns to a subset V of Y is the measure of $g^*(V)$, or in symbols, $g_* m(V) = m(g^*(V))$. Thus a fluid motion ϕ_t preserves the mass given by the measure m exactly when $\phi_{t*} m = m$, for all t (in the unsteady case m may be time dependent). Similarly, incompressibility is expressed by $\phi_{t*}\nu = \nu$, for all t.

From a strictly kinematic point of view there is little difference between preserving a mass m or preserving the volume ν. This is consequence of a theorem of Oxtoby and Ulam [21] (Moser [19] in the smooth category). If the mass density ρ is a reasonable function and for simplicity we assume that the total mass equals the total volume, $\int \rho\, d\nu = \int d\nu$, then the theorems give a homeomorphism (diffeomorphism) h of B to itself which pushes forward m to ν, $h_* m = \nu$. Thus if ϕ_t preserves m then by changing coordinates by h, the fluid motion $h_*(\phi_t)$ preserves ν.

Elementary measure theory and integration is covered in most textbooks in Analysis, [12] is a measure theory standard.

2.6. DIFFERENTIABILITY

The laws of mechanics are expressed using derivatives, and indeed, the development of mechanics went hand in hand with the invention of calculus. The continuity of the fluid motion arose from the need to preserve the continuum structure of the fluid body. But the assumption that the fluid region is locally Euclidean gives more than just a local topology, there is also the algebraic structure of Euclidean space as a vector space. The assumption of differentiability of fluid motion means that on small enough scales the fluid map looks like a linear map of a vector space to itself. This is made precise by Taylor's theorem and its converse which says that a map f is differentiable exactly when it is linear up to an error of specified type. More precisely, in Euclidean space the matrix A is the first derivative of the map f at the point $\mathbf{p}$ exactly when

$$f(\mathbf{p} + \mathbf{h}) = f(\mathbf{p}) + A\mathbf{h} + o(|\mathbf{h}|) , \tag{1}$$

with $|\mathbf{h}|$ the usual Euclidean norm and $o(|\mathbf{h}|)$ representing a quantity $r(\mathbf{h})$ that goes to zero sufficiently fast so that $\frac{r(\mathbf{h})}{|\mathbf{h}|} \to 0$ as $\mathbf{h} \to 0$.

The matrix A is, of course, the usual Jacobian matrix, for which there are numerous notations; its (i, j)-th entry is $\partial f_i / \partial x_j$. When f is real valued, i.e. has only one component, then the derivative matrix is a row vector representing the gradient, ∇f. It is common in Fluid Mechanics to also use ∇ in the multidimensional case, and so we shall denote the full derivative as ∇f, or $\nabla f(\mathbf{p})$. A function f is *continuously differentiable* if a matrix A satisfying (1) exists at each point $\mathbf{p}$ and the matrices vary continuously with the evaluation point. In this case one writes $f \in C^1$.

Higher derivatives can be defined using the analog of (1). A function is C^2 when it locally looks like a quadratic polynomial up to an error of $o(|\mathbf{h}|^2)$. The notation for the analog of (1) grows complicated, but in the simplest case of a real valued function α, C^2 means that

$$\alpha(\mathbf{p}+\mathbf{h}) = \alpha(\mathbf{p}) + \nabla\alpha \cdot \mathbf{h} + \mathbf{h}^T \cdot H(\alpha) \cdot \mathbf{h} + o(|\mathbf{h}|^2) \ ,$$

where $H(\alpha)$ is the Hessian matrix of α, $(\frac{\partial^2 \alpha}{\partial x_i \partial x_j})$, and the Hessian must vary continuously with the evaluation point. Continuing, a function is C^k if it locally looks like a degree k polynomial up to order $o(|\mathbf{h}|^k)$. If a function is C^k for all k, it is called C^∞. A condition that is even stronger than C^∞ is the requirement that the Taylor series converges in a neighborhood of each point. Such functions are called *real analytic* and their class is denoted C^ω.

Exercise 2.4 Let the real valued function α be defined by $\alpha(0) = 0$ and for $x \neq 0$, $\alpha(x) = \exp(-1/x^2)$. Show that all orders of derivatives at 0 are equal to 0 and thus the Taylor's series about 0 is identically zero. However, the function α is not identically zero in any neighborhood of the origin, and so α is an example of a C^∞ function that is not C^ω.

On a manifold, a function is C^k if it is C^k in local coordinates in the domain and the range, and we need to treat the derivative as a linear transformation rather than a matrix. The matrix represents the linear transformation in particular coordinates, but it is the linear transformation that has coordinate free meaning.

In formulating differentiability from the structural point of view, the objects in the category are C^∞-manifolds. These are manifolds in which C^∞-compatible local coordinates can be defined. The morphisms are C^∞-maps, and the isomorphisms are called *diffeomorphisms*, A diffeomorphism is a C^∞-bijection whose inverse is also C^∞. This category is the subject of Differential Topology. Henceforth we assume that fluid maps ϕ_t are C^∞-diffeomorphisms for each t, and further, that the family itself is C^∞ in t. It is worth noting that C^3 is enough differentiability to do virtually all of Fluid Mechanics, but we shall not deal with that level of technicalities here.

Introductions to Differential Topology are given in [17] and [24], and [11], [6], and [14] are standard texts.

3. Vector Fields and the Tangent Bundle

At this point we have developed all the basic kinematic assumptions for the simplest models of fluid flow. Forces enter into the models by their action on velocity fields, the prototypical example of a vector field. Vector fields as a mathematical structure are the basis of many additional structures and operations, and in this section we discuss some of those that are valuable in describing fluid motions. Principal among these are the tangent bundle and other structural bundles that live "above" the fluid body. The fluid motion and its derivatives induce an action on these bundles that transport the various structures. The nature of this transport is the key to describing geometric and topological aspects of the fluid motion.

3.1. VECTOR FIELDS AND THE VELOCITY FIELD

Given a fluid motion ϕ_t, its *velocity field* $\mathbf{u}$ at the point $\mathbf{p}$ at a time t is the instantaneous velocity of the fluid particle that occupies that point at that time. Note that this is the particle that started at $\phi_{-t}(\mathbf{p})$ at time zero, and so

$$\mathbf{u}(\mathbf{p}, t) = \frac{\partial \phi}{\partial t}(\phi_{-t}(\mathbf{p}), t) \ ,$$

or

$$\mathbf{u}(\phi_t(\mathbf{p}), t) = \frac{\partial \phi}{\partial t}(\mathbf{p}, t) \ . \tag{2}$$

This is sometimes called the *advection equation.* Given a velocity field $\mathbf{u}$, its fluid motion ϕ_t is obtained by solving the differential equation given by (2). It is easy to check that a fluid motion is steady as defined in Section 2.4 if and only if its velocity field is time independent, $\mathbf{u}(\mathbf{p}, t) = \mathbf{u}(\mathbf{p})$, and the motion is T-periodic exactly when the velocity field is T-periodic, $\mathbf{u}(\mathbf{p}, t + T) = \mathbf{u}(\mathbf{p}, t)$.

Exercise 3.1 The planar velocity field $\mathbf{u}(\mathbf{p}) = (-p_2, p_1)$ represents uniform counter-clockwise fluid motion. Show that its flow is $\phi_t(p_1, p_2) = (p_1 \cos(t) - p_2 \sin(t),\ p_1 \sin(t) + p_2 \cos(t))$, and that in this case the group property $\phi_t \phi_s = \phi_{s+t}$ is a reflection of standard trigonometric identities.

The velocity field at a fixed time is an example of a *vector field,* i.e. a rule that assigns a vector to each point in the fluid region. Other common examples of vector fields are the magnetic field and the vorticity field. We will adopt the fairly standard mathematical convention that the terminology "vector field" refers to a *time independent* field. Through equation (2) mathematical flows and vector fields go hand in hand and it is usual to pass freely from one to the other without comment.

In certain circumstances the time parameterization of a flow is not important, and one wants to think of the trajectories as just curved lines in space. In this case it is common in the sciences to call the curve constituting the trajectory a *field line*, as in a magnetic field line and a vortex (field) line. The mathematical name for a way to fill a manifold with a collection of curves is a *one-dimensional foliation*, and the individual curves are called the *leaves*. Note that the nonzero scalar multiple of a given nonvanishing vector field, $\alpha\,\mathbf{u}$, gives the same foliation as the original field, but gives rise to a different flow.

3.2. THE TANGENT BUNDLE

There is a common and quite understandable confusion that occurs early in a student's mathematical training; is the base of a vector at the origin or at the point along a curve to which it is tangent? This usually leads to the mathematical vector space called $\mathbb{R}^3$ being somehow conceptually distinct in the student's mind from the physical notion of velocity vectors.

The resolution of this unsatisfactory state is to attach a copy of the vector space $\mathbb{R}^3$ to each point of three-dimensional space and then one can accommodate all possible vectors being based at all possible points. The resulting object is the *tangent bundle* of three space, and is denoted $T\mathbb{R}^3$. It is equal to $\mathbb{R}^3 \times \mathbb{R}^3$, with the first factor holding base points and the second the tangent vectors. The collection of vectors attached to a single basepoint $\mathbf{p}$ is called the *tangent space* or *fiber* at that point and is written $T_{\mathbf{p}}\mathbb{R}^3$.

A vector field assigns a vector to each basepoint in $\mathbb{R}^3$. Thus properly speaking a vector field is a map $\mathbf{U} : \mathbb{R}^3 \to T\mathbb{R}^3$. Amongst all such maps vector fields have the distinguishing property that they map a basepoint to a vector based at that point. This is symbolically expressed by defining the projection $\pi : T\mathbb{R}^3 \to \mathbb{R}^3$ which takes a vector to its basepoint $\pi(\mathbf{p}, \vec{v}) = \mathbf{p}$. A vector field then satisfies $\pi \circ \mathbf{U} = \mathrm{id}$. Since it picks out just one vector "above" each point, a vector field is called a *cross section* of the tangent bundle.

The velocity field studied in Exercise 3.1 is formally written as $\mathbf{U}(p_1, p_2) = (p_1, p_2, -p_2, p_1)$. It is clear why in practise the first two components are usually suppressed, but it is equally clear that they need to be there in the proper understanding of a vector field. A compromise is to use the base point as a subscript, and so, for example, $\mathbf{v}_{\mathbf{p}}$ represents a vector based at $\mathbf{p}$.

For a fluid region B contained in $\mathbb{R}^3$, its tangent bundle is denoted TB and is also a Cartesian product $TB = B \times \mathbb{R}^3$. For a 3-dimensional manifold the collection of tangent vectors based at a single point is always the vector space $\mathbb{R}^3$, and locally any tangent bundle looks like a product, but in general the entire bundle is not a product. The simplest example is in one dimension lower and is provided by the two sphere, S^2. Now if in fact, $TS^2 = S^2 \times \mathbb{R}^2$, then for each point $\mathbf{p}$ on the sphere, we pick the vector $(\mathbf{p}, (1,0))$, yielding a nonvanishing vector field on the sphere. This contradicts the "hairy ball" theorem (see, for example, [11]).

3.3. TRANSPORT OF VECTOR FIELDS

The next step is to understand how vectors and vector fields are moved by the fluid, or more generally transported by a diffeomorphism. Recall that Taylor's theorem

$$f(\mathbf{p}+\mathbf{h}) = f(\mathbf{p}) + \nabla f(\mathbf{p})\mathbf{h} + o(|\mathbf{h}|)$$

says that up to an error, on small scales the function f acts like the linear map $\nabla f(\mathbf{p})$. In the sciences it is common to use $\delta\mathbf{p}$ in place of $\mathbf{h}$, representing a differential or infinitesimal displacement in the actual space. However, in Differential Topology the linear action of the derivative is lifted from the manifold up to the attached tangent vectors. Thus the function f induces a map on the tangent bundle with the map itself transporting basepoints and the derivative transporting tangent vectors. This induced map is called the *tangent map* and is written as $Tf(\mathbf{p}, \mathbf{v}_\mathbf{p}) = (f(\mathbf{p}), \nabla f(\mathbf{p})\mathbf{v}_\mathbf{p})$.

The transport of vectors can also be viewed through the chain rule. Let us focus on a vector $\mathbf{v}_\mathbf{p}$ and say it is a vector in a magnetic field and that $\gamma(s)$ is a parameterization of the field line through $\mathbf{p}$, with $\gamma(0) = \mathbf{p}$ and $\gamma'(0) = \mathbf{v}_\mathbf{p}$. Now we transform the entire field by the diffeomorphism f. The field line through $\mathbf{p}$ is transformed to a curve parameterized as $f(\gamma(s))$ and so the vector $\mathbf{v}_\mathbf{p}$ is sent to the corresponding tangent vector of the image,

$$\left.\frac{df(\gamma(t))}{dt}\right|_{t=0} = \nabla f(p)\gamma'(0) = \nabla f(p)\mathbf{v}_\mathbf{p} \ .$$

Thus we see that the appropriate transport of the vector $\mathbf{v}_\mathbf{p}$ is its image under the linear transformation ∇f. Note that under the tangent map a tangent vector is not rigidly transported, but is acted on by the linear map that best reflects the stretching, shearing, and rotation that occurs near the point $\mathbf{p}$ as it is transformed by f.

Since the tangent map can act on all tangent vectors at once, a diffeomorphism $f : M \to N$ induces a tangent map $Tf : TM \to TN$. It also acts on vector fields, pushing forward a vector field $\mathbf{v}$ defined on M to one defined on N, denoted $f_*(\mathbf{v})$. It is also common to use the pull back $f^*(\mathbf{v})$, defined by pushing forward by f^{-1}. A vector field $\mathbf{v}$ is said to be *invariant* under the action of a diffeomorphism f if $f_*(\mathbf{v}) = \mathbf{v}$. Given a fluid motion ϕ_t and a vector field $\mathbf{v}$ (say a magnetic field), then its *push forward* or *transport* under the flow is $\phi_{t*}(\mathbf{v})$. The vector field is *invariant under the flow* or *frozen into the fluid* if $\phi_{t*}(\mathbf{v}) = \mathbf{v}$ for all times t.

As noted above, vector fields and flows go hand in hand. Since a diffeomorphism can be viewed as a change of coordinates one would expect that pushing forward a vector field and then constructing its flow should give the same result as pushing forward the flow of the original vector field. More precisely, if $\mathbf{u}$ has flow ϕ_t and $\mathbf{v}$ has flow ψ_t, we have $f_*\mathbf{u} = \mathbf{v}$ if and only if $f_*\phi_t = \psi_t$, which by definition says that $f \circ \phi_t \circ f^{-1} = \psi_t$.

3.4. LIE DERIVATIVES

As a general notion, the Lie derivative of a structure with respect to the vector field $\mathbf{u}$ measures the rate of change of the structure as it is transported by the flow of $\mathbf{u}$. Assume now that $\mathbf{u}$ and its corresponding flow ϕ_t are steady and that $\mathbf{v}$ is also time independent, then the Lie derivative of $\mathbf{v}$ with respect to $\mathbf{u}$ is

$$\mathcal{L}_{\mathbf{u}}\mathbf{v} = \left.\frac{d(\phi_t^*\mathbf{v})}{dt}\right|_{t=0} .$$

The Lie derivative is sometimes called "the fisherman's derivative" since it corresponds to sitting at one point and measuring the rate of change as the transported vector field goes by. In Euclidean space a computation yields

$$\mathcal{L}_{\mathbf{u}}\mathbf{v} = (\mathbf{u}\cdot\nabla)\mathbf{v} - (\mathbf{v}\cdot\nabla)\mathbf{u} . \tag{3}$$

Although the Lie derivative by definition only measures what is happening at time $t = 0$, it also captures other times as well. This is expressed in the formula

$$\frac{d}{dt}(\phi_t^*\mathbf{v}) = \phi_t^*(\mathcal{L}_{\mathbf{u}}\mathbf{v}) , \tag{4}$$

which says that at any time the derivative of the pull back of $\mathbf{v}$ is the pull back of the Lie derivative. Equation (4) immediately yields that $\mathcal{L}_{\mathbf{u}}\mathbf{v} = 0$ if and only if $\phi_t^*\mathbf{v} = \mathbf{v}$, and so the vanishing of the Lie derivative is a differential condition that implies that $\mathbf{v}$ is frozen into the flow of $\mathbf{u}$.

The Lie derivative $\mathcal{L}_{\mathbf{u}}\mathbf{v}$ is sometimes written as the Lie bracket $[\mathbf{u},\mathbf{v}]$ and it has many marvellous algebraic and analytic properties. We mention just two here. The first is that $[\mathbf{u},\mathbf{v}] = -[\mathbf{v},\mathbf{u}]$, and so $\mathcal{L}_{\mathbf{u}}\mathbf{v} = 0$ implies $\mathcal{L}_{\mathbf{v}}\mathbf{u} = 0$. In Fluid Mechanics one usually thinks of a fluid flow with velocity field $\mathbf{u}$ and a different kind of physical object, say a magnetic field $\mathbf{v}$, as being transported in the flow. But both are vector fields and can be used to generate flows. If $\mathbf{v}$ is frozen in the flow of $\mathbf{u}$, then we can turn $\mathbf{v}$ into a flow and $\mathbf{u}$ will be frozen into that. Another nice property is that when $[\mathbf{u},\mathbf{v}] = 0$, the corresponding flows commute, i.e. $\phi_t \circ \psi_s = \psi_s \circ \phi_t$ for all t and s.

We can also define the Lie derivative of a time independent scalar field α as

$$\mathcal{L}_{\mathbf{u}}\alpha = \left.\frac{d(\phi_t^*\alpha)}{dt}\right|_{t=0} .$$

As with vector fields, $\mathcal{L}_{\mathbf{u}}\alpha = 0$ means that α is frozen in, i.e. constant on trajectories of the flow. Since by definition $\phi_t^*\alpha(\mathbf{p}) = \alpha(\phi_t(\mathbf{p}))$, we see that in Euclidean space, the Lie derivative of a steady scalar field is the same as its material derivative and is computed as $D\alpha/Dt = (\mathbf{u}\cdot\nabla)\alpha$.

Example 3.2 The prototypical fluid mechanical example of a frozen in field is the vorticity field $\boldsymbol{\omega} = \text{curl}(\mathbf{u})$ for a steady, incompressible, constant density, Euler flow. In Euclidean space the vector field satisfies the equation

$$(\mathbf{u}\cdot\nabla)\mathbf{u} = -\nabla P , \tag{5}$$

where we use a capital P for the pressure and assume that the density is one. Letting $\Phi = \frac{1}{2}(\mathbf{u}\cdot\mathbf{u}) + P$ be the Bernoulli function, standard vector identities turn (5) into

$$\nabla(\Phi) = \mathbf{u}\times\boldsymbol{\omega} . \tag{6}$$

Dotting this by $\mathbf{u}$ gives $\mathcal{L}_{\mathbf{u}}\Phi = 0$ and so Φ is constant on flow lines. Taking the curl of (6) gives

$$0 = \text{curl}(\mathbf{u}\times\boldsymbol{\omega}) , \tag{7}$$

and vector identities with (3) yield that $\mathcal{L}_{\mathbf{u}}\boldsymbol{\omega} = 0$. More can be obtained by assuming that $\mathbf{u}\times\boldsymbol{\omega} \neq 0$, except at perhaps a finite number of points. Via (6) this implies that except for a finite number of exceptions any level set S of Φ (called a Lamb surface) is a two-manifold. Further, (6) says that $\boldsymbol{\omega}$ as well as $\mathbf{u}$ are tangent to S. Thus restricted to S, since $\mathcal{L}_{\mathbf{u}}\boldsymbol{\omega} = 0$, the flow and the (artificial) flow made from $\boldsymbol{\omega}$ commute. We can now invoke a classical theorem of Liouville which says that S has to be a two-torus or a topological cylinder and the flow of $\mathbf{u}$ on it has constant direction and magnitude (perhaps after a change of coordinates). This is an outline of a basic piece of the Bernoulli-Lamb-Arnol'd theorem. See [4] and [10] for more details.

For an unsteady vector field $\mathbf{u}_t$, the Lie derivative $\mathcal{L}_{\mathbf{u}_t}$ is by definition a family of derivatives, one for each t. To compute a member of the family, one freezes a time and then computes the Lie derivative with respect to that vector field, thus using the time t streamlines and *not* the unsteady flow of $\mathbf{u}_t$. If $\mathbf{v}$ is steady, then (4) with ϕ_t the unsteady fluid motion of $\mathbf{u}_t$ still holds and $\mathbf{v}$ is frozen in if and only if $\mathcal{L}_{\mathbf{u}_t}\mathbf{v} = 0$ for all t. If $\mathbf{v}_t$ is also time dependent then

$$\frac{d\phi_t^*\mathbf{v}_t}{dt} = \phi_t^*(\frac{\partial \mathbf{v}_t}{\partial t}) + \phi_t^*(\mathcal{L}_{\mathbf{u}_t}\mathbf{v}_t) = \phi_t^*(\frac{\partial \mathbf{v}_t}{\partial t} + \mathcal{L}_{\mathbf{u}_t}\mathbf{v}_t) .$$

Thus since ϕ_t is a diffeomorphism, the condition for $\mathbf{v}_t$ to be frozen in, $\phi_{t*}v_t = v_0$, can be written

$$\frac{\partial \mathbf{v}_t}{\partial t} + \mathcal{L}_{\mathbf{u}_t}\mathbf{v}_t = 0 \tag{8}$$

for all t.

For more information on the Lie derivative see [25], [3], or [1].

4. Geometry, Metrics, and Connections

4.1. THE NEED FOR ADDITIONAL STRUCTURE

As a fluid flows, subregions of fluid are deformed by the surrounding fluid. The forces involved in these deformation are, in fact, what determine the equations that characterize fluid motions. Since the fluid maps are diffeomorphisms, all topological properties of the evolving subregions stay the same. The nature of the deformation lies in changing angles and lengths, and is therefore geometric. Thus we need a geometric structure on the flow region.

There are several other ways in which geometric considerations can be seen entering into mechanics. Most simply, the magnitude of a velocity vector is required for the kinetic energy. In addition, we have seen that a velocity vector lies in the tangent bundle, and so the acceleration (the velocity of the velocity) lies in the tangent bundle of the tangent bundle. Thus a force vector and the acceleration live in different mathematical objects, and there is no way to equate them as required by Newton's second law.

The acceleration of the fluid is the rate of change of the velocity field along a trajectory, and is thus a special case of what in Fluid Mechanics is called the material derivative. For simplicity, let $\mathbf{u}$ and $\mathbf{v}$ be steady. The derivative we require is the instantaneous rate of change of one vector field in the direction of another. In Mathematics this is called the directional derivative of $\mathbf{v}$ in the direction of $\mathbf{u}$ and is defined in Euclidean space by

$$\nabla_{\mathbf{u}}\mathbf{v}(\mathbf{p}) = \left.\frac{d(\mathbf{v}(\phi_t(\mathbf{p}))}{dt}\right|_{t=0} .$$

The chain rule and the advection equation (2) then yield that $\nabla_{\mathbf{u}}\mathbf{v} = (\nabla\mathbf{v})\cdot\mathbf{u}$, which is more commonly written in Fluid Mechanics as $(\mathbf{u}\cdot\nabla)\mathbf{v}$ where $\nabla\mathbf{v}$ is the derivative matrix of $\mathbf{v}$, sometimes called the velocity gradient 2-tensor, and has components $\frac{\partial v_i}{\partial x_j}$. To uncover the implicit assumptions in this calculation let us return to the definition of the derivative,

$$\nabla_{\mathbf{u}}\mathbf{v}(\mathbf{p}) = \lim_{t\to 0}\frac{\mathbf{v}(\phi_t(\mathbf{p})) - \mathbf{v}(\mathbf{p})}{t} . \tag{9}$$

Thus computing the derivative requires the subtraction of $\mathbf{v}(\phi_t(\mathbf{p}))$ and $\mathbf{v_p}$. But note that the vector $\mathbf{v}(\phi_t(\mathbf{p}))$ lives in the tangent space attached to point $\phi_t(\mathbf{p})$ while $\mathbf{v_p}$ is attached to the point $\mathbf{p}$, and thus they lie in different vector spaces and their difference has no meaning. When we compute this limit in Euclidean space we implicitly identify the two tangent spaces by rigidly transporting one onto the other, but on a manifold there is no natural way to do this without some additional structure.

So we see that there are, in fact, two geometric structures required to proceed with the fluid model. The first is a way to measure geometric quantities like angles and lengths, and the second is a way to compare the geometry at different points in the fluid body. These needs are fulfilled by a Riemannian metric and parallel transport, respectively. The infinitesimal version of parallel transport is the differentiation of one vector field with respect to another, and it is this process that is most fundamental and has been entitled a *linear connection.*

An introduction to Differential Geometry is given in [6], the chapter in [18] is short and sweet, and the redoubtable [23] is comprehensive and comprehensible.

4.2. CONNECTIONS AND COVARIANT DERIVATIVES

We now assume that our fluid region has a linear connection. A connection defines a way of taking derivatives, a process. It is rather algebraic in character, taking a pair of vectors fields $\mathbf{u}$ and $\mathbf{v}$ and returning a third denoted $\nabla_{\mathbf{u}}\mathbf{v}$. It is required to satisfy properties that make it behave like the familiar directional derivative in Euclidean space.

- It is linear in the $\mathbf{u}$ slot with respect to multiplication by scalar fields, $\nabla_{\alpha\mathbf{u}+\hat{\alpha}\hat{\mathbf{u}}}\mathbf{v} = \alpha\nabla_{\mathbf{u}}\mathbf{v} + \hat{\alpha}\nabla_{\hat{\mathbf{u}}}\mathbf{v}$.
- It is linear in the $\mathbf{v}$ slot with respect to multiplication by constants, $\nabla_{\mathbf{u}}(r\mathbf{v} + \hat{r}\hat{\mathbf{v}}) = r\nabla_{\mathbf{u}}\mathbf{v} + \hat{r}\nabla\hat{\mathbf{v}}$.
- When one multiplies by a scalar field in the $\mathbf{v}$ slot it must satisfy the product rule, $\nabla_{\mathbf{u}}(\alpha\mathbf{v}) = \alpha\nabla_{\mathbf{u}}\mathbf{v} + (\mathcal{L}_{\mathbf{u}}\alpha)\mathbf{v}$.

The connection itself is indicated by the symbol ∇, and $\nabla_{\mathbf{u}}\mathbf{v}$ is called the *covariant derivative* of $\mathbf{v}$ in the direction of $\mathbf{u}$. The standard connection on Euclidean space is the usual convective or directional derivative, $\nabla_{\mathbf{u}}\mathbf{v} = (\mathbf{u}\cdot\nabla)\mathbf{v}$.

In order to use the connection to compare different points we first connect them with a curve, which we parameterize as $\gamma(s)$, with s not necessarily the arc length. For simplicity we assume that γ is the integral curve of a vector field $\mathbf{u}$, i.e. the derivative of the curve $\gamma'(s)$ at the point $\gamma(s)$ is the value of the vector field at that point, $\gamma'(s) = \mathbf{u}(\gamma(s))$. Physically this curve could correspond to a fluid trajectory or perhaps a field line. We define the *covariant derivative* of the vector field $\mathbf{v}$ along γ as

$$\frac{D\mathbf{v}}{Ds}(s) = \nabla_{\mathbf{u}}\mathbf{v}(\gamma(s)) .$$

Note that $\frac{D\mathbf{v}}{Ds}$ is a vector field which is defined along γ. The vector field $\mathbf{v}$ is said to be parallel along γ if $\frac{D\mathbf{v}}{Ds} = 0$ on the whole curve. If γ starts at the point $\mathbf{p}$ and $\mathbf{w_p}$ is a tangent vector, there is a unique way to continue $\mathbf{w_p}$ to a vector field that is parallel along γ. In this way any tangent vector at $\mathbf{p}$ can be *parallel transported* to a tangent vector based at $\gamma(s)$. This process defines the parallel transport maps $\mathcal{P}_s$ from the tangent space at $\mathbf{p}$ to the tangent space at $\gamma(s)$.

Although we have assumed the existence of a connection and used it to define parallel transport note that our discussion is consistent in the sense that using $\mathcal{P}$ as the analog of implicit rigid translation in (9) does correctly compute the derivative because

$$\nabla_{\mathbf{u}}\mathbf{v}(\mathbf{p}) = \left.\frac{d\mathcal{P}_s^*\mathbf{v}}{ds}\right|_{s=0} = \lim_{s\to 0}\frac{\mathcal{P}_s^*\mathbf{v} - \mathbf{v}(\mathbf{p})}{s} . \tag{10}$$

It is important to note that the parallel transport of tangent vectors between two points usually depends on the curve we choose between them. In fact the only situation in which all parallel transport is independent of path is when there is no curvature. In spite of this, the limit in (10) is independent of the choice of curve, and in fact, it may be used to define $\nabla_{\mathbf{w_p}}\mathbf{v}$ for a given vector $\mathbf{w_p}$, since the limit will not depend on how $\mathbf{w_p}$ is extended to a vector field on the whole fluid region.

Given a connection, the acceleration of fluid particles in the steady case is $\nabla_{\mathbf{u}}\mathbf{u}$ which is, as required, a vector field in the tangent bundle and not in the tangent tangent bundle.

4.3. RIEMANNIAN METRICS

We also need to quantify such geometric notions as lengths, angles and volumes. In Euclidean space this is done via the usual inner product $\vec{u}\cdot\vec{v} = \sum u_i v_i$, which defines a length as $\|\vec{v}\| = (\vec{v}\cdot\vec{v})^{\frac{1}{2}}$ and an angle using $\frac{\vec{v}\cdot\vec{u}}{\|\vec{v}\|\|\vec{u}\|}$. In general, an *inner product* on a vector space is a rule that takes two vectors and returns a number. A usual notation is $\langle\vec{u},\vec{v}\rangle$ or $\iota(\vec{u},\vec{v})$. It is further required that an inner product be linear in each argument, symmetric $\iota(\vec{u},\vec{v}) = \iota(\vec{v},\vec{u})$, and positive definite $\vec{v}\neq 0$ implies $\iota(\vec{v},\vec{v}) > 0$. A linear isomorphism L (i.e. a linear bijection) induces a pull back on a inner product as $L^*(\iota)(\vec{u},\vec{v}) = \iota(L(\vec{u}),L(\vec{v}))$. The isomorphism is said to be an *isometry* between the inner products $\hat{\iota}$ and ι if it preserves the inner product, $L^*(\iota) = \hat{\iota}$.

On a curved manifold the way of measuring lengths and angles vary from point to point and so the geometry is specified by a *Riemannian metric* or *metric tensor* which is a family of inner products, with one defined on the tangent space to each point. The metric itself is denoted g, the particular inner product at the point $\mathbf{p}$ is denoted $g_{\mathbf{p}}$ (but the subscript is often suppressed), and its value on a pair of vectors $\mathbf{v_p}$ and $\mathbf{u_p}$ based at $\mathbf{p}$ is $g_{\mathbf{p}}(\mathbf{v_p}, \mathbf{u_p})$. A metric is often indicated by its components in a basis as g_{ij}, or as a line element $ds^2 = g_{11}\, dx_1^2 + g_{22}\, dx_2^2 + \ldots$. Recall from Section 3.3 that a diffeomorphism $h : \hat{M} \to M$ induces a linear isomorphism on tangent spaces via the tangent maps, $Th_{\mathbf{p}} : T\hat{M}_{\mathbf{p}} \to TM_{h(\mathbf{p})}$. If M has metric g, then each of these $Th_{\mathbf{p}}$ induce a pull back of $g_{\mathbf{p}}$, which in turns creates a pull back of the entire metric which is denoted h^*g. If $\hat{M}$ has a metric $\hat{g}$, then h is called an *isometry* if Th is an isometry on every tangent space, i.e. if $h^*g = \hat{g}$. Isometries are the isomorphisms in the category of Riemannian manifolds.

The Riemannian metric immediately allows to define the speed, or magnitude of a velocity vector as $\|\mathbf{u_p}\|_g = g_{\mathbf{p}}(\mathbf{u_p}, \mathbf{u_p})^{1/2}$ and the length of the curve $\gamma : [a, b] \to B$ as

$$\int_a^b \|\gamma'(s)\|_g \, ds \; . \tag{11}$$

Parallel transport as defined by the connection was introduced as the analog of rigid translation in Euclidean space. In particular, it should preserve lengths and angles, and so it should preserve the inner products on tangent spaces. Thus we now require our connection to be *compatible with the metric* in the sense that the parallel transport maps, $\mathcal{P}_s$, must induce an isometry of the tangent spaces at points along the curve.

There are actually many connections that are compatible with a given metric. To get a unique connection, we also require that the connection be compatible with the Lie derivative in the same way it is in Euclidean space

$$\mathcal{L}_{\mathbf{u}}\mathbf{v} = \nabla_{\mathbf{u}}\mathbf{v} - \nabla_{\mathbf{v}}\mathbf{u} \; .$$

If this condition holds the connection is said to be *symmetric* or *torsion free*. The fundamental theorem of Riemannian geometry states that there is a unique symmetric connection (called the *Riemannian* or *Levi-Civita* connection) that is compatible with a given Riemannian metric. Henceforth we assume that our fluid region has a Riemannian metric g and ∇ is its connection.

4.4. VOLUME FORMS

In Euclidean space we get the volume of a box from the product of its side lengths. A Riemannian metric determines a way of computing length and so it also provides a way to compute a volume. As with the inner product we begin with the notion of a linear volume on $\mathbb{R}^3$. Given three vectors, the Euclidean volume of the parallelepiped they span is the determinant of the matrix whose columns are the three vectors. The appropriate notion of volume generalizes this situation. A volume element σ is a map that takes three vectors and returns a number, $\sigma(\vec{u}, \vec{v}, \vec{w})$. Since a volume element is supposed to act like the usual determinant, it is required to be linear in all three arguments and interchanging two vectors must change the sign. A linear isomorphism L induces a pull back on a volume element as $L^*(\sigma)(\vec{u}, \vec{v}, \vec{w}) = \sigma(L(\vec{u}), L(\vec{v}), L(\vec{w}))$.

Exercise 4.1 If σ and $\widehat{\sigma}$ are volume elements on $\mathbb{R}^3$, show that there is a number r so that $\widehat{\sigma} = r\,\sigma$, i.e. there is a single r that works for all choices of triples of input vectors.

A family of volume elements on a manifold, one for each tangent space, is called a *volume form*. A diffeomorphism, f, induces a pull back, $f^*\chi$, of a volume form χ in a manner completely analogous to the pull back of an inner product. The diffeomorphism *preserves the volume form* if $f^*\chi = \chi$.

To construct a volume form that is compatible with the metric first note that as a consequence of the exercise, given an inner product on $\mathbb{R}^3$ and an orthonormal basis b there is a unique volume element which gives it volume 1. Switching exactly two elements in the chosen orthonormal basis yields a basis $\hat{b}$ with volume -1. Every other orthonormal basis will be assigned a volume of 1 or -1, and the two classes are distinguished by whether the basis in question can be continuously moved to b or to $\hat{b}$. A Riemannian manifold has an inner product on each tangent space, and so we may choose an orthonormal basis on each tangent space and thus obtain a compatible volume element. Since parallel transport is an isometry on tangent spaces, it takes orthonormal bases to orthonormal bases. If our volume elements are to fit together in a nice family, the volume of an orthonormal basis should not be changed by parallel transport. If the volume elements can be chosen so this is possible, the manifold is called *orientable*. The simplest example of a non-orientable manifold is the two-dimensional Möbious band with the metric induced from how it lives in $\mathbb{R}^3$. As one traverses the core circle, an orthonormal basis will come back with one vector flipped, resulting in a basis assigned the opposite area it had initially.

We now assume that the fluid region is orientable, and for definiteness we pick a orthonormal basis b on one tangent space and a volume element which assigns it a volume 1. Then we can find a family of volume elements, one on each tangent space, which give all parallel translates of b the volume 1. The resulting object is called a *Riemannian volume form*, and we denote it μ. Note that switching a pair of elements of the original basis b would result in the negative of the volume form we have chosen. Applying Exercise

4.1 on each tangent space shows that for any volume form χ there is a scalar field r called the *density* with $\chi = r\mu$.

In (11) we computed the length of a curve by integrating the norm of its tangent vectors as defined by the metric. In the same way we can compute the volume of a region by integrating the value of the volume form on its tangent vectors. This is done most succinctly by using the pull back under its parameterization. We begin with the case where U is a region in $\mathbb{R}^3$ with the Euclidean form μ_E. Since μ_E should give the usual volumes, if χ is a volume form on U with density r we define

$$\int_U \chi = \int_U r\mu_E = \int_U r(x,y,z)\, dx\, dy\, dz \ .$$

Now if V is a region in a manifold that fits into a single coordinate chart it has a parameterization in terms of a subset of Euclidean space. This is a diffeomorphism h from some region U in $\mathbb{R}^3$ onto V. If χ is a volume form on the manifold define

$$\int_V \chi = \int_U h^*\chi \ ,$$

where to be very careful we must insist that the choice of h was such that $\int_U h^*(\mu) > 0$ for the Riemannian volume form μ on V, or equivalently that $h^*(\mu) = r\,\mu_E$ for a positive function r . If the region V is large, chop it up into smaller pieces that fit into charts and add the integrals.

Volumes and measures are closely related and in many cases one can pass from one to the other. If μ a Riemannian volume form, the volume form χ is *positive* if its density r is a positive function. Given a positive volume form we can define a measure via integration $m(V) = \int_V \chi$ on nice sets V and on nastier sets using limits of nice sets. In a steady fluid motion if ρ is the mass density, then $\chi = \rho\mu$ is the *mass form* and the flow is mass preserving if that form is preserved, $\phi_t^*(\chi) = \chi$. This implies that the corresponding measure is preserved, $\phi_t^*(m) = m$. One advantage which volume forms have over measures is that they are linear objects and so can be added, subtracted, differentiated, etc.

Recall that in Euclidean space the Jacobian $J(f)$ of a diffeomorphism f is the determinant of the Jacobian matrix, i.e. $\det(\nabla f)$. Since the determinant of a linear transformation is equal to the (signed) volume of the image of the unit cube, and the derivative (the Jacobian matrix) is the best local linear approximation of the map, we see that the Jacobian computes the local change in volume under the map f. It follows fairly easily that for a diffeomorphism f, $f_*(\mu_E) = J(f)\mu_E$. Thus, in particular, a fluid motion is incompressible exactly when $J(\phi_t) = 1$ for all times and at all points. Similarly, a flow preserves a mass with density ρ if and only if $\rho\mu_E = \phi_t^*(\rho\mu_E) = \phi_t^*(\rho)\phi_t^*(\mu_E) = \phi_t^*(\rho)J(\phi_t)\mu_E$, which reduces to a scalar equation $\phi_t^*(\rho) = J(\phi_t)\rho$. The same derivation works if ϕ_t is unsteady and ρ is time dependent in which case the scalar equation is written in the familiar form $\rho = J\rho_0$.

On the general Riemannian manifold case one *defines* the Jacobian (with respect to μ) of the diffeomorphism h as the scalar field $J(h)$ that satisfies $h_*(\mu) = J(h)\mu$, which is to say that $J(h)$ is the density of $h_*(\mu)$. This is a common strategy in coordinate free definitions, a property of an object in Euclidean space is taken as the coordinate free definition.

4.5. THE LIE, COVARIANT AND MATERIAL DERIVATIVES

Both the Lie derivative and the covariant derivative measure a rate of change of one vector field with respect to another. In this subsection we compare them and remark on their generalizations. We put the definitions side by side

$$\nabla_{\mathbf{u}}\mathbf{v} = \left.\frac{d\mathcal{P}_t^*(\mathbf{v})}{dt}\right|_{t=0} \quad \text{and} \quad \mathcal{L}_{\mathbf{u}}\mathbf{v} = \left.\frac{d\phi_t^*(\mathbf{v})}{dt}\right|_{t=0}, \qquad (12)$$

and note that the only difference is the way in which the vector field is pulled back. In the Lie case one pulls back the advected vector field using the derivative of the flow and so the Lie derivative measures the deformations of the advected vector field. In contrast, in the covariant derivative one pulls back using an isometry, and so this derivative does not measure the deformations during advection, but rather just how the vector field is changing with respect to the metric. If $\mathcal{L}_{\mathbf{u}}\mathbf{v} = 0$, it means that the vector field $\mathbf{v}$ is transported to itself, i.e. $\phi_{t*}\mathbf{v} = \mathbf{v}$, which is a strictly topological notion, and indeed the Lie derivative requires only Differential Topology. On the other hand, $\nabla_{\mathbf{u}}\mathbf{v} = 0$ along a trajectory means that the vector field is constant along the trajectory *as measured by the metric*, and indeed $\nabla_{\mathbf{u}}\mathbf{v} = 0$ is used to define the geometric notion of parallel transport.

In analogy with (12) we may compute the Lie and covariant derivative of any kind of structure that can be pulled back and subtracted. Such structures include Riemannian metrics and volume forms. These are examples of *contravariant k-tensors* which are families of multilinear maps, one for each tangent space, which take k tangent vectors as input and return a number at each point, so as a global object they return a scalar field. Note that what most mathematicians call a contravariant tensor is called a covariant tensor by most engineers and physicists, and vice versa. Only one kind of tensor arises here so we drop the potentially confusing adjective; a Riemannian metric is a 2-tensor and a volume form is a 3-tensor. In general, if τ

is a tensor, define

$$\nabla_{\mathbf{u}}\tau = \left.\frac{d\mathcal{P}_t^*(\tau)}{dt}\right|_{t=0} \quad \text{and} \quad \mathcal{L}_{\mathbf{u}}\tau = \left.\frac{d\phi_t^*(\tau)}{dt}\right|_{t=0} . \tag{13}$$

In this case the Lie and covariant derivatives have the same dynamical meaning as in the vector field case. The condition $\mathcal{L}_{\mathbf{u}}\tau = 0$ says that $\phi_t^*(\tau) = \tau$ and so τ is frozen into the flow. On the hand, $\nabla_{\mathbf{u}}\tau = 0$, means that from the point of view of the metric, τ is not changing along trajectories.

Example 4.2 The Riemannian connection associated with a metric g was required to have the property that each $\mathcal{P}_s$ is an isometry. This says that $\mathcal{P}_t^* g = g$, or that $\nabla_{\mathbf{u}} g = 0$ for any $\mathbf{u}$. In contrast, $\mathcal{L}_{\mathbf{u}} g = 0$ is a very rare situation. It says that the metric is invariant under the flow, and so each ϕ_t is an isometry. Such a $\mathbf{u}$ is called a *Killing field* for the metric and in Euclidean space with the usual metric any flow of a Killing field is either a rigid rotation or a translation, or composition of these.

A scalar field α is a 0-tensor and in this case the two derivatives are equal, $\mathcal{L}_{\mathbf{u}}\alpha = \nabla_{\mathbf{u}}\alpha$. This is because the pull backs are the same; an advected scalar field does not feel the local deformations during the evolution.

The Lie and covariant derivatives behave differently when transported by a diffeomorphism h. The Lie derivative satisfies $h^*(\mathcal{L}_{\mathbf{u}}\mathbf{v}) = \mathcal{L}_{h^*\mathbf{u}} h^*\mathbf{v}$. This is sometimes called the "naturalness" of the Lie derivative, and expresses the fact that the definition of the Lie derivative requires only Differential Topology and so is preserved by a diffeomorphism. Thus the same definition works before and after the action of h, or in symbols $h^*(\mathcal{L}) = \mathcal{L}$. On the other hand, the covariant derivative depends on the metric, and so $h^*(\nabla)$ is the Riemannian connection $\hat{\nabla}$ of the pulled back metric $h^*(g)$, and so $h^*(\nabla_{\mathbf{u}}\mathbf{v}) = \hat{\nabla}_{h^*\mathbf{u}} h^*\mathbf{v}$.

In Euclidean space the *material derivative* with respect to $\mathbf{u}$ is the operator $\frac{D}{Dt} = \frac{\partial}{\partial t} + (\mathbf{u} \cdot \nabla)$ and it computes the rate of change of say a time dependent vector field $\mathbf{v}$ along the flow as

$$\frac{D\mathbf{v}}{Dt}(\phi_t(\mathbf{p})) = \frac{\partial \mathbf{v}(\phi_t(\mathbf{p}), t)}{\partial t} . \tag{14}$$

The obvious generalization using the covariant derivative $\frac{D}{Dt} = \frac{\partial}{\partial t} + \nabla_{\mathbf{u}}$ maintains this meaning. Thus, for example, the acceleration of a fluid particle under the perhaps unsteady velocity field $\mathbf{u}$ is given as usual by $\frac{D\mathbf{u}}{Dt}$, and for a perhaps time dependent scalar field the familiar condition $\frac{D\alpha}{Dt} = 0$ says that the scalar field is passively transported.

4.6. DIV, GRAD, CURL AND THE LAPLACIAN

We may now give coordinate free definitions of the standard operations on velocity fields. In Euclidean space these can be built out of the velocity gradient tensor, $\nabla \mathbf{u} = (\partial u_i / \partial x_j)$. Its generalization is the *covariant derivative* of $\mathbf{u}$, also denoted $\nabla \mathbf{u}$. In the previous subsections we took covariant derivatives in the direction of $\mathbf{u}$. To define the derivative of $\mathbf{u}$ itself recall that a connection is required to be linear over scalar fields in the direction in which one is differentiating, i.e. $\nabla_{\mathbf{v}} \mathbf{u}$ is linear in the $\mathbf{v}$ slot. Thus $\nabla \mathbf{u}$ can be defined as the linear transformation on each tangent space which inputs a vector $\mathbf{v}_{\mathbf{p}}$ and outputs the rate of change of $\mathbf{u}$ in that direction $\nabla_{\mathbf{v}_{\mathbf{p}}} \mathbf{u}$. Thus $\nabla \mathbf{u}(\mathbf{v})$ means $\nabla_{\mathbf{v}} \mathbf{u}$.

In a similar vein, given a scalar field α its derivative is the linear functional that inputs a vector $\mathbf{v}_{\mathbf{p}}$ and outputs the rate of change of α in that direction $\nabla_{\mathbf{v}_{\mathbf{p}}} \alpha$. Since the notation $\nabla \alpha$ has a commonly accepted meaning as a vector field, the linear functional is called $d\alpha$, and so $d\alpha(\mathbf{v}_{\mathbf{p}}) = \nabla_{\mathbf{v}_{\mathbf{p}}} \alpha$. The *gradient vector field* $\nabla \alpha$ is defined using the metric as the unique vector field that satisfies $d\alpha(\mathbf{v}_{\mathbf{p}}) = g(\nabla \alpha, \mathbf{v}_{\mathbf{p}})$ for all vectors $\mathbf{v}_{\mathbf{p}}$. This means that $\nabla \alpha$ is the direction of maximum increase of α as measured by the metric, in accord with the usual situation in Euclidean space where $\nabla_{\mathbf{v}_{\mathbf{p}}} \alpha = \mathbf{v}_{\mathbf{p}} \cdot \nabla \alpha$.

Since $\nabla \mathbf{u}$ gives a linear transformation on each tangent space, its trace is independent of the choice of basis and so we may define $\operatorname{div}(\mathbf{u}) = \operatorname{trace}(\nabla \mathbf{u})$. As in the Euclidean case the divergence measures the infinitesimal rate of change of volumes as they are transported which is expressed by $\mathcal{L}_{\mathbf{u}} \mu = \operatorname{div}(\mathbf{u}) \mu$, where μ is the Riemannian volume form. Thus as usual, ϕ_t is incompressible exactly when $\operatorname{div}(\mathbf{u}) = 0$. If the perhaps time dependent density is ρ_t, then the mass form is $\rho_t \mu$, and conservation of mass says that $\phi_t^*(\rho_t \mu) = \rho_0 \mu$. Thus using the analog of (8) for tensors we have that

$$0 = \frac{\partial(\rho_t \mu)}{\partial t} + \mathcal{L}_{\mathbf{u}}(\rho_t \mu) \ .$$

Now Leibnitz' rule for the Lie derivative says that $\mathcal{L}_{\mathbf{u}}(\rho_t \mu) = (\mathcal{L}_{\mathbf{u}} \rho_t) \mu + \rho_t \mathcal{L}_{\mathbf{u}} \mu = (\nabla_{\mathbf{u}} \rho_t) \mu + \rho_t \operatorname{div}(\mathbf{u}) \mu$, using the definition of the divergence and the equality of the Lie and covariant derivatives for scalar fields. Since μ is time independent $\frac{\partial(\rho_t \mu)}{\partial t} = \frac{\partial \rho_t}{\partial t} \mu$. Equating the coefficients of μ and using the definition of the material derivative yields the *mass conservation equation* or *continuity equation*

$$\frac{D\rho_t}{Dt} + \rho_t \operatorname{div}(\mathbf{u}) = 0 \ . \qquad (15)$$

Note that the preceding derivation did not require $\mathbf{u}$ to be steady.

In Euclidean space the symmetric and skew symmetric parts of $\nabla \mathbf{u}$ yield the deformation and rotation tensor. To formulate the generalization we require a transpose. Recall that in $\mathbb{R}^3$, the transpose of a linear transformation A with respect to inner product ι is the unique linear transformation A^T that satisfies $\iota(A^T(\vec{v}), \vec{w}) = \iota(\vec{v}, A(\vec{w}))$. Thus working with the metric g we can define $(\nabla\mathbf{u})^T$ as satisfying $g((\nabla\mathbf{u})^T(\mathbf{v_p}), \mathbf{w_p}) = g(\mathbf{v_p}, \nabla\mathbf{u}(\mathbf{w_p}))$ on each tangent space.

We then define the symmetric part of $\nabla\mathbf{u}$ as $D(\mathbf{u}) = \frac{1}{2}(\nabla\mathbf{u} + (\nabla\mathbf{u})^T)$ and the skew symmetric part as $\Omega(\mathbf{u}) = \frac{1}{2}(\nabla\mathbf{u} - (\nabla\mathbf{u})^T)$, and so $\nabla\mathbf{u} = D(\mathbf{u}) + \Omega(\mathbf{u})$. Since D is symmetric, it has three orthogonal principal directions and D diagonalizes in that basis. The diagonal elements represent the infinitesimal deformation rates in each direction and so D is called the *deformation tensor* (although at this point it is formally a linear transformation). In Euclidean space Ω is related to $\boldsymbol{\omega} = \text{curl}(\mathbf{u})$ by the formula

$$\Omega(\vec{v}) = \frac{1}{2}\boldsymbol{\omega} \times \vec{v} \,, \tag{16}$$

for any vector $\vec{v}$, and so Ω is sometimes called the *rotation tensor.*

We can find our way to the covariant definition of the curl by dotting both sides of (16) by $\vec{w}$, yielding

$$\vec{w} \cdot \Omega(\vec{v}) = \frac{1}{2}\vec{w} \cdot (\boldsymbol{\omega} \times \vec{v}) = \frac{1}{2}\det(\vec{w}, \boldsymbol{\omega}, \vec{v}) \,, \tag{17}$$

where we have used the standard identity connecting the triple scalar product and the determinant of the matrix whose columns are the vectors. All of the terms in (17) have a covariant generalization, so in a now familiar move we define curl($\mathbf{u}$) as the unique vector field $\boldsymbol{\omega}$ satisfying

$$g(\vec{w}, \Omega(\vec{v})) = \frac{1}{2}\mu(\vec{w}, \boldsymbol{\omega}, \vec{v}) \,. \tag{18}$$

Since the metric quantifies deformation, one would expect a close relationship between the deformation tensor and the metric. Since the metric is a 2-tensor, we will change D from a linear transformation to a 2 tensor. There is a standard way to do this using the metric called lowering the indices or the $\flat$ operator. The $\widetilde{D(\mathbf{u})}$ which is associated with $D(\mathbf{u})$ is the unique 2-tensor $\widetilde{D(\mathbf{u})}$ with $\widetilde{D(\mathbf{u})}(\mathbf{w_p}, \mathbf{v_p}) = g(D(\mathbf{u})(\mathbf{w_p}), \mathbf{v_p})$. The relationship of the deformation tensor and the metric is expressed by $\mathcal{L}_{\mathbf{u}}g = 2\widetilde{D(\mathbf{u})}$, which says that $\widetilde{D(\mathbf{u})}$ exactly measures how the metric is deformed as it is advected by the flow. If we also turn Ω into a 2-tensor we find that the curl satisfies $\mu(\mathbf{w_p}, \text{curl}(\mathbf{u}), \mathbf{v_p}) = 2\widetilde{\Omega(\mathbf{u})}(\mathbf{w_p}, \mathbf{v_p})$.

The Laplacian of a scalar field is covariantly defined as $\Delta(\alpha) = \text{div}(\nabla\alpha)$. Since the first derivative of a vector field requires the use of a connection one might suspect that the Laplacian of a vector field would require yet more structure. This is fortunately not the case. We can use (13) to define the covariant derivative of $\nabla \mathbf{u}$ in a given direction. Treating the result as a function of the direction we get an object denoted $(\nabla\nabla)\mathbf{u}$ which takes two vectors as input and gives another as output. The Laplacian of $\mathbf{u}$ is the trace of this object, $\Delta\mathbf{u} = \mathrm{trace}((\nabla\nabla)\mathbf{u})$. To be clear on the type of trace we are taking, if we choose an orthonormal basis, e_i, with respect to the metric on each tangent space, then $\Delta\mathbf{u} = \sum_i (\nabla\nabla)\mathbf{u}(e_i, e_i)$.

The Laplacian we have just defined is sometimes called the *analyst's Laplacian.* The *topologist's Laplacian* is defined using the (negative of the) analog of the Euclidean space formula $\Delta\mathbf{u} = \nabla(\mathrm{div}(\mathbf{u})) - \mathrm{curl}\,\mathrm{curl}\,\mathbf{u}$. After a sign switch the two Laplacians differ by the Ricci curvature.

Differential forms are the other common way to give coordinate free definitions of the standard vector calculus notions. Both points of view are important and have their virtues: forms work well with integration and are directly connected to the underlying topology, but the covariant derivative is most naturally related to velocity fields. The two methods are intimately connected and we have, in fact, already encountered the 1-form $d\alpha$, the 2-form $\tilde{\Omega}$, and the three form μ. Forms were not explicitly discussed here only because space and time limitations demanded the most direct path to the goal. The reader is urged to consult [9], [25], [2], [3] or [1].

5. Equations

We now have the mathematical equipment to bring forces into the model and state the basic dynamical equations of Fluid Mechanics. This is familiar material for fluid mechanicians, but for completeness we give a brief summary. Assume that there are no external forces and so the only forces to consider are internal, the force that the fluid body exerts on a subregion across its boundary. The force per unit area on the bounding surface is the *stress* and its exact form is encapsulated in the existence and properties of the Cauchy stress tensor. The usual derivation of the basic dynamic equations in Euclidean space invokes Newton's laws to say that the rate of change of momentum of a patch of fluid is equal to the total surfaces forces on it. If it is assumed that the only stresses are normal to the bounding surface, one obtains Euler's equation

$$\rho(\frac{\partial \mathbf{u}}{\partial t} + \nabla_{\mathbf{u}}\mathbf{u}) = -\nabla P \ . \tag{E}$$

If tangential components of the stress are included one obtains the Navier-Stokes equation

$$\rho(\frac{\partial \mathbf{u}}{\partial t} + \nabla_{\mathbf{u}}\mathbf{u}) = -\nabla P + \bar{\mu}\Delta\mathbf{u} \ . \tag{NS}$$

In these equations P is the pressure, ρ is the mass density and $\bar{\mu}$ is the viscosity (not a volume form!). Note that all the operations involved in the equations have been covariantly defined. However, there are serious and subtle problems in trying to formulate the usual derivations of (NS) on a manifold. We refer the reader to [16] for a careful exposition and merely remark that the Laplacian in (NS) is the analyst's Laplacian.

Depending on the relative importance of viscosity in the fluid system under study either (E) or (NS) is adapted as the basic dynamical equation. The appropriate boundary conditions are slip and no slip, respectively. For a complete system in which all variables are determined additional equations must be added. In the most common situations it can be assumed that the viscosity is constant throughout the fluid, mass is conserved, and the fluid is incompressible. After inclusion of the mass conservation equation (15), incompressibility is equivalent to $\frac{D\rho}{Dt} = 0$, or that the density of a particle remains constant as it is transported. The simplest compressible systems use thermodynamic considerations to justify the assumption that the pressure and the density are functionally dependent.

There are three obligatory remarks to be made. The first is that all our mathematical modelling would be meaningless if it were not for the fact that the resulting models and equations give results that agree extremely well with experimental data. The second is that the existence-uniqueness theory of the Navier-Stokes and Euler equations is still far from being understood. The third is that what's in this paper just sets the stage; the real action is the understanding and prediction of fluid behavior.

One advantage of having defined all operations covariantly is that changing coordinates or regions with a diffeomorphism $h : B \to \hat{B}$ preserves the property of being a solution to a system of fluid equations. More precisely, if $\mathbf{u}$ satisfies a system on B with respect to the metric g, then $h_* \mathbf{u}$ satisfies the same system on $\hat{B}$. That's the good news. The bad news that the operations in the equations on $\hat{B}$ such as ∇ and Δ must be defined in terms of the metric $h_* g$ which may not be the metric which you care about. This begs a question:

Question 5.1 Is there a physical meaning to doing Fluid Mechanics with a general Riemannian metric?

We only hazard a few remarks. If there is a general physical interpretation of a curved metric, it cannot involve an intrinsic property of the fluid because everything in the fluid is advected, and the metric (at least as developed here) stays fixed on the manifold. There are few cases where it is clear that fluid flows over a curved space. One is the surface of the earth. Another is in very large scale fluid mechanical models in Cosmology where one can need to take into account the curvature of space-time. Also note that changing from Euclidean into curvilinear, non-orthogonal coordinates forces one to work with the push forward of the Euclidean metric. This is a rather special metric, however, being by definition isometric to the Euclidean one and therefore lacking curvature. The last remark comes from the philosophy of Mathematics described in the introduction: by understanding fluid mechanics in the most general context in which it makes sense, one gains new insights into the particular cases of interest.

Acknowledgements

Thanks to David Groisser for several very useful conversations during the preparation of this paper.

References

1. Abraham, R., Marsden, J. & Ratiu, T. (1988) *Manifolds, Tensor Analysis, and Applications.* Springer Verlag.
2. Aris, R. (1962) *Vectors, Tensors, and the Basic Equations of Fluid Mechanics.* Dover Publications.
3. Arnol'd, V. (1989) *Mathematical Methods in Classical Mechanics.* Springer Verlag.
4. Arnol'd, V. & Khesin, B. (1998) *Topological Methods in Hydrodynamics.* Springer Verlag.
5. Batchelor, G.K. (1967) *An Introduction to Fluid Mechanics.* Cambridge University Press.
6. Boothby, W. (1986) *An Introduction to Differentiable Manifolds and Riemannian Geometry.* Academic Press.
7. Chorin, A. & Marsden, J. (1993) *A Mathematical Introduction to Fluid Mechanics.* Springer Verlag.
8. Christenson, C. & Voxman, W. (1998) *Aspects of Topology.* BCS Associates.
9. Flanders, H. (1963) *Differential Forms with Applications to the Physical Sciences.* Academic Press.
10. Ghrist, R. & Komendarczyk, R. (2001) Topological features of inviscid flows. In *An Introduction to the Geometry and Topology of Fluid Flows* (ed. R.L. Ricca), this volume. NATO-ASI Series: Mathematics, Kluwer, Dordrecht, The Netherlands.
11. Guillemin, V. & Pollack, A. (1974) *Differential Topology.* Prentice-Hall.
12. Halmos, P. (1950) *Measure Theory.* Van Nostrand.
13. Halmos, P. (1960) *Naive Set Theory.* Van Nostrand.
14. Hirsch, M. (1976) *Differential Topology.* Springer Verlag.
15. MacLane, S. (1998) *Categories for the Working Mathematician.* Springer Verlag.
16. Marsden, J. & Hughs, T. (1994) *Mathematical Foundations of Elasticity.* Dover Publications.
17. Milnor, J. (1965) *Topology from the Differentiable Viewpoint.* University Press of Virginia.
18. Milnor, J. (1969) *Morse Theory.* Princeton University Press.
19. Moser, J. (1965) On the volume elements on a manifold. *Trans. Amer. Math. Soc.* **120**, 286–294.
20. Munkres, J. (1975) *Topology. A First Course.* Prentice-Hall.
21. Oxtoby, J. & Ulam, S. (1941) Measure-preserving homeomorphisms and metrical transitivity. *Ann. of Math.* **42**(2), 874–920.
22. Serrin, J. (1959) Mathematical Principles of Classical Fluid Mechanics. In *Handbuch der Physik.* Vol. VIII/1, Springer-Verlag.

23. Spivak, M. (1979) *A Comprehensive Introduction to Differential Geometry.* Vol. I–V. Publish or Perish, Inc..
24. Spivak, M. (1965) *Calculus on Manifolds.* W.A. Benjamin.
25. Tur, A. & Yanovsky, V. (1993) Invariants in dissipationless hydrodynamic media. *J. Fluid Mech.* **248**, 67–106.
26. Vilenkin, N. (1968) *Stories about Sets.* Academic Press.

III. GEOMETRY AND TOPOLOGY OF FLUID FLOWS

INTRODUCTION TO A GEOMETRICAL THEORY OF FLUID FLOWS AND DYNAMICAL SYSTEMS

TSUTOMU KAMBE
Institute of Dynamical Systems
Higashi-Yama 2-11-3, Meguro-ku
Tokyo 153-0043, Japan

E-mail: *kambe@gate01.com*

Abstract. Dynamical systems can be formulated on the basis of the Riemannian geometry and Lie algebra, provided that a dynamical system has a group symmtery, namely it is invariant under a group transformation, and further that the group manifold is endowed with a Riemannian metric. The basic ideas and tools are described, and application to various physical systems are considered: (i) free rotation of a rigid body; (ii) geodesic equation and KdV equation on the group of diffeomorphisms of a circle and its extended group; (iii) self-gravitating system of a finite number of point masses; (iv) geometrical formulation of hydrodynamics of an incompressible ideal fluid on the group of volume preserving diffeomorphisms, where the origin of Riemannian curvatures of the fluid flow is considered; (v) geodesics on a loop group and the local induction equation of a vortex filament. It is remarkable that the present geometrical formulations are successful for all the problems considered here and give insight into deep background common to the diverse physical systems. Further, the geometrical formulation opens a new approach to various dynamical systems.

1. Introduction

Various dynamical systems have often common geometrical structures and can be formulated on the basis of Riemannian geometry and Lie group theory. In this paper, the mathematical features are illustrated and physical aspects are exemplified by five systems. Before describing the details of the dynamical systems, mathematical concepts are presented and reviewed

R.L. Ricca (ed.), An Introduction to the Geometry and Topology of Fluid Flows, 137–165.

briefly. In Sec. 2, an introductory review is given for flows and diffeomorphisms, Lie group and Lie algebra, and in Sec. 3 the theory of Riemannian differential geometry is reviewed and basic concepts are presented. Full accounts are found in the textbooks by Frankel [9], Arnold [3], and Abraham & Marsden [1].

In the subsequent Secs. 4-8, five typical dynamical systems are reformulated according to the mathematical framework presented in the preceeding sections. The governing equation of each system is now obtained as a geodesic equation on a group manifold associated with the individual system. It is to be noted that, although the governing equations are derived already in physics, the present derivations are new and based on very general setup and concepts of metric, connection and Lie algebra in Riemannian differential geometry. See [16] for more details.

2. Flows, Diffeomorphisms and Lie Group

2.1. DIFFERENTIABLE MAP AND DIFFEOMORPHISMS

A *manifold* M^n is an n-dimensional topological space that is locally $\mathbb{R}^n$, namely, in terms of local coordinates, a point $p \in M^n$ is represented as $p = (x_p^1, \cdots, x_p^n)$. Let $F : M^n \to V^r$ be a map from a manifold M^n to another V^r, represented as $y = F(x)$ or $y = y(x)$, where $F^i(x)$ are differentiable functions of x^j $(i = 1, \cdots, r;\ j = 1, \cdots, n)$.

When $n = r$, we say that F is a *diffeomorphism*, provided that F is differentiable, one-to-one, onto, and in addition F^{-1} is differentiable. Such an F is a *differentiable homeomorphism* with a differentiable inverse F^{-1}. In the next section, the fluid flow is described to be a sequence of diffeomorphisms of particle configuration (of infinite dimension).

2.2. VECTOR FIELDS AND FLOWS

Given a flow of a fluid in $\mathbb{R}^3$, one can construct a 1-parameter family of maps, $\phi_t : \mathbb{R}^3 \to \mathbb{R}^3$, where ϕ_t takes a fluid particle located at p when $t = 0$ to the position of the same fluid particle $\phi_t(p) = x_t(p)$ at a later time t. The maps are the *Lagrangian* representation of motion of fluid particles. Associated with any such flow, we have a velocity,

$$v(p) = \frac{\mathrm{d}}{\mathrm{d}t}\, x_t(p) = \frac{\mathrm{d}}{\mathrm{d}t}\, \phi_t(p)\ .$$

Taking a smooth function $f : \mathbb{R}^3 \to \mathbb{R}$ and differentiating $f(x_t(p)) = f(\phi_t(p))$ with respect to t, we have

$$\frac{\mathrm{d}}{\mathrm{d}t} f \circ \phi_t(p) \equiv \frac{\mathrm{d}}{\mathrm{d}t} f(\phi_t(p)) = \frac{\mathrm{d}x^j}{\mathrm{d}t}\frac{\partial f}{\partial x^j} = v^j(p)\,\frac{\partial}{\partial x^j} f = v_p\, f\ , \tag{1}$$

where $v_p = v(p)$. Conversely, to each vector field $v(x) = (v^j)$ in $\mathbb{R}^3$, one may associate a flow $\{\phi_t\}$ having v as its velocity field. The map $\phi_t(p)$, with t a continuous parameter, can be determined by solving the system of ordinary differential equations, $\mathrm{d}x^j/\mathrm{d}t = v^j(x^1(t), x^2(t), x^3(t))$ with the initial condition, $x(0) = p$. Thus one finds the integral curves $\phi_t(p)$, called a flow *generated* by the vector field v. The map $\phi_t(p)$ is a diffeomorphism. This is assured by the property that each fluid particle is a physical entity. Hence it keeps its identity during the motion and two particles are assumed not to occupy the same point simultaneously.

In general, on a manifold M^n, we consider a vector v tangent to the parametrized curve ϕ_t at p. The vector field $v(x)$ at a point $x \in M^n$ is called a *tangent* vector, or a vector in the *tangent* space $T_x M$ to the manifold M.

Remark 2.1 Continuous distribution of fluid particles in a space has infinite degrees of freedom. In this context, the set of diffeomorphisms ϕ_t forms an *infinite-dimensional* manifold $D^{(\infty)}$ and a point $\eta = \phi_t \in D^{(\infty)}$ represents a configuration (as a whole) of all particles at a given time t.

A flow $\phi_t(p) = (x_t^j(p))$ on an n-dimensional manifold M^n is described by the system of ordinary differential equations,

$$\frac{\mathrm{d}x^j}{\mathrm{d}t} = v^j\left(x^1(t), \ldots, x^n(t)\right), \qquad (j = 1, \ldots, n) \tag{2}$$

with the initial condition, $\phi_0 = p$. The one-to-one correspondence between the tangent vector $v = (v^j)$ to M^n at p and the first-order differential operator $v^j(p)\partial/\partial x^j$, mediated by (1)–(2), implies $v = \sum_j v^j(x)\frac{\partial}{\partial x^j}$, which defines the vector field v as a differential operator [3]. Usually, in differential geometry, no distinction is made between a vector and its associated differential operator.

The tangent space to M^n at the point $x \in M^n$, written $T_x M$, is a vector space consisting of all tangent vectors to M^n at x. The basis vector $\boldsymbol{\partial}/\boldsymbol{\partial} x^j$ is simply written as $\boldsymbol{\partial}_j$. A tangent vector X is usually written as $X = X^j \boldsymbol{\partial}_j$, or $X_x = X^j(x)\boldsymbol{\partial}_j$.

In the case of *time-dependent* velocity field, an additional coordinate x^0 should be introduced, and the n equations (2) are replaced by the following $(n+1)$ equations,

$$\frac{\mathrm{d}x^j}{\mathrm{d}t} = X^j\left(x^0(t), x^1(t), \ldots, x^n(t)\right), \quad \text{with} \quad X^0 = 1 \tag{3}$$

for $j = 0, 1, \ldots, n$. Naturally the additional equation is reduced to $x^0 = t$. Correspondingly, the tangent vector in the time-dependent case, denoted by the *hat* symbol, is written as

$$\hat{X} = \hat{X}^i\,\partial_i = X^0\partial_0 + X^\alpha\partial_\alpha = \partial_t + X^\alpha\partial_\alpha\,, \tag{4}$$

where the index α denotes the spatial components ($\alpha = 1, \ldots, n$).

2.3. COVECTOR AND INNER PRODUCT

2.3.1. *Covector (1-form)*

We regard the differential of f defined by $df[v] = vf$ as a linear functional $M^n \to \mathrm{I\!R}$ for any vector $v \in M^n$. Given a vector $v = v^j\partial_j$, using (1), we have $df[v] = df[v^j\partial_j] = \sum_j v^j(x)\,\partial f/\partial x^j$. The most general linear functional, $\alpha : M^n \to R$, is expressed in coordinates as $\alpha = a_1 dx^1 + \ldots + a_n dx^n$. α is called *covector*, or *covariant vector*, or (differential) *1-form*. With $v = v^j\partial_j$, the 1-form $\alpha[v]$ takes the value, $\alpha[v] = \sum_{ij} a_i\, dx^i[v^j\partial_j] = \sum_j a_j\, v^j$.

2.3.2. *Inner (scalar) product*

Let the vector space E^n be endowed with an inner (scalar) product $\langle \cdot, \cdot \rangle$. For each pair of vectors $X, Y \in E^n$, the scalar product $\langle X, Y \rangle$ is a real number, and it is bilinear and symmetric with respect to X and Y. The inner product is also called the *metric* in Riemannian geometry. By writing $X = X^i\partial_i$ and $Y = Y^j\partial_j$, we have

$$\langle X, Y \rangle = g_{ij}\, X^i Y^j \ , \tag{5}$$

where $g_{ij} = g_{ji}$. The 1-form $\alpha[X]$ is defined in terms of the the inner product as $\alpha[X] = \langle A, X \rangle = g_{ij}\, A^i X^j = a_j X^j$, where $a_j = g_{ij} A^i$.

2.4. MAPPING OF VECTORS AND COVECTORS

Let $\phi : M^n \to V^r$ be a smooth map, and let $\phi_* : M_x \to V_y$ be the corresponding differential map. The map ϕ is represented by the function $y = F(x)$, where $x \in M^n$ and $y \in V^r$. Let $p(t)$ be a curve on M with $p(0) = p$ and $\dot{p}(0) = X$ (velocity vector), where $X \in T_pM^n$. Then, the differential map $\phi_* X$ is defined by $Y = \phi_* X = F_* X = d/dt(F(p(t)))$. Hence, the vector X is mapped to Y, whose component, by definition, is given by $Y^k = (\phi_* X)^k = (\partial F^k/\partial x^j)X^j = (\partial y^k/\partial x^j)X^j$. The transformation ϕ_* is linear with respect to the scalar coefficient X^j, written as

$$Y = Y^k\frac{\partial}{\partial y^k} = \phi_* X = \phi_*\Big[X^j\frac{\partial}{\partial x^j}\Big] = X^j\,\phi_*\Big[\frac{\partial}{\partial x^j}\Big] = X^j\frac{\partial y^k}{\partial x^j}\frac{\partial}{\partial y^k} \ . \tag{6}$$

This is called a *push-forward* transformation of the velocity vector X to the vector Y (the velocity vector of the image curve at $F(p)$) and is written as $Y = JX$, where the transformation matrix is the Jacobian.

Corresponding to the *push-forward* ϕ_*, one can define the *pull-back* ϕ^*, which is the linear transformation taking covectors at y to covectors at x, *i.e.* $\phi^* : V_y^* \to M_x^*$. Suppose that a vector X at $x \in M$ is transformed to $Y = \phi_*(X)$ at $y = \phi(x) \in V$, then the pull-back ϕ^* of a covector α is defined, using the push-forward $\phi_*(X)$, by

$$(\phi^* \alpha)[X] \equiv \alpha[\phi_*(X)] = \alpha[Y] , \tag{7}$$

for any covector $\alpha = A_i \, dy^i$. The definition (7) is understood as the *invariance* of the pull-back transformation. In fact, the value of the covector $\alpha = Y_i \, dy^i$ with $Y = \phi_* X$ (in V) is equal to the value of the pull-back covector $\phi^* \alpha$ with the original vector X (in M). See also the next section. Based on this invariance, the general pull-back formula is defined for the integral of a form α by

$$\int_{\phi(\sigma)} \alpha = \int_\sigma \phi^* \alpha , \tag{8}$$

where $\phi : \sigma \subset M \to \phi(\sigma) \subset V$.

2.5. LIE GROUP, LIE ALGEBRA AND LIE DERIVATIVE

2.5.1. *Lie group and invariant fields*

A *Lie group* is a group which is a differentiable manifold, and the group operations of multiplication and inverse are also differentiable. The identity is denoted by e. A Lie group always has two families of diffeomorphisms, theleft and right translations. Given a tangent vector X_e to G at the identity e, one may right-translate or left-translate X_e to every point of G by

$$X_g^{\mathrm{R}} = R_{g*} X_e = X_e \circ g \qquad \text{(right translation)}, \tag{9}$$

and $X_g^{\mathrm{L}} = L_{g*} X_e = g_* \, X_e$ (left translation). It is said that a vector field X on G is *right invariant* (or *left invariant*) if it is invariant under all right-translations (or left-translations), namely if $R_{h*} X_g^{\mathrm{R}} = X_{gh}^{\mathrm{R}}$ (or $L_{h*} X_g^{\mathrm{L}} = X_{hg}^{\mathrm{L}}$, respectively). It should be evident that the transformation (9) gives a right-invariant field generated by X_e at e.

For two right-invariant tangent vectors X_g^{R} and Y_g^{R}, the metric (5) is called *right-invariant* if

$$\langle X_g^{\mathrm{R}}, Y_g^{\mathrm{R}} \rangle|_g = \langle X_e, Y_e \rangle|_e .$$

The right-translation of the vector field $X(x)$ (at the identity e) by the map g is denoted by $X(g(x)) \equiv X \circ g(x)$.

2.5.2. *Adjoint operator and Lie bracket*

Every pair of vector fields defines a new vector field called the *Lie bracket* $[\,\cdot\,,\cdot\,]$. More precisely, the tangent space $T_e G$ at the identity e of a Lie group

G, equipped with the bracket operation $[\cdot, \cdot]$ of bilinear skew-symmetric pairing, is called the *Lie algebra* $\mathbf{g}$ of the group G, if the bracket satisfies the Jacobi identity :

$$[[X, Y], Z] + [[Y, Z], X] + [[Z, X], Y] = 0 , \tag{10}$$

for any triplet of $X, Y, Z \in \mathbf{g} = T_e G$. Provided that an element $\xi \in G$ acts as a linear transformation on $\forall Y \in \mathbf{g}$ (*i.e.* its own Lie algebra element) in the following way: $Ad_\xi Y \equiv L_{\xi *} \circ R_{\xi^{-1} *} Y = \xi Y \xi^{-1}$, then the operator Ad_G is called the *adjoint* representation. Consider a curve $\xi_t : t \mapsto G$ with the tangent $\dot{\xi}_0 = X$. The derivative of $Ad_{\xi_t} Y$ with respect to t is a linear transformation from Y to $ad_X Y$:

$$ad_X Y \equiv \frac{\mathrm{d}}{\mathrm{d}t} \xi_t Y \xi_t^{-1} = [X, Y] . \tag{11}$$

The Lie bracket is skew-symmetric: $[X, Y] = -[Y, X]$, as shown below. The bracket operation is usually called the *commutator*.

Example 2.2 Consider the rotation group $G = SO(3)$, of which an element A is represented by a 3×3 orthogonal matrix of $\det A = 1$. Let $\xi(t)$ be a curve issuing from e with the velocity $\mathbf{a}$ on $SO(3)$. Then one has $\xi(t) = e + t\,\mathbf{a} + O(t^2)$ for an infinitesimal time parameter t, where $\mathbf{a} = \dot{\xi}(0)$ is an element of the algera $\mathbf{g} = \mathbf{so}(3)$ and represented by a skew-symmetric matrix due to the orthogonality property and $\det \xi(t) = 1$. The algebra element $\mathbf{a}$ is called a *generator* of $\xi(t)$. Then, for $\mathbf{a}, \forall \mathbf{b} \in \mathbf{so}(3)$, the operator $ad_{\mathbf{a}} : \mathbf{g} \to \mathbf{g}$ is given by

$$ad_{\mathbf{a}}\, \mathbf{b} = [\mathbf{a}, \mathbf{b}] = \mathbf{ab} - \mathbf{ba} , \tag{12}$$

[4], where $[\mathbf{a}, \mathbf{b}] = \mathbf{ab} - \mathbf{ba}$ is the commutator which defines a third algebra element denoted by $\mathbf{c}$. The well-known representation of a 3×3 skew-symmetric matrix $\mathbf{a}$ in terms of a three-component vector $\hat{\mathbf{a}}$ leads to the rule of cross product,

$$\hat{\mathbf{c}} = \hat{\mathbf{a}} \times \hat{\mathbf{b}} . \tag{13}$$

2.5.3. *Lie derivative and Lagrange derivative*

Suppose that we are given a vector field $X = X^i \partial_i$ on a manifold M^n. With every such vector field, we associate a flow, or one-parameter group of *diffeomorphisms* ξ_t: $M \to M$, for which $\xi_0 = id$ and $(\mathrm{d}/\mathrm{d}t)\xi_t \mathbf{x}|_{t=0} = X(\mathbf{x})$. In addition, a first-order differential operator $\mathcal{L}_X$ is defined as

$$\mathcal{L}_X f(\mathbf{x}) \equiv X^i (\partial / \partial x_i)\, f(\mathbf{x}) = \frac{\mathrm{d}}{\mathrm{d}t} f(\xi_t \mathbf{x}) = \frac{\mathrm{d}}{\mathrm{d}t} \xi_t^* f(\mathbf{x}) . \tag{14}$$

The operator $\mathcal{L}_X$ is the *Lie derivative*, sometimes called fisherman's derivative [4]. This refers to the rightmost expression $(\mathrm{d}/\mathrm{d}t)\,\xi_t^* f(\mathbf{x})$. In fluid dynamics however, the same derivative is called the *Lagrange* derivative, which refers to the middle expression,

$$\frac{\mathrm{D}f}{\mathrm{D}t} = \frac{\mathrm{d}}{\mathrm{d}t} f(\xi_t \mathbf{x}) \; .$$

This is understood as denoting the time derivative observed by the fluid particle at $\xi_t\mathbf{x}$ moving with the flow.

In order to consider the derivative of vectors, suppose that we are given another vector field $Y = Y^i \partial_i$, and denote the flow generated by Y as η_s with $\eta_0 = id$ (the id is used here instead of e in order to emphasize that this is a *map*). The commutator of $\mathcal{L}_X$ and $\mathcal{L}_Y$ defines the *Poisson bracket*, $[\,\mathcal{L}_X, \mathcal{L}_Y\,] = \mathcal{L}_X\,\mathcal{L}_Y - \mathcal{L}_Y\,\mathcal{L}_X = \mathcal{L}_{\{X,Y\}}$ [4].

Now, let us consider the Lie derivative of a vector field $Y(\mathbf{x})$. The flow is transporting the vector $Y(\mathbf{x})$ in front of the fisherman sitting at a fixed point p. After an infinitesimal time t, the point $\xi_t^{-1}p$ will arrive at p and the man will see the vector Y transported forward by the map ξ_t from there, that is $\xi_t\, Y(\xi_t^{-1}p)$ (precisely, $\xi_{t*}\, Y(\xi_t^{-1}p)$). Thus we have

$$\mathcal{L}_X\, Y = ad_X\, Y = [\,X, Y\,] = \frac{\mathrm{d}}{\mathrm{d}t}\, \xi_t\, Y\, \xi_t^{-1} = \frac{\partial}{\partial t}\frac{\partial}{\partial s}\, \xi_t\, \eta_s\, \xi_t^{-1} \, , \tag{15}$$

where $[X, Y]$ is the *Lie bracket*. According to the definition (15), the Lie derivative of a vector field Y is given as $\mathcal{L}_X Y = -\{X, Y\}$, where

$$\{\,X, Y\,\} = \{\,X, Y\,\}^k \partial_k = \sum_j \left(X^j \frac{\partial Y^k}{\partial x^j} - Y^j \frac{\partial X^k}{\partial x^j} \right) \partial_k \; . \tag{16}$$

In summary, we have

$$\mathcal{L}_X Y = [\,X, Y\,] = -\,\{X, Y\} = -\,\mathcal{L}_{\{X,Y\}} = -\,[\,\mathcal{L}_X, \mathcal{L}_Y\,] \; . \tag{17}$$

Note the minus sign in front of $\{X, Y\}$ in (17), which is characteristic of the right-invariant field, likewise the diffeomorphisms we are considering (see also Secs. 5 and 7). This is in contrast with the left-invariant field (see Sec. 4) associated with the commutator (12). From the point of view of fluid dynamics, the Lagrange derivative would be given by $(\mathrm{D}/\mathrm{D}t)\, Y^k(\xi_t x)\, \partial_k = X^j(x)\,(\partial Y^k(x)/\partial x^j)\,\partial_k$. As a particular case, we note that the coordinate bases commute. In fact, by substituting $X = \partial_\alpha$, $Y = \partial_\beta$, we have

$$[\,\partial_\alpha, \, \partial_\beta\,] = \partial_\alpha\, \partial_\beta - \partial_\beta\, \partial_\alpha = 0 \; . \tag{18}$$

Remark 2.3 A vector field Y defined along the integral curve of X is said to be *invariant* if $Y(\xi_t x) = \xi_{t*}Y(x)$. Then, the equation (15) reduces to

$$0 = \mathcal{L}_X\, Y(x) = -\sum_i \Big(\frac{\mathrm{D}Y^i}{\mathrm{D}t} - Y^j \frac{\partial X^i}{\partial x^j} \Big)\, \partial_i\,. \tag{19}$$

Note the minus sign in front of $\mathrm{D}Y/\mathrm{D}t$, suggesting that appropriate time evolution is given by $-\mathcal{L}_X Y(x)$. In fluid dynamics, the equation (19) is often called an equation of *frozen field.* The Jacobi field Y ($= J$ in §3.9) satisfies this equation (see §7.2.4). Its solution is given by the Cauchy's solution, $Y^\alpha(t) = Y^j(0)\partial y^\alpha/\partial x^j$, which is nothing but the equation (6).

2.6. DIFFEOMORPHISM OF A CIRCLE S^1

Diffeomorphism of the manifold S^1 (circle of periodicity T) is represented by a map $g\, :\, x \in S^1 \to g(x) \in S^1$ and $g(x+T) = g(x)$. Such maps constitute a group $D(S^1)$ of diffeomorphisms with the composition law:

$$h = g \circ f\,, \qquad i.e.\ \ h(x) = g(f(x)) \in D(S^1)\ ,$$

for $f, g \in D(S^1)$. The diffeomorphism is a pointwise map of infinite degrees of freedom. In §5, the diffeomorphism is assumed to be orientation-preserving in the sense that $g'(x) > 0$, where the prime denotes $\partial/\partial x$.

Associated with a flow $\xi_t(x)$ that is a continuous sequence of diffeomorphisms with the parameter t, its tangent field at t is defined by

$$\dot{\xi}_t(x) = \frac{\mathrm{d}}{\mathrm{d}t}\,\xi_t(x) = \lim_{\tau\to 0} \frac{\xi_\tau(x) - 1}{\tau}\, \xi_t(x) = u(x) \circ \xi_t(x)\,,$$

a right-invariant form. The tangent field at the identity is given by $u(x) = \mathrm{d}\xi_t(x)/\mathrm{d}t\,|_{t=0}$. Put it in another way, an element X in the *tangent space* $T_x D(S^1)$ is represented as $X = u(x)\,\partial_x$. The *Lie bracket* (commutator) is

$$[X, Y] = -(\,u\,v' - v\,u'\,)\,\partial_x\ . \tag{20}$$

for $X = u(x)\partial_x$, $Y = v(x)\partial_x \in T_x D(S^1)$. See (17).

3. Riemannian Geometry

3.1. RIEMANNIAN METRIC

On a Riemannian manifold M^n, a positive definite inner product $\langle\cdot\,,\cdot\rangle$ is defined on the tangent space $T_x M^n$ at $x = (x^1, \cdots, x^n) \in M$ with differentiability. The *Riemannian metric* is defined by $\langle X, Y\rangle(x) = g_{ij}\, X^i(x) Y^j(x)$, for $X = X^i(x)\partial_i$, $Y = Y^j(x)\partial_j \in TM^n$, where the metric tensor, $g_{ij}(x) =$

$\langle \partial_i, \partial_j \rangle = g_{ji}(x)$, is symmetric and differentiable with respect to x^i. This bilinear quadratic form is called the *first fundamental form*. The line element $\mathrm{d}s$ is given by $\mathrm{d}s^2 = g_{ij}\, \mathrm{d}x^i\, \mathrm{d}x^j$.

The inner product is *non-degenerate*, for $\forall Y \in TM$, if

$$\langle X, Y \rangle = 0 \qquad \text{only when} \quad X = 0 \, . \tag{21}$$

3.2. EXAMPLES OF METRIC TENSOR

3.2.1. *Finite dimension*

Consider a dynamical system of N degrees of freedom in a *gravitational* field with the potential $V(\bar{q})$, the kinetic energy $T = (1/2)\, a_{ij}\, \dot{q}^i\, \dot{q}^j$, and the Lagrangian $L(\bar{q}, \dot{\bar{q}})$, where $\bar{q} = (q^i), (i = 1, \ldots, N)$. The *Eisenhart* metric is defined by introducing two additional coordinates q^0 and q^{n+1} where $dq^0 = dt,\ dq^{N+1} = (1 - L(\bar{q}, \dot{\bar{q}})\,)$, and the metric tensor is $g_{ij}(Q) = g^E_{ij}(Q)$ (for $i, j = 0, \cdots, N+1$) with $Q = (q^0, \bar{q}, q^{N+1})$, $\underline{\underline{a}} = (\, a_{ij}\,)$, and

$$g^E = \begin{pmatrix} -2V(\bar{q}) & \underline{0} & 1 \\ \underline{0}^T & \underline{\underline{a}} & \underline{0}^T \\ 1 & \underline{0} & 0 \end{pmatrix} . \tag{22}$$

3.2.2. *Infinite dimension*

A metric on the group $D(S^1)$ of diffeomorphisms (§2.7) is defined for the right-invariant tangent field $U_\xi(x) = u \circ \xi(x)$ and $V_\xi(x) = v \circ \xi(x)$ in the following invariant way (also see §7.1.2) :

$$\langle U, V \rangle_\xi = \int_{S^1} (U_\xi \circ \xi^{-1},\, V_\xi \circ \xi^{-1})_x\, \mathrm{d}x = \int_{S^1} u(x)\, v(x)\, \mathrm{d}x = \langle X, Y \rangle \, , \tag{23}$$

where $X = u(x)\, \partial_x,\ Y = v(x)\, \partial_x \in T_x D(S^1)$.

3.3. RIEMANNIAN CONNECTION AND COVARIANT DERIVATIVE

3.3.1. *Connection*

An affine connection is an operator ∇ satisfying the relations (i)-(iii), for any vector field X, Y and a function $f(x)$ with $a,\ b \in \mathbb{R}^1$:

Definition 3.1

$$\left.\begin{array}{lrcl} \text{(i)} & \nabla_\xi\, (aX + bY) & = & a\nabla_\xi X + b\nabla_\xi Y \, , \\ \text{(ii)} & \nabla_{a\xi + b\eta} X & = & a\nabla_\xi X + b\nabla_\eta X \, , \\ \text{(iii)} & \nabla_\xi\, f(x) X & = & f(x) \nabla_\xi X + \xi(f) X \, , \end{array}\right\} \tag{24}$$

for any vector ξ and η at x. The connection $\nabla_X Y$ is also called the *covariant derivative* of the vector field $Y(x)$ in the direction of the vector X. Applying the representations $X = X^i\,\partial_i$ and $Y = Y^j\,\partial_j$, we have

$$\nabla_X Y \;=\; \nabla_{X^i\partial_i} Y^j\partial_j \;=\; X^i\nabla_{\partial_i}(Y^j\partial_j) = (X^i\partial_i Y^j)\partial_j + X^i Y^j\,\Gamma^k_{ij}\,\partial_k \quad (25)$$

by using (i)-(iii), where Γ^k_{ij} is the Christoffel symbol defined by

$$\nabla_{\partial_i}\partial_j \;=\; \Gamma^k_{ij}\partial_k\;. \quad (26)$$

Therefore,

$$(\nabla_X Y)^i = \left[\frac{\partial Y^i}{\partial x^j} + \Gamma^i_{jk}Y^k\right] X^j\;. \quad (27)$$

Most dynamical systems are *time-dependent* and every tangent vector is written in the form $\hat{X} = \hat{X}^i\,\partial_i = \partial_t + X^\alpha\partial_\alpha$, see (4). Correspondingly, the connection should be written as

$$\nabla_{\hat{X}}\hat{Y} \;=\; \nabla^i_{\hat{X}}\partial_i\hat{Y}^j\partial_j = \nabla_{\partial_t}\hat{Y}^j\partial_j + X^\alpha\nabla_{\partial_\alpha}(\hat{Y}^j\partial_j)\;,$$

where $\hat{Y} = \partial_t + Y^\alpha\partial_\alpha$ and α denotes indices of spatial components ($\alpha = 1,\ldots,n$). For the spatial part $Y = Y^\alpha\partial_\alpha$ (with $X = X^\alpha\partial_\alpha$), we have

$$\nabla_{\hat{X}}Y = \frac{\partial}{\partial t}Y + \nabla_X Y\,, \quad (\text{where}\quad \nabla_{\partial_t}Y = \partial_t Y \quad \text{and} \quad \nabla_{\partial_t}\partial_\alpha = 0). \quad (28)$$

3.3.2. *Riemannian connection*

There is a unique connection ∇ on a Riemannian manifold called *Riemannian* connection or *Levi-Civita* connection that satisfies

$$\text{(i)}\quad \nabla_X Y - \nabla_Y X \;= [X, Y] \qquad \text{(torsion free)} \quad (29)$$

$$\text{(ii)}\quad Z\,\langle X, Y\rangle \;= \langle\nabla_Z X, Y\rangle + \langle X, \nabla_Z Y\rangle \quad \text{(compatibility)} \quad (30)$$

for vector fields X, Y and Z. The torsion-free requires the symmetry, $\Gamma^k_{ij} = \Gamma^k_{ji}$, with respsect to i and j. See the next section for the condition (ii).

Due to the above two properties, the Riemannian connection satisfies the following identity,

$$\begin{aligned} 2\langle\nabla_X Y, Z\rangle \;=\; & X\langle Y, Z\rangle + Y\langle Z, X\rangle - Z\langle X, Y\rangle \\ & +\langle[X,Y], Z\rangle - \langle[Y,Z], X\rangle + \langle[Z,X], Y\rangle\;, \end{aligned} \quad (31)$$

where $X\langle\cdot\,,\cdot\rangle = X^j\partial_j\,\langle\cdot\,,\cdot\rangle$. The last equation defines the connection ∇ in terms of the inner product $\langle\cdot\,,\cdot\rangle$ and the commutator $[\cdot\,,\cdot]$.

In most dynamical systems studied below, the metrics are defined invariant (with respect to either right or left translation). In such cases, the first three terms on the right hand side of (31) vanishes identically:

$$2\langle \nabla_X Y, Z\rangle = \langle [X,Y],Z\rangle - \langle [Y,Z],X\rangle + \langle [Z,X],Y\rangle \, . \tag{32}$$

3.3.3. *Covariant derivative along a parametrized curve*

Consider a curve $x(t)$ on M passing through p whose tangent at p is given by $T = \mathrm{d}x/\mathrm{d}t = \dot{x}$, and let Y be a tangent vector field defined along the curve. According to (27), the covariant derivative $\nabla_T Y$ is also written as

$$\nabla_T Y = \frac{\boldsymbol{\nabla} Y}{\mathrm{d}t} = \left[\frac{\partial Y^i}{\partial x^k} + \Gamma^i_{kj} Y^j\right] \dot{x}^k \boldsymbol{\partial}_i = \left[\frac{\mathrm{d}}{\mathrm{d}t} Y^i + \Gamma^i_{kj} \dot{x}^k Y^j\right] \boldsymbol{\partial}_i \, . \tag{33}$$

The second expression $\boldsymbol{\nabla} Y/dt$ emphasizes the derivative along the curve $x(t)$ parametrized with t.

On the manifold M endowed with the connection $\boldsymbol{\nabla}$, one can consider *parallel* displacement of a vector Y along a parametrized curve $x(t)$, which is defined by vanishing covariant derivative: $\boldsymbol{\nabla} Y/\mathrm{d}t = \nabla_T Y = 0$. Thus, Y is translated parallel along the curve $x(t)$ when $\dot{x}^k \, (\partial Y^i/\partial x^k) + \Gamma^i_{kj} \dot{x}^k Y^j = 0$. It is readily seen from (30) that the scalar product is invariant, $T\langle X, Y\rangle = 0$, for the vector fields translated parallel along the curve:

$$\langle X, Y \rangle = \text{constant} \quad \text{(under parallel translation)}.$$

3.3.4. *Curvature tensor*

The vector field given by

$$\begin{aligned} R(X, Y)\, Z \;&=\; \nabla_X(\nabla_Y Z) - \nabla_Y(\nabla_X Z) - \nabla_{[X,Y]}\, Z & (34)\\ &\equiv\; (R^i_{jkl} Z^j X^k Y^l)\, \partial_i & (35) \end{aligned}$$

is called the curvature transformation for a pair of vector fields X, Y, where R^i_{jkl} is the *Riemannian curvature tensor.* Then we have

$$\langle W, R(X, Y)\, Z\rangle = R^\alpha_{jkl} \langle \partial_i, \partial_\alpha\rangle\, W^i Z^j X^k Y^l = R_{ijkl}\, W^i Z^j X^k Y^l \, , \tag{36}$$

where $R_{ijkl} = g_{i\alpha} R^\alpha_{jkl}$ and $g_{i\alpha} = \langle \partial_i, \partial_\alpha\rangle$. From its definition, one may write $R(X, Y) = [\nabla_X, \nabla_Y] - \nabla_{[X,Y]}$, which clearly shows the antisymmetry $R(X, Y) = -R(Y, X)$. Correspondingly, we have $R^i_{jkl} = -R^i_{jlk}$. The curvature tensor $\langle \partial_\delta, R(\partial_\alpha, \partial_\beta)\, \partial_\gamma\rangle = R^i_{\gamma\alpha\beta}\langle \partial_\delta, \partial_i\rangle \equiv R_{\delta\gamma\alpha\beta}$ is anti-symmetric with respect to γ and δ as well as α and β.

3.3.5. *Induced connection and second fundamental form*

Let V^m be a submanifold of a Riemannian maniforl M^n with the metric g_{ij}. Let us consider the restriction of the Riemannian metric g_{ij} to the tangent vectors to V. This action induces the Riemannian metric for the subspace V. An arbitrary vector field Z in M can be decomposed into two orthogonal components: $Z = Z_V + Z_N$, where $Z_V = \mathtt{P}\{Z\}$ is the projected component to V and $Z_N = \mathtt{Q}\{Z\}$ the component orthogonal to V [18, 16].

Let $\hat{\nabla}$ be the Riemannian connection for M^n, and define a new connection $\bar{\nabla}$ for V^m $(m < n)$ as follows. Consider a vector X tangent to V and a vector field Z, with Z_N not necessarily zero. Then the $\bar{\nabla}$ is defined by

$$\bar{\nabla}_X Z \equiv \hat{\nabla}_X Z - \mathtt{Q}\{\hat{\nabla}_X Z\} = \mathtt{P}\{\hat{\nabla}_X Z\}\ , \tag{37}$$

where the right hand side is the projection of $\hat{\nabla}_X Z$ onto the tangent space of V. It can be checked that the operator $\bar{\nabla}$ satisfies the properties (24).

Now, *the second fundamental form* $S(X,Y)$ is defined by

$$\hat{\nabla}_X Y = \bar{\nabla}_X Y + S(X,Y), \tag{38}$$

which is called *Gauss's formula.* It is not difficult to see that the function $S(X,Y) \equiv \hat{\nabla}_X Y - \bar{\nabla}_X Y$ satisfies the following symmetry relation,

$$S(X,Y) = \mathtt{Q}\{\hat{\nabla}_X Y\} = \mathtt{Q}\{\hat{\nabla}_Y X\} = S(Y,X)\ . \tag{39}$$

This is a Riemannian generalization of the *Weingarten equation.*

Corresponding to $\hat{\nabla}$ and $\bar{\nabla}$, we have two kinds of curvature tensors, $\hat{R}(X,Y)Z$ and $\bar{R}(X,Y)Z$. Using the definition (34) and the above relations (38) and (39), one can show the following *Gauss-Codazzi equation*:

$$\begin{aligned}\langle W, \hat{R}(X,Y)Z\rangle &= \langle W, \bar{R}(X,Y)Z\rangle + \langle S(X,Z), S(Y,W)\rangle \\ &\quad -\langle S(X,W), S(Y,Z)\rangle\ ,\end{aligned} \tag{40}$$

where X, Y, Z, W are tangent to V.

3.4. GEODESIC EQUATION

3.4.1. *Local coordinates representation*

A parametrized curve $\gamma(s)$ is a *goedesic* if its tangent $T = \mathrm{d}\gamma/\mathrm{d}s = \dot{\gamma}$ is displaced in parallel :

$$\frac{\boldsymbol{\nabla} T}{\mathrm{d}s} = \boldsymbol{\nabla}_T T = \frac{\boldsymbol{\nabla}}{\mathrm{d}s}\left(\frac{\mathrm{d}\gamma}{\mathrm{d}s}\right) = 0\ .$$

In coordinates $\gamma = (x^i)$, we have $T = T^i\,\partial_i = \mathrm{d}\gamma/\mathrm{d}s = (\mathrm{d}x^i/\mathrm{d}s)\partial_i$, and

$$\frac{\boldsymbol{\nabla} T}{\mathrm{d}s} = \boldsymbol{\nabla}_T T = \left[\frac{\mathrm{d}T^i}{\mathrm{d}s} + \Gamma^i_{jk}\,T^j\,T^k\right]\partial_i = 0\ . \tag{41}$$

Thus we obtain the *geodesic equation* :

$$\frac{\mathrm{d}^2 x^i}{\mathrm{d}s^2} + \Gamma^i_{jk} \frac{\mathrm{d}x^j}{\mathrm{d}s} \frac{\mathrm{d}x^k}{\mathrm{d}s} = 0 \ . \tag{42}$$

3.4.2. *Group-theoretic representation*

On the Riemannian manifold of invariant metric, another formulation of the geodesic equation is possible, because most dynamical systems considered below are equipped with invariant metrics (with respect to either right or left translation). In such cases, the following derivation would be useful.

In terms of the adjoint operator $ad_X\, Z = [X, Z]$ introduced in (11), let us define the *coadjoint* operator by

$$\langle ad^*_X\, Y, Z\rangle = \langle Y, ad_X\, Z\rangle = \langle Y, [X, Z]\rangle \ . \tag{43}$$

Then the equation (32) is transformed to

$$2\langle \nabla_X Y, Z\rangle = \langle ad_X Y, Z\rangle - \langle ad^*_Y X, Z\rangle - \langle ad^*_X Y, Z\rangle \ .$$

The non-degeneracy of the inner product (21) leads to

$$\nabla_X\, Y = \frac{1}{2}\, (ad_X\, Y - ad^*_X Y - ad^*_Y X) \ . \tag{44}$$

Thus, another form of the geodesic equation is given by $\nabla_X X = -ad^*_X X = 0$, since $ad_X X = [X, X] = 0$. In particular, the geodesic equation of a time-dependent problem is represented as

$$\nabla_{\hat{X}}\, X = \partial_t\, X + \nabla_X\, X = \partial_t\, X - ad^*_X\, X = 0 \ , \tag{45}$$

for the spatial part X. It should be noted that this is valid for the left-invariant field such as the rotation group considered in §2.5.2 and §4. There is a $\pm$ difference in the sign of the relation between the commutator of the Lie algebra and the Poisson bracket whether the vector fields are left-invariant or right-invariant, as shown in §2.5 [4].

In the case of the right-invariant field, it was shown in §2.5.3 that $[X, Y] = \mathcal{L}_X Y = -\{(\mathrm{d}Y^i/\mathrm{d}t) - Y^j(\partial X^i/\partial x^j)\}\, \partial_i$. When the time evolution of such system is concerned, this requires that

$$\nabla^{(\mathrm{R})}_X\, Y \equiv -\, \nabla_X\, Y$$

should be used instead of (44), and that the time-dependent geodesic equation takes the form (instead of (45)),

$$\partial_t\, X + \nabla^{(\mathrm{R})}_X\, X = \partial_t\, X + ad^*_X\, X = 0 \ . \tag{46}$$

3.5. JACOBI EQUATION

Let $C_0 : \gamma_0(s)$ be a geodesic curve with $s \in [0, L]$ the arc-length, and C_α : $\gamma(s, \alpha)$ a varied geodesic curve where $\alpha \in (-1, +1)$ is a variation parameter and $\gamma_0(s) = \gamma(s, 0)$, s being the arc-length for $\alpha = 0$. Because $\gamma(s, \alpha)$ is a geodesic, we have $\nabla(\partial_s\gamma)/\partial s = 0$ for all α. The function $\gamma(s, \alpha)$ is a differentiable map $\gamma : U \subset \mathrm{I\!R}^2 \to M^n$ with the property $[\partial/\partial s, \partial/\partial\alpha] = 0$ (see (18)). In this circumstance, the following two identities hold:

$$\left(\frac{\nabla}{\partial s}\right) \partial_\alpha\gamma = \left(\frac{\nabla}{\partial \alpha}\right) \partial_s\gamma \, , \qquad \frac{\nabla}{\partial s}\left(\frac{\nabla Z}{\partial \alpha}\right) - \frac{\nabla}{\partial \alpha}\left(\frac{\nabla Z}{\partial s}\right) = R(\partial_s\gamma, \partial_\alpha\gamma)\, Z \, ,$$

(see [9]), where $\partial_s\gamma = \partial\gamma/\partial s$ and $\partial_\alpha\gamma = \partial\gamma/\partial\alpha$. Along the reference geodesic $\gamma_0(s)$, let us use the notation $T = \partial_s\gamma$ for the tangent to the geodesic and $J = \partial_\alpha\gamma$ ($\alpha = 0$) for the variation vector. Using $\nabla T/\partial s = 0$ and the above identities with $Z = T$, we have

$$0 = \frac{\nabla}{\partial \alpha}\frac{\nabla T}{\partial s} = \frac{\nabla}{\partial s}\frac{\nabla T}{\partial \alpha} - R(T, J)\, T = \frac{\nabla}{\partial s}\frac{\nabla}{\partial s} J + R(J, T)\, T \, . \tag{47}$$

Thus we have obtained the *Jacobi equation* for the geodesic variation J

$$\frac{\nabla}{\partial s}\frac{\nabla}{\partial s} J + R(J, T)\, T = 0 \, . \tag{48}$$

The variation vector field J is called *Jacobi field.* Defining $\| J \|^2 \equiv \langle J, J \rangle$, differentiating it twice with respect to s and using (48) and (30), we obtain

$$\frac{\mathrm{d}^2}{\mathrm{d}s^2} \frac{\| J \|^2}{2} = \| \nabla_T J \|^2 - K(T, J) \, , \tag{49}$$

where $\nabla_T J = \nabla J/\partial s$, and

$$K(T, J) \equiv \langle J, \, R(J, T)T \rangle = R_{ijkl}\, J^i\, T^j\, J^k T^l \tag{50}$$

is a *sectional curvature* associated with the two-dimensional section spanned by J and T. Writing $J = \| J \| \, \mathbf{e}_J$, the equation (49) is transformed to

$$\frac{\mathrm{d}^2}{\mathrm{d}s^2} \| J \| = \Big(\| \nabla_T \mathbf{e}_J \|^2 - K(T, \mathbf{e}_J)\Big) \| J \| \, , \tag{51}$$

where $\| \mathbf{e}_J \| = 1$. The equation (48) provides the link between the stability of geodesic curves and the Riemannian curvature and one of the bases for geometrical description of dynamical systems considered later.

4. Free Rotation of a Rigid Body

We now consider a physical problem, that is, an application of the geometrical theory formulated in the previous two sections to one of the simplest dynamical system: free rotation of a rigid body without action of external torque. The basic idea is that the governing equation is the geodesic equation over the manifold space of a group of transformations $SO(3)$ (a Lie group) and that it describes the motion of the physical problem. We begin with this simplest system in order to illustrate the underlying geometrical ideas. See Arnold (1966) [2], Kambe (1998) [15], and Suzuki *et al.* (1998) [23].

In the mechanics of rigid bodies, *free* rotation is described by the Euler's equation: $J_1\,(\mathrm{d}\Omega_1/\mathrm{d}t)-(J_2-J_3)\Omega_2\Omega_3=0,\ J_2\,(\mathrm{d}\Omega_2/\mathrm{d}t)-(J_3-J_1)\Omega_3\Omega_1=0,\ J_3\,(\mathrm{d}\Omega_3/\mathrm{d}t)-(J_1-J_2)\Omega_1\Omega_2=0,$ in the body frame (*i.e.* the moving frame of reference), where $\Omega=(\Omega^1,\Omega^2,\Omega^3)$ is the angular velocity vector in the principal frame (x^1,x^2,x^3) of the body's *moment of inertia* $J=(J_{\alpha\beta})=\mathrm{diag}(J_1,J_2,J_3)$, J_α being the diagonal elements.

Stability is considered by deriving the Jacobi equation for the geodesic variation vector J.

4.1. ROTATION AND THE GROUP $SO(3)$

Rotation of a rigid body is regarded as a sequence of transformations of the body axes (*e.g.* the principal axes), which are a sequence of elements of the special orthogonal transformation group: $SO(3)$. By a transformation matrix $A\in SO(3)$, a point $\mathbf{x}$ in the body is mapped to $\mathbf{x}'=A\mathbf{x}$ where $\det A=1$. The group $G=SO(3)$ is a (smooth) Lie group of dimension three and consists of all orientation-preserving rotations (*i.e.* $\det A=+1$).

An element g of the group G corresponds to a position of the body arrived at g by its motion from the initial state (the identity e). A motion of the body is described by a curve $C:t\to g_t$ on the manifold $SO(3)$, where $g_t=A(t)$ and $A(t)\in SO(3)$ [3, 2]. An infinitesimal transformation Ω is defined by $\delta A=\Omega\,\delta t$ for an infinitesimal time increment δt at g_t, where Ω is shown to be anti-symmetric.

The vector Ω is a *tangent* vector at the identity e of the group G and is an element of the Lie algebra $\mathbf{so}(3)$. The space of such vectors is denoted by $T_eG=\mathbf{so}(3)$. The commutator in this algebra is given by the vector product in $\mathrm{I\!R}^3$ (see (12) and (13)) :

$$[X,\,Y]=X\times Y\,,\qquad\qquad \text{for } X,\,Y\in T_eG=\mathbf{so}(3)\,. \tag{52}$$

The kinetic energy K of the body motion is given by the scalar product of the angular velocity and the angular momentum multiplied by $\frac{1}{2}$, and it does not depend on the coordinate change of left-translation mentioned above. Hence the energy K gives a left-invariant Riemannian metric on the group: $2K=(J\Omega,\,\Omega)$, where J is a diagonal matrix with positive elements

$J_\alpha(>0)$. Now one can define the metric $<\cdot,\cdot>$ on T_eG by

$$< X, Y > \equiv (J\,X, Y) \qquad \text{for } X, Y \in T_eG\,. \tag{53}$$

The group G is a Riemannian manifold with the left-invariant metric (53).

4.2. GEOMETRY OF THE RIGID BODY MOTION

Let us consider the geodesic equation on the manifold $SO(3)$. We have already introduced the commutator (52) and the inner product (53) together with the moment of inertia J in diagonal form. Further, the metric is left-invariant, that is, the metric (53) is conserved by the left-translation on the Lie group $SO(3)$. In such a case of invariant metric, the connection satifies the equation (32), and in terms of the operators ad and ad^*, we have the expression (44) of $\nabla_X Y$ for X, $Y \in \mathbf{so}(3)$. By the definition (11), we have $ad_X Y = [X, Y] = X \times Y$. Then the $ad^*_X\, Y$ satisfies

$$\langle ad^*_X\, Y, Z\rangle = \langle Y, ad_X\, Z\rangle = (JY, X \times Z)_{\mathbb{R}^3} = (JY \times X, Z)_{\mathbb{R}^3}\,.$$

Hence, the non-degeneracy of the metric leads to $ad^*_X\, Y = J^{-1}(JY \times X)$. Thus it is found that [23]

$$\nabla_X Y = \frac{1}{2} J^{-1}\Big(J(X \times Y) - (JX) \times Y - (JY) \times X \Big)\,,$$

which is equal to $(1/2)\, J^{-1}(\tilde{K} X \times Y)$, where $\tilde{K}$ is a diagonal matrix with the diagonal elements, $\tilde{K}_\alpha \equiv -J_\alpha + J_\beta + J_\gamma$ for $(\alpha, \beta, \gamma) = (1,2,3)$ or its cyclic permutation (all $\tilde{K}_\alpha > 0$).

In the case of the left-invariant field, it is sufficient to give expressions at e, and the tangent vector at e is the angular velocity Ω. The geodesic equation of the time-dependent problem is given by (45) with $X = \Omega$

$$\frac{\mathrm{d}}{\mathrm{d}t}\Omega - J^{-1}\Big((J\Omega) \times \Omega \Big) = 0\,, \tag{54}$$

which is also written as $(\mathrm{d}/\mathrm{d}t)J\Omega = (J\Omega) \times \Omega$. This is nothing but the Euler's equation mentioned in the beginning. Equation of the *Jacobi* field J along the geodesic generated by X is derived by Suzuki *et al.* [23]. Further, stability analysis is made in terms of the sectional curvatures. The results are consistent with what is known in classical mechanics.

5. Geodesic Equation on $\mathbf{D}(\mathbf{S}^1)$ and KdV Equation on $\hat{\mathbf{D}}(\mathbf{S}^1)$

Now we consider a second example of application of the geometrical theory, which is given by a dynamical system of continuous mapping along a circle; this is a fundamental problem in physics. The geodesic is a curve over the manifold of orientation-preserving diffeomorphisms of a circle S^1 (with a periodicity T) noted in §2.7. Two systems are considered below: the first one is the geodesic equation over a group of diffeomorphisms $D(S^1)$, describing a simple diffeomorphic flow over S^1, and the second one is the KdV equation, which is the geodesic equation on an extended group $\hat{D}(S^1)$. The manifold S^1 is spatially one-dimensional, but its diffeomorphism has infinite degrees of freedom because the pointwise mapping describes arbitrary (but orientation-preserving) deformation along the circle. Thus we consider two infinite-dimensional Lie groups $D(S^1)$ and $\hat{D}(S^1)$ with an infinite-dimensional algebra called the Virasoro algebra [7].

5.1. DISPERSIONLESS KDV

Consider a group $D(S^1)$ of diffeomorphisms of a circle S^1 of periodicity $T = 1$, equipped with a *right-invariant* metric (see §3.2). Because of this metric invariance, the Riemannian connection ∇ is given by the expression (44): $\nabla_X Y = \frac{1}{2}(ad_X Y - ad^*_Y X - ad^*_X Y)$. More precisely, using the tangent fields $X = u(x)\partial_x$, $Y = v(x)\partial_x$, $Z = w(x)\partial_x \in TD(S^1)$ of the diffeomorphisms (see §2.6) and the definition (20), we have, $ad_X Y = [X, Y] = -(u\,v' - v\,u')\,\partial_x$, where $u' = \partial_x u = u_x$. Then the $ad^*_X Y$ is given by

$$\langle ad^*_X Y, Z\rangle = \langle Y, ad_X Z\rangle = -\int_{S^1} v(uw' - wu')\,\mathrm{d}x = \int_{S^1}(uv' + 2vu')w\,\mathrm{d}x\ ,$$

where the definition of metric (23) is used and integration by parts is performed with the periodic boundary conditions $u(x+1) = u(x)$, etc. . Hence,

$$ad^*_X Y = (\,u\,v' + 2v\,u'\,)\,\partial_x\ .$$

It is found from (44) that the Riemannian connection on the group manifold $D(S^1)$ is given by

$$\nabla_X Y = -\,(\,2uv' + vu'\,)\,\partial_x\ .$$

Remember that the present system is right-invariant (§2.7, §3.2). Then, $\nabla^{(R)}_X Y = -\nabla_X Y$ must be used for the time evolution, and the geodesic equation is given by (46) :

$$\partial_t X + \nabla^{(R)}_X X = 0\ .$$

Thus it is found that the geodesic equation on the manifold $D(S^1)$ is

$$u_t + 3\,u\,u_x = 0\ . \tag{55}$$

Compared with the KdV equation of the form $\bar{u}_t + \bar{u}\bar{u}_x + \kappa\bar{u}_{xxx} = 0$, there is no third-order dispersion term $\kappa\bar{u}_{xxx}$, where κ is a constant and $\bar{u} = 3u$.

The third-order derivative term is only introduced by the central extension considered in the next section. The flow described by (55) can be called a *simple diffeomorphic* flow.

5.2. CENTRAL EXTENSION

An extention of the group $D(S^1)$, denoted by the hat $\hat{}$, is defined by $\hat{f} = (f, a)$, $\hat{g} = (g, b) \in \hat{D}(S^1)$, for $f, g \in D(S^1)$ and $a, b \in \mathbb{R}$, where $\hat{D}(S^1) = D(S^1) \oplus \mathbb{R}$. The group operation is given by [17]

$$\hat{g} \circ \hat{f} \quad = \quad (g \circ f,\ a + b + B(g, f)) \ , \tag{56}$$

where $B(g, f) = (1/2) \int_{S^1} \ln \partial_x (g \circ f) \, \mathrm{d} \ln \partial_x f$. Consider a subgroup $\hat{D}_0$ defined by

$$\hat{D}_0 \equiv \{\hat{f}_a : \hat{f}_a = (f_0,\ a),\ |\ f_0(x) = x,\ a \in \mathbb{R}\} \ .$$

For any two elements $\hat{f}_a, \hat{f}_b \in \hat{D}_0$, it can be shown that: i) $\hat{f}_a \circ \hat{f}_b = \hat{f}_{a+b}$ (invariant subgroup); and ii) $\hat{g} \circ \hat{f}_a = \hat{f}_a \circ \hat{g}$ (commutative with any element $\hat{g} \in \hat{D}$). Thus, it is found that $\hat{D}_0$ is a *center* of the extended group $\hat{D}$.

5.3. KDV EQUATION AS A GEODESIC EQUATION ON $\hat{D}(S^1)$

Let us consider the geodesic equation on the extended manifold $\hat{D}(S^1)$ [20]. Associated with a flow defined by $\hat{f}_t(x) = (f_t(x),\ \alpha t + a)$, the tangent field is written as $\hat{u} = (\, u(x)\partial_x,\ \alpha)$. For any two tangent fields $\hat{u}\,, \hat{v} \in T\hat{D}(S^1)$, where

$$\hat{u} = (u(x,t),\ \alpha), \quad \hat{v} = (v(x,t),\ \beta) \ ,$$

the *metric* is defined by

$$\langle\, \hat{u},\ \hat{v}\,\rangle = \int u(x)\, v(x)\, dx + \alpha\, \beta \ , \tag{57}$$

and the *commutator* is given by

$$[\, \hat{u},\ \hat{v}\,] = -(\, u\, \partial_x v - v\, \partial_x u\,,\ c(u, v)) \tag{58}$$

where

$$c(u, v) = \int \partial_x^2 u\, \partial_x v\, dx \ = \ -c(v, u) \ , \tag{59}$$

is the Gelfand-Fuchs cocycle [16, 11]. Following the procedure of §5.1, the covariant derivative is derived for $\hat{u}$ and $\hat{v}$,

$$\nabla^{\mathrm{R}}_{\hat{u}}\, \hat{v} = -\, \nabla_{\hat{u}}\, \hat{v} = \left(w\, \partial_x,\ \frac{1}{2} \int_{S^1} u_{xx} v_x \, \mathrm{d}x \right) \ ,$$

where

$$w = 2uv_x + vu_x + \frac{1}{2}(\alpha v_{xxx} + \beta u_{xxx}) .$$

The geodesic equation is written as $\partial\hat{u}/\partial t + \nabla^{\mathrm{R}}_{\hat{u}}\hat{u} = 0$, which leads to the following two equations:

$$\begin{aligned} u_t + 3uu_x + \alpha u_{xxx} &= 0 ; \\ \partial_t \alpha &= 0 . \end{aligned} \tag{60}$$

Thus, the *KdV equation* is derived, where the constant α is called a *central charge.* The extended algebra is called the *Virasoro algebra* [7].

5.4. SECTIONAL CURVATURES OF KDV SYSTEM

The geometrical theory leads to a relationship between the stability of geodesic curves on a Riemannian manifold and its curvature. The link is expressed by the Jacobi equation for geodesic variation J, derived in §3.5. An evolution equation for the norm $\| J \|$ is given by Eq. (49). The term $K(J,T)$ on the right-hand side is the sectional curvature associated with the two-dimensional section spanned by J and T. If $K(J,T)$ is negative, the right-hand side of (49) becomes positive. Then exponential growth of the magnitude $\| J \|$ is predicted, which is understood that the geodesics are *unstable.* In this context, the sectional curvature of the KdV system is estimated [15, 17] for the section spanned by the two tangent vectors, $\hat{u} = (\, u(x,t),\, \alpha)$ and $\hat{v} = (\, v(x,t),\, \alpha)$. For the sinusoidal fields $\hat{u}_n = (a_n \sin nx\, \partial_x,\, \alpha)$ and $\hat{v}_n = (a_n \cos nx\, \partial_x,\, \alpha)$, it is found that both of the sectional curvatures $K(\hat{v}_1,\, \hat{u}_n)$ and $K(\hat{v}_1,\, \hat{v}_n)$ are *positive* for $n \geq 3$. Thus, most sectional curvatures are positive, however there are some sections which are not positive definite.

6. Geometrical Theory of a Hamiltonian System

A self-gravitating system of N point masses is one of the typical systems studied in classical analytical dynamics. The differential geometric formulation is applied to this system of finite degrees of freedom, as a third example. A simplest non-trivial case is the Hénon-Heiles system, a two-degree-of-freedom Hamiltonian system, which is well known to be a chaotic system. With the present framework, stability of the trajectories of the dynamical system is studied by calculating the Riemannian curvatures of the manifold. This leads to the geometric characterization of *Hamiltonian chaos* [6].

It has recently been discovered that the phenomenon of phase transitions is related at a deep level to a change in the topology of configuration space of the system. Fluctuations of the configuration-space curvature exhibit a singular behavior at the phase transition.

In this section, only the former chaos analysis is presented. As to the latter aspect of geometrical theory of phase transition, we simply refer to the work done by Casetti, Pettini and Cohen (2000) [5].

6.1. FORMULATION

Consider a dynamical system described by the Lagrangian function

$$L(\bar{q},\dot{\bar{q}}) = T - V = \frac{1}{2}a_{ij}(\bar{q})\dot{q}^i\dot{q}^j - V(\bar{q}) ,$$

where $\bar{q} = (q^1, \ldots, q^N)$ and $\dot{\bar{q}}$ are the coordinates and velocities respectively, and $V(\bar{q})$ is the potential of self-gravitation. In the Eisenhart metric g^E given in §3.2.1, two extra coordinates q^0 and q^{N+1} are introduced. The arc-length $\mathrm{d}s$ is represented by $\mathrm{d}s^2 = g^E_{ij}(Q)\mathrm{d}Q^i\mathrm{d}Q^j$, where the metric tensor g^E is given by Eq.(22) and $Q = (q^0, \bar{q}, q^{N+1})$.

Using the covariant derivative $\nabla Y/\mathrm{d}s$ defined by (33), the geodesic equation is represented by $(\nabla/\mathrm{d}s)\mathrm{d}Q/\mathrm{d}s = 0$, which is given as $(\mathrm{d}/\mathrm{d}t)^2 q^i = -\partial V/\partial q^i$ $(i = 1, \ldots, N)$. This equation is the Newton's equation of motion. Thus it is found that the geometric machinery works for the present dynamical system too. The link between the stability of trajectories and the geometrical characterization of the manifold $(M(\bar{q}) \times \mathrm{I\!R}^2, g^E)$ is expressed by the Jacobi equation (48).

6.2. GEOMETRICAL ANALYSIS OF TRANSITION TO CHAOS IN THE HÉNON-HEILES SYSTEM

Information of dynamical behaviours, either regular or chaotic, are included in this geometrical characterization. In order to see this, the previous formulation is applied to a two-degree-of-freedom system described by $L = \frac{1}{2}\left((\dot{q_1})^2 + (\dot{q_2})^2\right) - V(q_1, q_2)$. The equation for the norm of geodesic separation $\| J \|$ is Eq. (49), where the curvature $K(J, \dot{Q})$ is given by

$$K(J,\dot{Q}) = \frac{\partial^2 V}{\partial q_1^2}J_1^2 + 2\frac{\partial^2 V}{\partial q_1 \partial q_2}J_1 J_2 + \frac{\partial^2 V}{\partial q_2^2}J_2^2 .$$

The authors define the integral of negative curvature value $\hat{K}_- = \{\hat{K} : \hat{K}(\dot{Q}, Q) < 0\}$ by $\langle \hat{K}_- \rangle =$ (space average). The quantity $\langle \hat{K}_- \rangle$ was estimated at different energy values E.

In the Hénon-Heiles [13] model, it is shown that the transition from order to chaos is quantitatively described by measuring on a Poincaré section the ratio σ of the area covered by the regular trajectories divided by the total area accessible to the motion. For low energies E the whole area is practically covered by regular orbits and hence the ratio σ is 1. As E increases, σ begins to decrease from 1, and drops rapidly to very small values. It is shown [6] that for low energies E the integral of the negative curvture $\langle \hat{K}_- \rangle$ is almost zero, but that, at the same value of $E \approx 0.1$) at which σ begins to decrease, the value $\langle \hat{K}_- \rangle$ also starts to increase. Hence the exact coincidence between the critical energy level at which σ decreases below 1 and the one at which $\langle \hat{K}_- \rangle$ increases above 0 is the onset of a sharp increase of chaotic domains, detected by the increase of the negative curvature integral $\langle \hat{K}_- \rangle$.

7. Flows of an Inviscid Incompressible Fluid

Motion of fluid particles of an inviscid incompressible fluid on a bounded domain is described on the basis of a geometrical framework. Fluid flows are well described already by conventional fluid dynamics. However, fluid flows may be equivalently described by the geodesics on the manifold of all diffeomorphic mappings. This formulation is based on Riemannian geometry and Lie group theory, originally developed by Arnold [2]. The present approach reveals new aspects which are not studied in conventional fluid dynamics. For example, Riemannian curvatures are found to be mostly negative, that seems to be related to the mixing and ergodicity of fluid motion.

7.1. BASIC CONCEPTS

7.1.1. *Lagrangian and Eulerian descriptions*

We consider flows of an inviscid incompressible fluid on a manifold M (which is the flow region): $\mathbb{R}^2$ (or T^2), or $\mathbb{R}^3$ (or T^3). The motivation of the geometrical analysis is the observation that Euler's equation of motion is a geodesic equation on a group of volume-preserving diffeomorphisms with the metric defined by the kinetic energy. The set of all volume-preserving diffeomorphisms of M constitutes a group manifold $D_\mu(M)$, of which an element $g \in D_\mu(M)$ is a map $g : M \to M$.

Suppose that a curve $t \mapsto g_t(x)$ denotes a fluid flow, then a point $x \in M$ is mapped to the point $g_t(x)$ at time t. This is a *Lagrangian* description of flows. The tangent vector (velocity field) at the time t is represented by

$$U_t(x) = \dot{g}_t(x) = \lim_{\Delta t \to 0} \frac{1}{\Delta t}(g_{\Delta t} - 1) \circ g_t(x) = u \circ g_t(x) = u(g_t(x)) \,.$$

By definition, the tangent field $U_t(x) = U_{g_t}$ is *right-invariant* (see §7.2). Right translation with g_t^{-1} yields the velocity field at $e = g_0$:

$$u_t(x) = \dot{g}_t \circ g_t^{-1}(x) \,, \qquad u_t \in T_e D^s_\mu(M) \,.$$

At the identity e (i.e. the initial manifold M_0), the velocity field $u_t(x)$ satisfies the divergence-free property, i.e. $\operatorname{div} u_t = 0$. The suffix t denotes that the tangent vector u_t is time-dependent. In terms of fluid-dynamics $u_t(x)$ represents an *Eulerian* velocity field. In mathematical language u_t is an element of the Lie algebra, $T_e D_\mu(M)$. It is useful to consider the manifold $D^s_\mu(M)$, which is a subgroup of volume-preserving diffeomorphisms (of M) of Sobolev class H^s, where $s > n/2 + 1$ ($n = \dim M$)[1].

An arbitrary vector field $v(x) \in T_e M$ can be decomposed into L^2-orthogonal components of divergence-free part $\bar{v}$ and gradient part [18, 16]

$$v = \bar{v} + \operatorname{grad} f \; . \tag{61}$$

7.1.2. *Right invariant fields*

Consider the tangent field $U_\eta \in T_\eta D^s_\mu(M)$ at any $\eta \in D^s_\mu(M)$, and suppose that U_η is *right-invariant*, that is $U_\eta(x) = U_e \circ \eta(x)$ for $U_e \in T_e D^s_\mu(M)$. Correspondingly, the right-invariant L^2-metric is defined by

$$\langle U, V \rangle_\eta \equiv \int_M \left(U_\eta \circ \eta^{-1}, V_\eta \circ \eta^{-1}\right)_x \mathrm{d}\mu(x) = \langle U_e, V_e \rangle_e \; , \tag{62}$$

where $U_\eta, V_\eta \in T_\eta D^s_\mu(M)$, $\mathrm{d}\mu$ is the volume form, and $(\,\cdot\,,\,\cdot\,)_x$ is the scalar product at each point $x \in M$. Applying a right translation to the middle side of (62) by η, we have $\langle U, V \rangle_\eta = \int_M (U_\eta(x), V_\eta(x))_{\eta(x)} \, \mathrm{d}\mu(x)$. This equality is verified by the commutability of d and η^* and the change of variables with the volume-preserving property $\eta^* \mu(\eta^{-1} x) = \mu(x)$. Thus the present L^2-metric (62) is *isometric* with respect to the right translation by any $\eta \in D^s_\mu(M)$.

The metric (62) induces on $D^s(M)$ and $D^s_\mu(M)$ smooth Riemannian connections $\hat{\nabla}$ and $\bar{\nabla} = \mathsf{P}\hat{\nabla}$, respectively. For any right invariant vector fields $U_\eta, V_\eta \in T_\eta D^s_\mu(M)$, we have the *right-invariant* connection,

$$(\hat{\nabla}_{U_\eta} V_\eta)_\eta = (\nabla_{U_e} V_e)_e \circ \eta \; , \tag{63}$$

for $\eta \in D^s_\mu(M)$, where ∇ is the covariant derivative on M (manifold of Eulerian description). Similarly, we have $\bar{\nabla}_{U_\eta} V_\eta = \mathsf{P}[\nabla_{U_e} V_e] \circ \eta$. The symbol P is the projection operator to the divergence-free part. The difference between $\hat{\nabla}$ and $\bar{\nabla}$ is given by the second fundamental form $S(U_\eta, V_\eta)$:

$$S(U_\eta, V_\eta) = \hat{\nabla}_{U_\eta} V_\eta - \bar{\nabla}_{U_\eta} V_\eta = \mathsf{Q}_\eta [\, \hat{\nabla}_{U_\eta} V_\eta] \tag{64}$$

[1]The group manifold $D^s_\mu(M)$ is a weak Riemannian submanifold of the group $D^s(M)$ of all Sobolev H^s-diffeomorphisms of M [18].

(see (39)). For tangent fields $U, V, W, Z \in T_\eta D^s_\mu(M)$, the curvature tensor $\hat{R}$ on $D^s(M)$ and $\bar{R}$ on $D^s_\mu(M)$ are also defined in the right-invariant way:

$$(\hat{R}(U,V)\,W)_\eta \;=\; (R(U\circ\eta^{-1}, V\circ\eta^{-1})\,W\circ\eta^{-1})\circ\eta\,, \tag{65}$$

and $(\bar{R}(U,V)\,W)_\eta = \mathsf{P}_\eta[(\hat{R}(U,V)\,W)_\eta]$, where R is the curvature tensor on M for $u,\ v,\ w, \in T_e D^s_\mu(M)$: $R(u,\ v)\,w = \nabla_u(\nabla_v\,w) - \nabla_v(\nabla_u\,w) - \nabla_{[u,v]}\,w$. The curvature tensors on $D^s(M)$ and $D^s_\mu(M)$ are related by the Gauss-Codazzi equation (40):

$$\begin{aligned}\left\langle \hat{R}(U,\,V)W,\,Z\right\rangle_{L^2} \;&=\; \left\langle \bar{R}(U,\,V)W,\,Z\right\rangle + \left\langle S(U,\,W),\,S(V,\,Z)\right\rangle \\ &\quad -\left\langle S(U,\,Z),\,S(V,\,W)\right\rangle\,. \end{aligned}\tag{66}$$

7.2. FORMULATION OF HYDRODYNAMICS

7.2.1. *Hydrodynamic connection*

Let us consider the geodesic equation on the manifold $D^s_\mu(M)$, the group of volume-preserving diffeomorphisms of M, that is the governing equation of an inviscid incompressible fluid. This is a mathematical derivation of the hydrodynamic equation of fluid flows, for which the Eulerian representation is given on the manifold M at the identity.

Because of the right-invariance of the metric on $TD^s_\mu(M)$, the Riemannian connection satisfies the expression like (32). Using the symbol $\bar{\nabla}$ for the connection, we obtain $2\langle\,\bar{\nabla}_u v, w\,\rangle = \langle\,[u,v], w\,\rangle - \langle\,[v,w], u\,\rangle + \langle\,[w,u], v\,\rangle$, for $u,\ v,\ w \in T_e D^s_\mu(M)$. The commutator of the present problem is

$$[\,u, v\,](s) = \bar{\nabla}_u v - \bar{\nabla}_v u \tag{67}$$

(see (29)), where the right-hand side is divergence-free too.

Introducing the adjoint operator $ad_v\,w = [v, w]$ and the coadjoint operator by $\langle ad^*_v\,u, w\rangle = \langle u, ad_v\,w\rangle = \langle u, [v, w]\rangle$, we obtain

$$\bar{\nabla}_u\,v = \mathsf{P}\left[\frac{1}{2}\left([u,\ v] - ad^*_u\,v - ad^*_v\,u\right) + \mathrm{grad}f\right]\,,$$

where f is a scalar function of x. The function f is naturally introduced because of the divergence-free of the tangent vector w. In fact, $\langle\,\mathrm{grad}\,f,\ w\,\rangle = 0$. Further, we have $\mathrm{div}\,\bar{\nabla}_u v = 0$.

Because of the right-invariance of the diffeomorphisms, the commutator is $[u,\ v] = \mathcal{L}_u v = -\mathsf{P}\{(\mathrm{d}v^i/\mathrm{d}t) - v^j(\partial u^i/\partial x^j)\}\,\partial_i$. Therefore the time evolution is described by $\nabla^{(\mathrm{R})}_u\,v = -\,\bar{\nabla}_u\,v$ (see §3.4.2).

7.2.2. *Formulas in* $I\!R^3$

For $\mathbf{u}$, $\mathbf{v} \in T_e D_\mu(\mathrm{I\!R}^3)$ (using bold-face letters for vectors in the space $\mathrm{I\!R}^3$),

$$-\,ad_{\mathbf{u}}\,\mathbf{v} \;\equiv\; -[\,\mathbf{u},\,\mathbf{v}\,] = [\mathcal{L}_{\mathbf{u}},\,\mathcal{L}_{\mathbf{v}}] = (\mathbf{u}\cdot\boldsymbol{\nabla})\mathbf{v} - (\mathbf{v}\cdot\boldsymbol{\nabla})\mathbf{u}\;,$$

where $\boldsymbol{\nabla} = (\partial_1, \partial_2, \partial_3)$, $\mathbf{u}\cdot\boldsymbol{\nabla} = u_1\partial_1 + u_2\partial_2 + u_3\partial_3$ and $\nabla\cdot\mathbf{u} = 0$, $\nabla\cdot\mathbf{v} = 0$. The relation $\langle ad_{\mathbf{u}}\,\cdot\,,\,\mathbf{v}\rangle = \langle\,\cdot\,,\,ad^*_{\mathbf{u}}\,\mathbf{v}\rangle$ by integration by parts yields

$$-\,ad^*_{\mathbf{u}}\,\mathbf{v} \;=\; -(\mathbf{u}\cdot\boldsymbol{\nabla})\mathbf{v} - v^k\boldsymbol{\nabla}u^k = \mathbf{u}\times(\boldsymbol{\nabla}\times\mathbf{v}) - \boldsymbol{\nabla}h - \boldsymbol{\nabla}f_{uv}\;,$$

for a function f_{uv}, where $h = u^k\,v^k = \mathbf{u}\cdot\mathbf{v}$. Thus, we obtain

$$\begin{aligned}\bar{\nabla}^{(\mathrm{R})}_{\mathbf{u}}\,\mathbf{v} = -\,\bar{\nabla}_{\mathbf{u}}\,\mathbf{v} \;&=\; \mathsf{P}[(1/2)(\,[\mathbf{u},\,\mathbf{v}] - ad^*_{\mathbf{u}}\mathbf{v} - ad^*_{\mathbf{v}}\mathbf{u}\,) + \mathrm{grad}\,f\,]\\ &=\; \mathsf{P}\,[(\mathbf{u}\cdot\nabla)\mathbf{v} + \boldsymbol{\nabla}\,p]\;,\end{aligned}$$

where $p = (1/2)h + f + f_{uv} + f_{vu}$ and $\mathrm{div}\,\bar{\nabla}_{\mathbf{u}}\mathbf{v} = 0$. In particular setting $\mathbf{v} = \mathbf{u}$, we have

$$\bar{\nabla}^{(\mathrm{R})}_{\mathbf{u}}\,\mathbf{u} = \mathsf{P}\,[(\mathbf{u}\cdot\boldsymbol{\nabla})\mathbf{u} + \boldsymbol{\nabla}\,p]\;,\quad \text{with}\quad \boldsymbol{\nabla}\cdot\mathbf{u} = 0\;.$$

7.2.3. *Geodesic equation*

The geodesic equation in the right-invariant time-dependent problem must be considered according to the formulation of §3.4.2. Using (63), a geodesic is a curve g_t whose tangent is translated parallel along itself:

$$0 = \mathsf{P}\Big[\,\Big(\hat{\nabla}^{(\mathrm{R})}_{\dot{g}_t}\,\dot{g}_t\Big)_{g_t}\,\Big] = \mathsf{P}\Big[\,\partial_t\,u + \nabla^{(\mathrm{R})}_u\,u + \mathrm{grad}\,p\Big]\circ g_t\;, \tag{68}$$

$(U_e = u)$, where $\mathrm{div}\,u = 0$. The Euler equation of motion of an incompressible fluid in $\mathrm{I\!R}^3$ is obtained by right translation g_t^{-1} to $e = g_0$:

$$\partial_t\,\mathbf{u} + (\mathbf{u}\cdot\boldsymbol{\nabla})\mathbf{u} + \boldsymbol{\nabla}\,p = 0\;,\qquad \boldsymbol{\nabla}\cdot\mathbf{u} = 0\;. \tag{69}$$

The Jacobi equation is given by (51), where $T = \partial_t g$ and $J = \partial_\alpha g = \|\,J\,\|$ $\mathbf{e}_J$, and $K(T,\mathbf{e}_J) = \langle\bar{R}(T,\mathbf{e}_J)\,\mathbf{e}_J,\,T\rangle$ is the sectional curvature in the two dimensional section spanned by the tangent T and the Jacobi vector J.

7.2.4. *Jacobi field as frozen field*

Let us consider the Jacobi field from a different point of view. According to the above definitions of T and J, the equation $(\boldsymbol{\nabla}/\partial s)\,\partial_\alpha\gamma = (\boldsymbol{\nabla}/\partial\alpha)\,\partial_s\gamma$ of §3.5 is written as $\bar{\nabla}_T J = \bar{\nabla}_J T$ (see (33)), and further the Lie derivative vanishes: $\mathcal{L}_T\,J = [\,T,\,J\,] = \bar{\nabla}_T J - \bar{\nabla}_J T = 0$ (see (19)). It can be verified that the torsion-free condition is valid not only with the connection $\bar{\nabla}$, but with the connection $\hat{\nabla}$ as well. Thus, we have $\hat{\nabla}_T J = \hat{\nabla}_J T$. In the time-dependent problem in $\mathbb{R}^3$ this is re-written as

$$\partial_t \mathbf{J} + (\mathbf{u} \cdot \boldsymbol{\nabla}) \mathbf{J} = (\mathbf{J} \cdot \boldsymbol{\nabla}) \mathbf{u} \ ,$$

which is equivalent to $\partial_t \mathbf{J} + \boldsymbol{\nabla} \times (\mathbf{J} \times \mathbf{u}) = 0$, due to a well-known vector identity and because $\boldsymbol{\nabla} \cdot \mathbf{u} = 0$ and $\boldsymbol{\nabla} \cdot \mathbf{J} = 0$. This equation is usually called the equation of *frozen* field, since it describes that the vector field $\mathbf{J}$ is carried along with the flow $\mathbf{u}$ as if $\mathbf{J}$ was frozen into the carrier fluid.

7.2.5. *Interpretation of the Riemannian curvature*
(i) Time evolution of Jacobi field
A Jacobi field $J(t)$ is uniquely determined by its value $J(0)$ and the value of $\nabla_T J$ at $t = 0$ on the geodesic g_t. Provided that $J(0) = 0$ and $(\nabla_T J)_{t=0} = a_0$, it can be shown [14] that

$$\frac{J(t)}{a_0} = t - \frac{t^3}{6} \frac{K(T, J)}{\| J \|^2} + O(t^4) \qquad (t > 0) \ , \tag{70}$$

where $\lim_{t \to 0} K(T, J)/ \| J \|^2 = K(T_0, \mathbf{e}_J) \equiv K_J(0)$, and $T_0 = T(0)$. The time development of the Jaocbi vector is controlled by the curvature $K(T_0, \mathbf{e}_J)$ and in particular by its sign. An L^2-distance between two states starting at the common e with different initial velocities v_1 and v_2 in a domain D is defined by $d(v_1, v_2 : t) = \left(\int_D |g_t(\mathbf{x} : v_1) - g_t(\mathbf{x} : v_1)|^2 \mathrm{d}\mathbf{x} \right)^{1/2}$. It is shown by Hattori and Kambe [12] that the *distance* d is proportional to the Jacobi field J at small t.

(ii) Second fundamental form
It is conjectured by Kambe [16] how the fluid motion acquires a curvature and what the curvature is. Here we consider it briefly.

The *curvature* of the closed submanifold $D_\mu^s(M)$ is given by $\langle \bar{R}(U, V)W, Z \rangle$ in the equation (66). In particular, the sectional curvature of the section spanned by the tangent vectors $X, Y \in T_\eta D_\mu^s(M)$ is given by

$$\bar{K}(X, Y)^{D_\mu} = \left\langle R(X, Y)Y, X \right\rangle^M + \left\langle S(X, X), S(Y, Y) \right\rangle - \| S(X, Y) \|^2 \ .$$

Even when the manifold M is flat, i.e. the curvature $\left\langle R(X, Y)Y, X \right\rangle^M$ vanishes, the sectional curvature $\bar{K}(X, Y)^{D_\mu}$ of the closed submanifold $D_\mu^s(M)$ does not necessarily vanish due to the second and third terms (defined as $\bar{K}_S(X, Y)$) associated with $S(X, Y)$, etc.. Namely the curvature of a fluid motion in this case *originates* from the $\bar{K}_S$ part, $\bar{K}_S(X, Y) = \left\langle S(X, X), S(Y, Y) \right\rangle_{L^2} - \| S(X, Y) \|^2$, for a flow of an incompressible (inviscid) fluid. Thus it is found that the restriction to be volume-preserving gives rise to the above curvature. Further it is shown that the curvature is related to the 'grad' part of the connection $\hat{\nabla}_X Y$, which is normal to $TD^s_\mu(M)$ and related to the pressure function.

7.2.6. *Space-periodic flows in a cubic space: Fourier representation*
Explicit forms are given for space-periodic flows in a cube by Fourier representation, i.e. for flows on the flat 3-torus $M = T^3 = \mathbb{R}^3/(2\pi Z)^3$ [19], [12]). Such periodic fields are represented by the Fourier forms. The metric, covariant derivative and commutator are explicitly given. Then the geodesic equation and curvature tensor are derived [19]. It is to be noted that the formulas derived reduce to those of Arnold [2] when two-dimensionality is imposed. As an application, a flow with Beltrami property is considered [19]. Negative sectional curvatures are found, and this is considered to be the three-dimensional conterpart of the Arnold's two-dimensional result [2].

8. Motion of a Vortex Filament

The dynamics of an isolated thin vortex filament with small curvature, embedded in an ideal incompressible fluid, is known to be well-approximated by the *local induction equation* (LIE) [21]. It is assumed here that a vortex filament is spatially periodic and given by a time-dependent C^∞-curve $\mathbf{x}(s, t)$ in $\mathbb{R}^3$ with t being the time and $s \in S^1$ the length parameter.

This system is shown to be characterized with the rotation group $G = SO(3)$ associated with the S^1 manifold pointwise. The group $G(S^1)$ of the smooth mappings, $g : s \in S^1 \mapsto g(s) \in G = SO(3)$, equipped with the point-wise composition law, $g''(s) = g'(s) \circ g(s)$ for $g, g', g'' \in G$, is an infinite-dimensional Lie group, *i.e.* the *loop group.* The corresponding loop algebra gives rise to the Landau-Lifshitz equation as the geodesic equation (§8.2). Further, its central extension results in the so-called Kac-Moody algebra [7], and corresponding geodesic equation is found to be the equation obtained by Fukumoto and Miyazaki (1991) [10]. See Kambe [16] for details.

8.1. LOCAL INDUCTION EQUATION

Suppose that the motion of a vortex filament is governed by the LIE,

$$\partial_t \mathbf{x} = \partial_s \mathbf{x} \times \partial_s^2 \mathbf{x} \ . \tag{71}$$

The Hamiltonian of this system is given by $H = \int \kappa^2(s)\, \mathrm{d}s$.

The motion of the curve $\mathbf{x}(s,t)$ is a map $t \mapsto \phi_t$, that is, $x_0(s) \mapsto x_t(s) = \phi_t \circ x_0(s)$, where $x_t(s) = \mathbf{x}(s,t)$. Following the motion, the tangent vector $T_t(s)$ on the curve is left-translated, i.e. $T_t(s) = \partial x_t(s)/\partial s = \phi_t \circ \partial x_0(s)/\partial s = (\phi_t)_* T_0(s)$. Note that $\partial_s T_t = \kappa(s) T_t(s)$. Differentiating the equation (71) with respect to s, one obtains

$$\partial_t T = T \times \partial_s^2 T = -\partial_s^2 T \times T, \tag{72}$$

for the unit tangent vector field T. The equation of the form $\partial_t X = \Omega \times X$ describes "rotation" of the vector X with the angular velocity Ω. Hence, the factor $-\partial_s^2 T$ ($= \Omega$, say) in the above equation is interpreted as the angular velocity at s pointwise, that is, the $\Omega = -T''$ is an element of the Lie algebra $\mathbf{so(3)}$. See §2.5 and §4.1 for the commutator of the $\mathbf{so(3)}$. Thus, it is found that, using the new symbol L for T, $L'' \in C^\infty(S^1, \mathbf{so(3)})$, where $C^\infty(S^1, \mathbf{so(3)}) = \mathcal{L}\mathbf{g} = \mathbf{so(3)}[S^1]$ is the *loop algebra* [7].

8.2. LOOP ALGEBRA AND LANDAU-LIFSHITZ EQUATION

The above dynamical system is reformulated in the following way. Let

$$X, Y, Z \in \mathcal{L}\mathbf{g} = \mathbf{so(3)}[S^1] = C^\infty(S^1, \mathbf{so(3)})$$

be the vector fields, and AX, AY, $AZ \in C^\infty(S^1, \mathbf{so(3)}^*)$ be the dual fields, where $A = -\partial_s^{-2}$. The left-invariant *metric* is defined by

$$\langle X, Y \rangle = \int_{S^1} (\partial_s^{-1} X, \partial_s^{-1} Y)_{\mathbb{R}^3}\, ds = \int_{S^1} (AX, Y)_{\mathbb{R}^3}\, ds .$$

The *commutator* is given by $[X, Y](s) = X(s) \times Y(s)$ at each s, pointwise. In the case of the invariant metric, the $ad_X^* Y$ satisfies

$$\begin{aligned} \langle ad_X^* Y, Z \rangle = \langle Y, ad_X Z \rangle &= \int_{S^1} (AY, X \times Z)_{\mathbb{R}^3}\, ds \\ &= \int_{S^1} \Big(AA^{-1}(AY \times X), Z \Big)_{\mathbb{R}^3} ds , \end{aligned}$$

which leads to $ad_X^* Y = A^{-1}(AY \times X) = -\partial_s^2 [AY, X]$. Thus it is found that the *connection* ∇ of the filament motion is given by [22]

$$\nabla_X Y = \frac{1}{2}\Big([X, Y] + \partial_s^2 [AX, Y] + \partial_s^2 [AY, X] \Big)$$

for $X(s), Y(s) \in \mathcal{L}\mathbf{g} = \mathbf{so(3)}[S^1]$, the loop algebra. The *geodesic equation* ($\partial_t X + \nabla_X X = 0$) on the loop group $LG = SO(3)[S^1]$ is given by

$$\partial_t X + \partial_s^2 (AX \times X) = 0 .$$

Applying the operator $A = -\partial_s^{-2}$, we get a corresponding equation in the dual space,

$$\partial_t L - (L \times L'') = 0 .$$

where $L = -AX$ and $X = L''$. Thus, we have recovered the equation (72) for the vector $T = L$. This type of equation is called the *Landau-Lifshitz* equation.

The *sectional curvature* is defined by $K(X,Y) = \langle R(X'',Y'')Y'', X''\rangle = \int_{S^1} f(s)\,ds$, for X'', $Y'' \in C^\infty(S^1, \mathbf{so(3)})$. The integrand $f(s)$ is explicitly given in terms of $X(s)$, $Y(s)$ and their derivatives. The curvature was estimated in two example cases [22].

In [16], the central extension of the algebra of the previous section is considerd in analogous to the KdV system (§5). The geodesic equation of the extended system is equivalent to the equation of Fukumoto and Miyazaki [10], shown to be a completely integrable system. The equation was originally derived for the motion of a thin vortex tube with an axial flow along it.

It was shown in §5 that the KdV equation is a geodesic equation on the diffeomorphism group of a circle S^1 with a central extension. Here it is verified that the motion of a vortex filament governed by the LIE equation is a geodesic on the loop group $LG = SO(3)[S^1]$ which is $SO(3)$-valued with pointwise multiplication. Further, the infinite-dimensional loop algebra $\mathcal{L}\mathbf{g}$ has non-trivial central extension equivalent to the Kac-Moody algebra [16].

9. Conclusion

A geometrical theory is develped for various dynamical systems of both finite and infinite degreees of freedom. Although those systems are already studied with the conventional methods of physics, the present formulation provides us an insight into the systems and adds new geometrical characterizations of the dynamical evolutions in terms of the geodesic equations, Jacobi fields, Lie algebras and Riemannian curvatures.

This paper originates from lecture notes prepared during the author's stay at the Isaac Newton Institute for Mathematical Sciences for the programme "Geometry and Topology of Fluid Flow" (September-December, 2000), benefitting greatly from discussions had with many participants to the programme. After that, the notes have been re-written, revised and largely expanded into the present paper.

Acknowledgements

The author, as one of the organizers of the programme 'Geometry and Topology of Fluid Flows', would like to express sincere thanks for the arrangement and help to the Isaac Newton Institute. The financial support from UK EPSRC Grant GR/K99015 is also kindly acknowledged. Further, the author is grateful to many participants for the discussions which contributed very much to the improvement and completion of the present article.

References

1. Abraham, R. & Marsden, J. (1978) *Foundations of Mechanics.* Benjamin, Cunnings.
2. Arnold, V.I. (1966) Sur la géométrie différentielle des groupes de Lie de dimension infinie et ses applications à l'hydrodynamique des fluides parfais. *Ann. Inst. Fourier* (Grenoble) **16**, 319–361.
3. Arnold, V.I. (1998) *Mathematical Methods of Classical Physics.* Springer.
4. Arnold, V.I. & Khesin, B.A. (1998) *Topological Methods in Hydrodynamics.* Springer.
5. Casetti, L., Pettini M. & Cohen, E.G.D. (2000) Geometric approach to Hamiltonian dynamics and statistical mechanics. *Physics Rep.* **337**, 237–341.
6. Cerruti-Sola M. & Pettini, M. (1996) Geometrical description of chaos in two-degrees-of-freedom Hamiltonian systems. *Phys. Rev. E* **53**, 179–188.
7. de Azcárraga, J.A. & Izquierdo, J.M. (1995) *Lie Groups, Lie Algebras, Cohomology and Some Applications in Physics.* Cambridge University Press.
8. Ebin, D.G. & Misiołek, G. (1999) The exponential map on D^s_μ. In *Proceedings of the Arnoldfest* (ed. E. Bierstone *et al.*), pp. 153–163. Fields Institute Communications **24**, Toronto.
9. Frankel, T. (1997) *The Geometry of Physics. (An Introduction).* Cambridge University Press.
10. Fukumoto Y. & Miyazaki, T. (1991) Three-dimensional distortions of a vortex filament with axial velocity. *J. Fluid Mech.* **222**, 369-416.
11. Gelfand, I.M. & Fuchs, D.B. (1968) The cohomology of the Lie algebra of vector fields on a circle. *Functs. Anal. Prirozhen.* **2**.
12. Hattori Y. & Kambe, T. (1994) Kinematical instability and line-stretching in relation to the geodesics of fluid motion. *Fluid Dyn. Res.* **13**, 97–117.
13. Hénon M. & Heiles, C. (1964) The applicability of the third integral of motion: some numerical experiments. *Astron. J.* **69**, 73–79.
14. Hicks, N.J. (1965) *Notes on Differential Geometry.* D. van Nostrand Co. Inc..
15. Kambe, T. (1998) Geometrical aspects in hydrodynamics and integrable systems. *Theor. Comput. Fluid Dynamics* **10**, 249–261.
16. Kambe, T. (2001) Geometry of dynamical systems and fluid flows. *INIMS Preprint* **NI01041-GTF**. Isaac Newton Institute, Cambridge.
17. Misiolek, G. (1997) Conjugate points in the Bott-Virasoro group and the KdV equation. *Proc. of AMS* **125**, 935–940.
18. Misiolek, G. (1993) Stability of flows of ideal fluids and the geometry of the group of diffeomorphisms. *Indiana Univ. Math. J.* **42**, 215–235.
19. Nakamura, F., Hattori Y. & Kambe, T. (1992) Geodesics and curvature of a group of diffeomorphisms and motion of an ideal fluid. *J. Physics A* **25**, L45–L50.
20. Ovsienko, V. Yu. & Khesin, B. A. (1987) Korteweg-de Vries superequation as an Euler equation. *Funct. Anal. Appl.* **21**, 329–331.
21. Saffman, P.G. (1992) *Vortex Dynamics.* Cambridge Univ. Press.
22. Suzuki, K., Ono, T. & Kambe, T. (1996) Riemannian geometrical analysis of the motion of a vortex filament: a system of $C^\infty(S^1, SO(3))$. *Phys. Rev. Lett.* **77**, 1679–1682.
23. Suzuki, K., Watanabe, Y. & Kambe, T. (1998) Geometrical analysis of free rotation of a rigid body. *J. Physics A: Math. Gen.* **31**, 6073–6080.

STREAMLINE PATTERNS AND THEIR BIFURCATIONS USING METHODS FROM DYNAMICAL SYSTEMS

MORTEN BRØNS
Department of Mathematics
Technical University of Denmark
DK-2800 Kongens Lyngby, Denmark

E-mail: *M.Brons@mat.dtu.dk*

Abstract. Patterns in viscous flow are studied with the aid of tools from low-dimensional nonlinear dynamics. We consider in particular change of patterns as parameters are varied, and show how normal form theory can be used to simplify the differential equations that describe the streamlines. The approach is used to obtain a basic qualitative description of separation and attachment close to a wall and to axisymmetric flow, and we illustrate how the ideas can be applied to a specific flow exhibiting secondary flow in the form of vortex breakdown. We describe how the possibility of chaotic streamlines in 3-dimensional flow complicates the classification of patterns in this case.

1. Introduction

The first step in a mathematical modelling of a fluid problem consists in writing down the governing equations for the flow, based on Newton's second law and the equation of continuity. Here we consider viscous flow only, and the relevant model is typically the Navier-Stokes equations,

$$\frac{\partial \mathbf{v}}{\partial t} + (\mathbf{v} \cdot \nabla)\mathbf{v} = -\frac{1}{\rho}\nabla p + \frac{\mu}{\rho}\Delta \mathbf{v}, \quad \nabla \cdot \mathbf{v} = 0, \tag{1}$$

where ρ is the density and μ is the dynamic viscosity of the fluid.

Performing one time integration of these equations with appropriate boundary conditions yields the velocity field $\mathbf{v}(\mathbf{x}, t)$ and the pressure field $p(\mathbf{x}, t)$. This is the *Eularian* point of view, where field quantities are given at specific points in space, as the fluid passes by.

R.L. Ricca (ed.), An Introduction to the Geometry and Topology of Fluid Flows, 167–182.

For many purposes this knowledge is sufficient, for example for obtaining forces and stresses on submerged bodies. However, if the motion of individual fluid particles is of interest, as for example in the study of mixing, one must switch to a *Lagrangian* description, and perform a further integration, since particles move according to

$$\dot{\mathbf{x}} = \mathbf{v}(\mathbf{x}, t), \tag{2}$$

which is a system of ordinary differential equations in dimension 2 or 3. The solution curves $\mathbf{x}(t)$ are the particle paths.

A further integration is also nedded to determine the instantaneous streamlines (at $t = t_0$ say), as they are defined as the solution curves to

$$\dot{\mathbf{x}} = \mathbf{v}(\mathbf{x}, t_0). \tag{3}$$

It is the purpose of the present paper to discuss how ideas from the qualitative theory of dynamical systems can be used to describe structures in fluids by analysing the features of (3). Indeed, a structure is qualitatively rather than quantitatively defined. For simplicity we will only consider steady flows where (2) and (3) are identical, and also primarily look at two-dimensional flows. As we shall see, several basic concepts in fluid mechanics are in fact naturally defined in dynamical systems terms.

Further results from this point of view on fluid mechanics can be found in the review papers [11, 23, 28]. For an introduction to modern nonlinear dynamics see e.g. [29].

2. Separation From a Wall

A basic problem in fluid mechanics is separation of the flow from a wall. We analyze the situation following [19, 22]. Let $\mathbf{x} = (x, y)$, the wall be $y = 0$, the flow region $y \geq 0$, and $\mathbf{v} = (u, v)$. Consider a Taylor expansion of the velocity field at the origin,

$$u = y(a_0 + a_1 x + a_2 y + \mathcal{O}_2), \quad v = y(b_0 + b_1 x + b_2 y + \mathcal{O}_2). \tag{4}$$

The common factor y accounts for the no-slip condition at the wall, and $\mathcal{O}_2$ denotes terms of second and higher order. To fulfil the continuity equation, we require

$$\nabla \cdot \mathbf{v} = b_0 + b_1 x + (a_1 + 2b_2)y + \mathcal{O}_2 = 0, \tag{5}$$

from which we get

$$b_0 = b_1 = 0, \quad b_2 = -\frac{a_1}{2}. \tag{6}$$

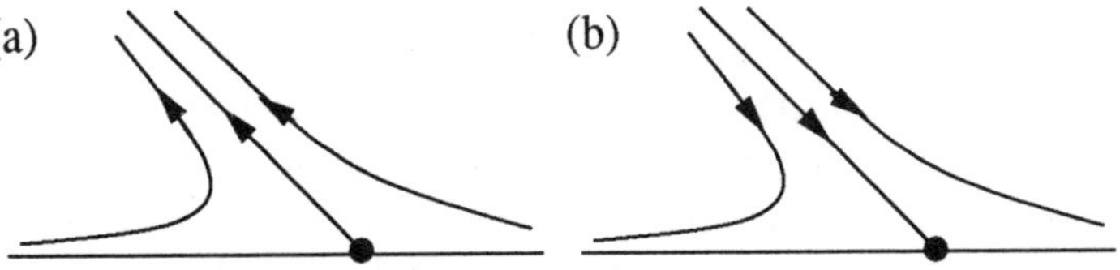

Figure 1. Hyperbolic critical points on the wall. (a): Separation, $a_1 < 0$. (b): Attachment, $a_1 > 0$.

To obtain the streamlines close to the origin, we may consider the differential equations

$$\dot{x} = \frac{u}{y} = a_0 + a_1 x + a_2 y + \mathcal{O}_2, \quad \dot{y} = \frac{v}{y} = -\frac{a_1}{2} y + \mathcal{O}_2, \tag{7}$$

as scaling by a common function does not change the streamlines (trajectories), but only the speed of the particle motion along the streamline.

If $a_0 = 0$, the origin is a critical point for (7), that is, a point where $\dot{x} = \dot{y} = 0$. Basic information on the streamlines in the vicinity of the critical point is found from the linearized system which is obtaind by dropping the $\mathcal{O}_2$-terms in (7). The linearized system has a system matrix which is the Jacobian of the full velocity field,

$$J = \begin{pmatrix} a_1 & a_2 \\ 0 & -\frac{a_1}{2} \end{pmatrix}. \tag{8}$$

The eigenvalues of J are $a_1, -\frac{a_1}{2}$. If $a_1 \neq 0$ these are of different sign, and the critical point is a saddle point with separatrices along the eigendirections. If $a_1 < 0$, the contracting eigendirection is $(1, 0)$ and the expanding eigendirection is $(2a_2, -3a_1)$. If $a_1 > 0$, the directions are exchanged. The two possible situations are shown in figure 1, the first corresponding to separation at the origin, the latter to attachment. In fluid mechanics, the off-wall separatrix is termed a *dividing streamline*.

It is a basic result in the theory of dynamical system that the local structure of the solution curves close to a critical point is determined solely by the linearization if the critical point is *hyperbolic*, i.e. if the eigenvalues of the linearization are off the imaginary axis. Furthermore, hyperbolic critical points are *structurally stable*, so the the qualitative structure of the trajectories is not changed if the velocity field is perturbed [29]. Hence, the fact that the critical point at the origin is a point of separation or attachment depends only on a_1, a_2, but the streamlines may be deformed depending on the higher-order terms, and in particular the dividing streamline may be curved. However, the eigendirection is tangent to the dividing streamline at the critical point.

For a specific flow, the expansion parameters a_i, b_i will depend on physical parameters such as boundary conditions and dimensionless parameters. If a structurally stable critical point is found for some specific combination of parameters, a small change of the parameters will result in a perturbation of the velocity field, but the structural stability ensures that no qualitative change of the streamlines occurs. Thus, a qualitative change – a *bifurcation* – of the local streamline pattern can occur only at parameter values where the critical points are non-hyperbolic.

In the present case, the critical point is non-hyperbolic if $a_1 = 0$, and the structure of the streamlines is not determined from the linearization alone. Possible bifurcations of patterns associated with this degeneracy are analyzed in [1, 16]. In the next section we consider bifurcation of structures in a different setting.

We end this section by noting that the decisive parameters a_1, a_2 have a direct physical interpretation. The wall shear stress is

$$\tau_w = \mu \left.\frac{\partial u}{\partial y}\right|_{y=0} = \mu(a_0 + a_1 x + \mathcal{O}_2), \tag{9}$$

so separation or attachment occurs when $\tau_w = 0$, and the type depends on the sign of the gradient, as

$$\frac{\partial \tau_w}{\partial x} = \mu a_1 \tag{10}$$

at the critical point. Furthermore, from eq. (1) we get that

$$\frac{\partial p}{\partial x} = \mu \Delta u = 2\mu a_2 \tag{11}$$

at the critical point, so the slope of the dividing streamline depends on the ratio of the the wall presure gradient and the wall stress gradient.

3. Bifurcation of Patterns in Axisymmetric Flow

Although three-dimensional, the description of axisymmetric flow can be reduced to two dimensions. Using cylindrical coordinates (r, θ, z) with corresponding velocity components (u, v, w) which only depend on r, z, a stream function $\psi(r, z)$ can be introduced such that

$$u = \frac{1}{r}\frac{\partial \psi}{\partial z}, \quad w = -\frac{1}{r}\frac{\partial \psi}{\partial r}. \tag{12}$$

Furthermore, ψ must be an even function of r, which makes it convenient to use $\rho = r^2/2$ as a radial variable. This gives the following differential equations for the streamlines

$$\dot{\rho} = \frac{\partial \psi}{\partial z}, \quad \dot{z} = -\frac{\partial \psi}{\partial \rho}, \quad \dot{\theta} = \frac{1}{r} v. \tag{13}$$

The two first equations can be solved independently, and constitute a Hamiltonian system with total energy ψ, whence the solution curves are level curves of ψ. These level curves are intersections of streamsurfaces with a meridional plane. Individual streamlines wind around the axis on the streamsurfaces. Here we concentrate on the patterns defined by ψ in a meridional plane.

As in the previous section, we will consider Taylor expansions of the velocity field, and consider now in general

$$\psi = \sum_{n+m=0}^{\infty} a_{nm}\rho^n z^m. \tag{14}$$

From the axisymmetry we have that the axis is a streamline, and assign the value $\psi = 0$ to the axis. For the Taylor coefficients in (14) this implies

$$a_{0m} = 0 \text{ for all } m. \tag{15}$$

This gives the following differential equations for the level curves of ψ,

$$\dot{\rho} = \rho(a_{11} + a_{21}\rho + 2a_{12}z + \cdots), \tag{16}$$
$$\dot{z} = -(a_{10} + 2a_{20}\rho + a_{11}z + 3a_{30}\rho^2 + 2a_{21}\rho z + a_{12}z^2 + \cdots). \tag{17}$$

These equations can also be interpreted as the equations for the streamlines of a truly two-dimensional flow, where ρ, z are cartesian coordinates, and the line $\rho = 0$ is a streamline. These equations occur in the study of patterns near free and viscous interfaces after a coordinate change which maps the interface to the straight line $\rho = 0$. See [4, 21].

If $a_{10} = 0$, the origin is a critical point, or in a fluid context, a stagnation point. Linearizing at the origin one obtains the Jacobian matrix

$$J = \begin{pmatrix} a_{11} & 0 \\ -2a_{20} & -a_{11} \end{pmatrix}. \tag{18}$$

If $a_{11} \neq 0$, the origin is a hyperbolic saddle with eigenvalues $a_{11}, -a_{11}$. The separatrix corresponding to $-a_{11}$ is the center axis, while the other separatrix is a dividing streamline given by

$$z = -\frac{a_{20}}{a_{11}}\rho + \mathcal{O}(\rho^2). \tag{19}$$

In terms of the original radial variable r, it follows that the dividing streamline is orthogonal to the center axis. Depending on the sign of a_{11}, the point is either a point of separation or attachment. See figure 2.

If $a_{10} = a_{11} = 0$, the origin is a non-hyperbolic critical point, since 0 is an eigenvalue of J of algebraic multiplicity two. Two subcases must be

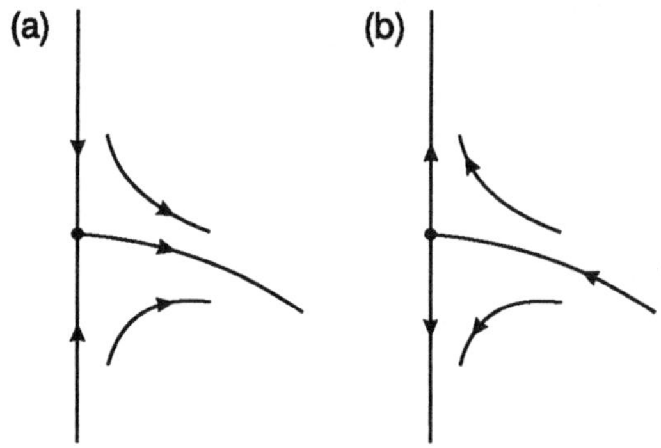

Figure 2. Hyperbolic critical point on the axis, shown in physical (r, z) coordinates. (a): Separation, $a_{11} > 0$. (b): Attachment, $a_{11} < 0$.

treated individually. If $a_{20} = 0$, J is the zero matrix and the geometric multiplicity of 0 is two. This case is treated in [4]. If $a_{20} \neq 0$, the algebraic multiplicity of 0 is one, and we treat this case now.

Finding the streamline pattern for the degenerate stream function with a non-hyperbolic critical point is itself of limited interest. Such a configuration cannot be realized physically, as it requires certain parameters to be exactly zero. Rather, we aim at finding all possible patterns that may occur when the degenerate streamfunction is perturbed. Some notions from singularity theory [15, 29] are useful:

Let $\psi_0(\rho, z)$ be a given stream function. An *unfolding* of ψ_0 is a parametrized familiy of functions $\psi(\rho, z; \lambda_1, \ldots, \lambda_k)$ such that $\psi(r, z; 0, \ldots, 0) = \psi_0(\rho, z)$. An unfolding is *versal* if all possible perturbations of ψ_0 are contained in the family. That is, for any function ψ' sufficently close to ψ_0, there is a member of the family ψ for which the qualitative structure of the level curves is same as for ψ'. A versal unfolding is *miniversal* if the number k of parameters is the lowest possible. The number is the *codimension* of ψ_0. A structurally stable stream function has codimension 0.

In this terminology, we are interested in finding a miniversal unfolding of the degenerate stream function. We start by introducing two small parameters by

$$\lambda_1 = a_{10}, \quad \lambda_2 = a_{11}, \tag{20}$$

and consider the two-parameter unfolding

$$\psi = \lambda_1 \rho + a_{20}\rho^2 + \lambda_2 \rho z + a_{30}\rho^3 + a_{21}\rho^2 z + a_{12}\rho z^2 + \cdots . \tag{21}$$

From this, we try to construct a miniversal unfolding by applying a number of simplifying coordinate transformations to obtain a normal form. We know that the linear terms of the differential equations do not suffice to describe the streamline and start by simplifying the quadratic terms of the differential equations, that is, the cubic terms of ψ.

As the differential equations for the level curves of ψ constitute a Hamiltonian system, we wish to preserve this property under the transformations. This is obtained by a *canonical transformation*, which is conveniently found from a *generating function*: Let the new coordinates be x, y. For any smooth function $S = S(y, \rho)$, consider the equations

$$x = \frac{\partial S}{\partial y}, \quad z = \frac{\partial S}{\partial \rho}. \tag{22}$$

If these can be solved for ρ, z, the solution defines a canonical transformation. See e.g. [14].

Here we try a generating function of the form

$$S = \rho y + \sum_{i+j+k+l=3} s_{ijkl}\rho^i y^j \lambda_1^k \lambda_2^l. \tag{23}$$

Solving eqns. (22) one finds

$$\begin{aligned} \rho &= x - 2s_{1200}xy - s_{2100}x^2 - s_{1101}x\lambda_2 - s_{1110}x\lambda_1 - 2s_{0210}y\lambda_1 \\ &\quad -s_{0111}\lambda_1\lambda_2 - s_{0120}\lambda_1^2 - 3s_{0300}y^2 - 2s_{0201}y\lambda_2 - s_{0102}\lambda_2^2 \\ &\quad +\mathcal{O}_3, \end{aligned} \tag{24}$$

$$\begin{aligned} z &= y + s_{1200}y^2 + s_{1101}y\lambda_2 + s_{1002}\lambda_2^2 + s_{1110}y\lambda_1 + s_{1011}\lambda_1\lambda_2 \\ &\quad +s_{1020}\lambda_1^2 + 2s_{2100}xy + 2s_{2001}x\lambda_2 + 2s_{2010}x\lambda_1 + 3s_{3000}x^2 \\ &\quad +\mathcal{O}_3, \end{aligned} \tag{25}$$

where $\mathcal{O}_3$ denotes terms in the variables of order 3 and higher.

For any choice of the s_{ijkl} the transformation is well-defined in a neighborhood of the origin, and we now try to choose these coefficients to make the transformed ψ as simple as possible. We want x to be a radial-like variable, and want the transformation to map the axis $\rho = 0$ to $x = 0$. This gives

$$s_{0210} = s_{0111} = s_{0120} = s_{0300} = s_{0201} = s_{0102} = 0, \tag{26}$$

and the transformed ψ is then

$$\begin{aligned} \psi &= \lambda_1 x + a_{20}x^2 + (a_{30} - 2a_{20}s_{2100})x^3 + (-4a_{20}s_{1200} + a_{21})x^2y \\ &\quad +(-s_{2100} - 2a_{20}s_{1110})x^2\lambda_1 - 2a_{20}s_{1101}\lambda_2x^2 + a_{12}xy^2 + \lambda_2xy \\ &\quad -2s_{1200}\lambda_1yx - s_{1110}x\lambda_1^2 - s_{1101}\lambda_2\lambda_1x + \mathcal{O}_4. \end{aligned} \tag{27}$$

The coefficients to $x^3, x^2y, x^2\lambda_1, x^2\lambda_2, x\lambda_1\lambda_2$ become zero if

$$s_{1200} = \frac{1}{4}\frac{a_{21}}{a_{20}}, \quad s_{2100} = \frac{1}{2}\frac{a_{30}}{a_{20}}, \quad s_{1110} = -\frac{1}{4}\frac{a_{30}}{a_{20}^2}, \quad s_{1101} = 0. \tag{28}$$

With this, we have

$$\psi = \mu_1 x + \mu_2 xy + a_{20}x^2 + a_{12}xy^2 + \mathcal{O}_4, \tag{29}$$

where

$$\mu_1 = \lambda_1 + \frac{a_{30}\lambda_1^2}{4a_{20}^2}, \quad \mu_2 = \lambda_2 - \frac{a_{21}\lambda_1}{2a_{20}}. \tag{30}$$

If $a_{12} \neq 0$, the single remaining cubic term in ψ is not degenerate. Assuming this, we can make further simplifications by the transformations

$$x \to \frac{a_{12}}{a_{20}}x, \quad y \to y - \frac{\mu_2}{2a_{12}}, \quad \psi \to \frac{a_{20}}{2a_{12}^2}\psi, \tag{31}$$

and finally get

$$\psi = -\mu x + \frac{1}{2}x^2 + \frac{1}{2}xy^2 + \mathcal{O}_4, \tag{32}$$

where

$$\mu = \frac{\mu_2^2}{8a_{12}^2} - \frac{\mu_1 a_{12}}{2a_{12}^2} \tag{33}$$

The stream function in (32) is a *normal form* of the unfolding (21). The analysis starts with considering the *truncated normal form* where the $\mathcal{O}_4$-term is dropped. This gives rise to the following differential equations for the iso-curves,

$$\dot{x} = xy, \quad \dot{y} = \mu - x - \frac{1}{2}y^2, \tag{34}$$

which are easily analyzed. There are critical points on the axis at $(x, y) = (0, \pm\sqrt{2\mu})$, that exist only for $\mu \geq 0$. There is also an off-axis critical point $(x, y) = (\mu, 0)$. The determinant of the Jacobian at this point is 2μ, so it is a saddle for $\mu < 0$ and a center for $\mu > 0$. In the original variables, only the region $r \geq 0$ is of physical significance. However, if a_{12}, a_{20} are of opposite sign, the transformation (31) maps the region $x \geq 0$ to $x \leq 0$, and hence the latter region is the relevant one. Summarizing, there are two possible bifurcation diagrams as μ is varied, as shown in figure 3. Physically, case (a) describes the creation of a re-circulation zone on the axis. Going from right to left, case (b) occurs in the merging of two re-circulation zones, where two stagnation points on the axis merge and "lift off" to a single critical point.

It is not very difficult to show that the $\mathcal{O}_4$-terms that were dropped to obtain (34) have no qualitative influence on the bifurcation diagrams. Hence, we have obtained a miniversal unfolding of the degenerate stream function with $a_{10} = a_{11} = 0$. The codimension is 1, since there is a single parameter μ in the normal form.

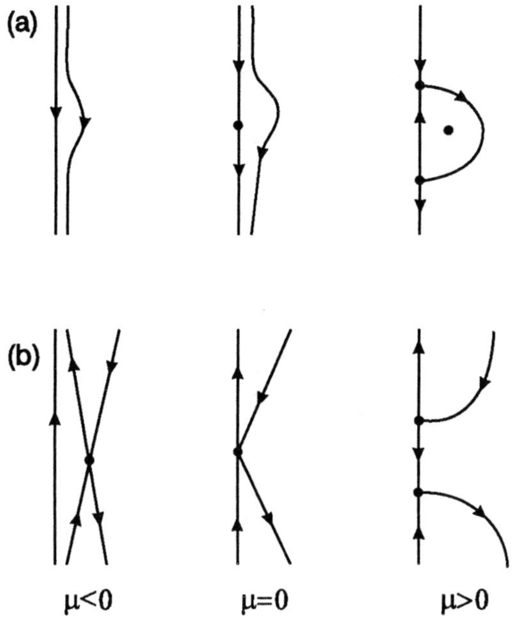

Figure 3. Bifurcation diagrams for the differential equations (34). (a): Streamlines in physical coordinates for $a_{20}/a_{12} > 0$. (b): Streamlines in physical coordinates for $a_{20}/a_{12} < 0$.

The analysis rests on the non-linear non-degeneracy condition $a_{12} \neq 0$. If this condition is violated, there are no third-order terms in (29), and terms of the fourth order must be considered. Proceeding as above, one can obtain an unfolding of this degeneracy, by applying further coordinate transformations of higher order. If $a_{13} \neq 0$, a normal form is

$$\psi = -\mu x + \frac{1}{2}x^2 + c_1 y + \frac{1}{3}xy^3 + \mathcal{O}_5 \tag{35}$$

which again is a versal unfolding of codimension 2 with the parameters (μ, c_1). For details, see [5]. The corresponding bifurcation diagram is shown in figure 4. It consists of three curves of codimension 1 bifurcations merging at the origin. The curve b_1 represents bubble creation bifurcations, figure 4(a), b_2 are bubble merging bifurcations, figure 3(b), and b_3 is the merging and disappearance of a saddle and a center off the axis. Bifurcations of the latter type are described in detail in [6]. Physically, the diagram describes possible interaction betweeen a small and a large re-circulation zone.

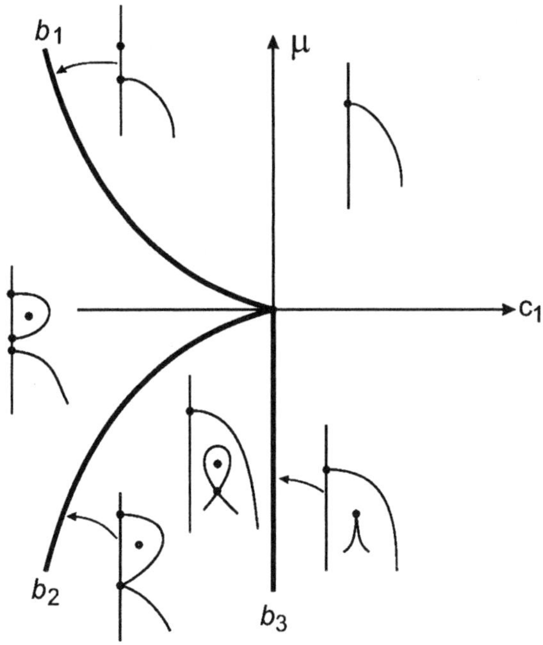

Figure 4. Codimension 2 bifurcation diagram of the normal form (35).

4. Bifurcation of Patterns in Vortex Breakdown

In this section we outline how the theory of the previous section can be applied to describe structures in a specific flow, and review some of the results of [7]. Consider a cylindrical container of height H of radius R filled with fluid. The bottom lid rotates with a constant angular velocity Ω_1 and the top with Ω_2. Three dimensionless parameters characterize the flow,

$$Re = \frac{\Omega_1 R^2}{\nu}, \quad h = \frac{H}{R}, \quad \gamma = \frac{\Omega_2}{\Omega_1}, \tag{36}$$

where ν is the kinematic viscosity of the fluid. The basic case is $\gamma = 0$, for which numerous experimental and numerical investigations exist. The basic flow is a vortex aligned with the axis, but Escudier [12] showed experimentally that depending on Re, h a number of recirculating zones may exist close to the axis in the steady regime, and that the flow is highly axisymmetric. These zones are reminiscent of the vortex breakdown of B-type [20], which makes the flow a useful tool for investigating this important phenomenon.

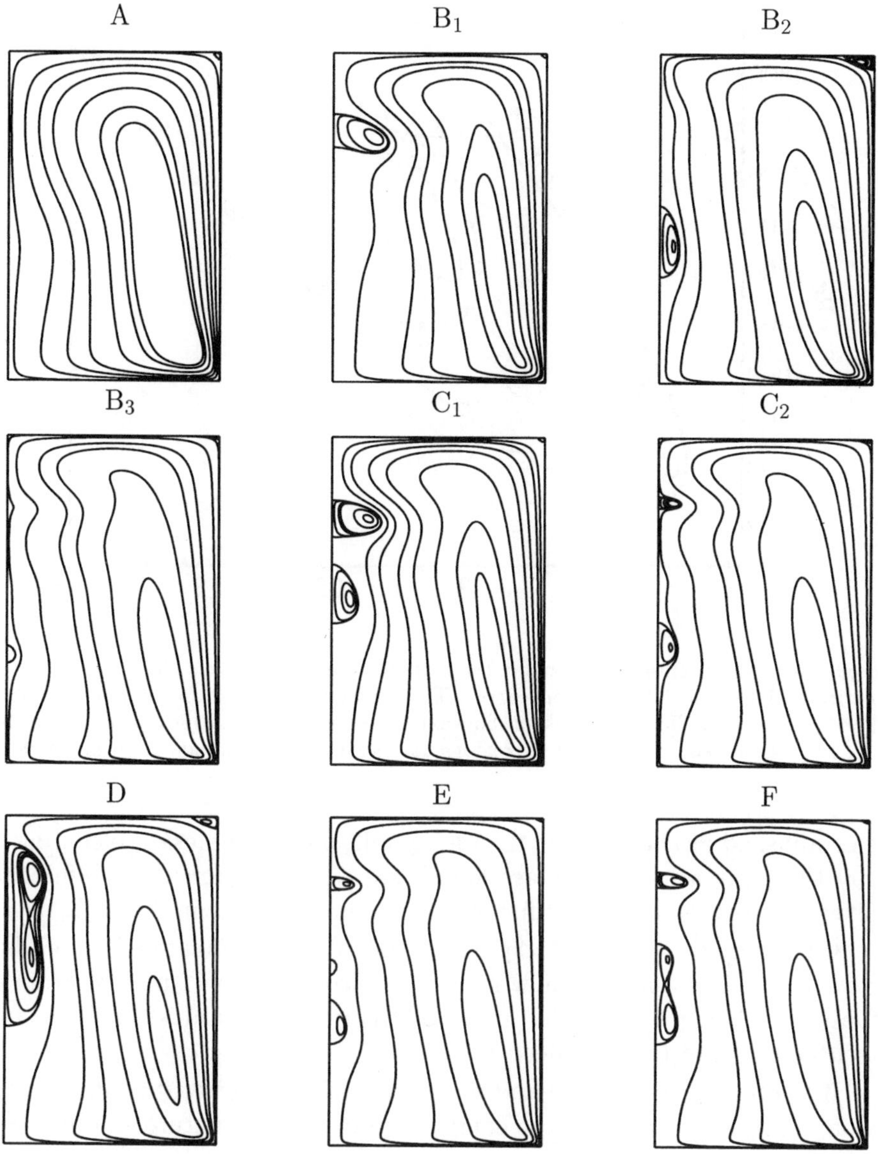

Figure 5. Numerically obtained flow topologies (iso-curves of ψ) for steady flow in a cylinder with rotating cover. Only the right half of a meridional plane is shown. The vertical line to the left is the axis.

Several axisymmetrical numerical studies, e.g. [13, 25], have reproduced the experimental flow structures with high accuracy. Sample computational results with different flow topologies are shown in figure 5.

On the basis of numerical simulations, a bifurcation diagram for $\gamma = 0$

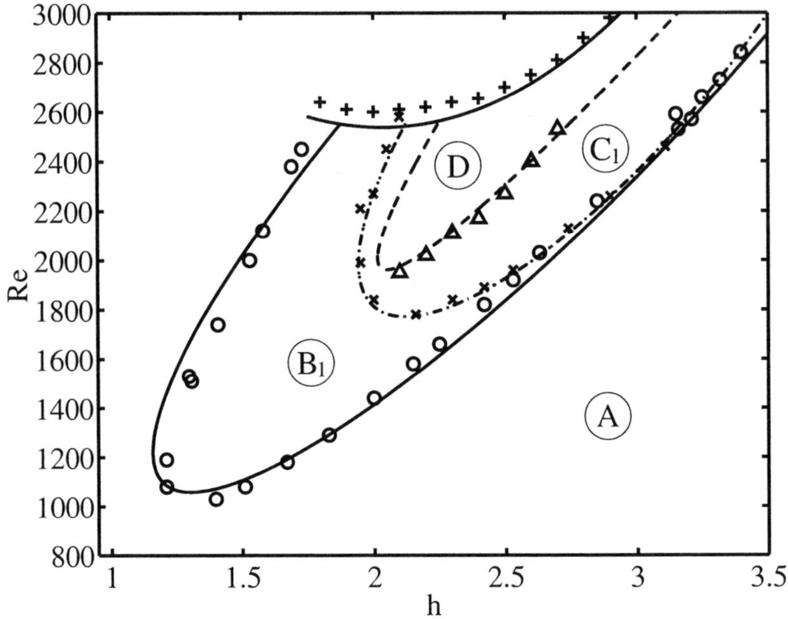

Figure 6. Bifurcation diagram for the cylinder flow with $\gamma = 0$. The curves are numerically determined bifurcation curves. The top curve is the stability limit, where the flow becomes periodic. The markers show experimental determinations of the curves from [12, 24]. The labels of the different regions correspond to the classification in figure 5.

is established as shown in figure 6 for details on the numerics, including the way the bifurcation curves are found, see [7]. There are three bifurcation curves, corresponding to codimension 1 bifurcations as in figure 3.

Changing to $\gamma = 0.01$, figure 7, two of the bifurcation curves have moved together, and they now meet in two codimension 2 points. From the analysis in § 3, we know that three bifurcation curves must emanate from such points, and this is indeed the case.

If the top lid is removed and replaced by a free surface, the flow changes profoundly, and a much more complex bifurcation structure emerges [8].

The analysis indicates a general approach to identifying pattern changes: In a problem with k parameters, one should identify parameter combinations associated with degeneracies of codimension k. These are *organizing centres* for the bifurcation diagram, since bifurcation curves or surfaces of lower codimension terminate here, and with knowledge of the unfoldings of these singularities one has a basic grip of the entire bifurcation diagram. Specifically, increasing γ further one can expect the two codimension 2 points in figure 7 to merge and form a codimension 3 point. The

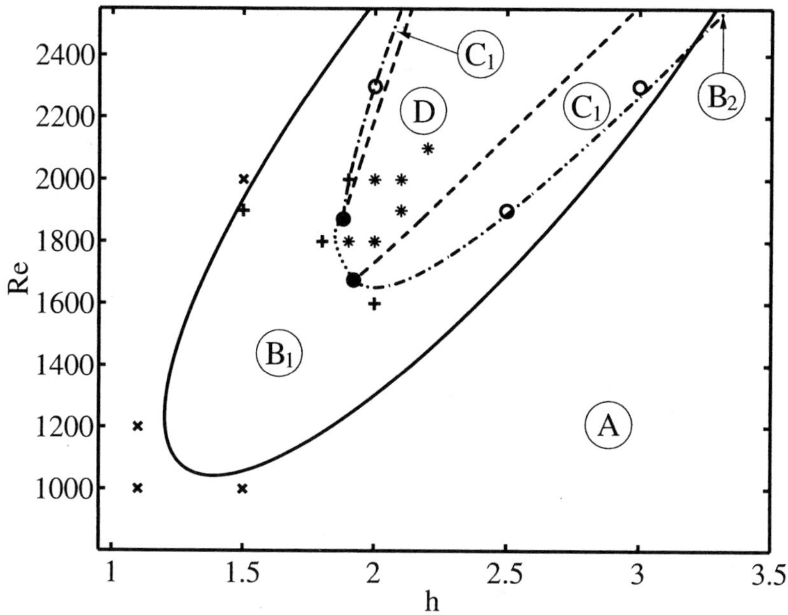

Figure 7. Bifurcation diagram for the cylinder flow for $\gamma = 0.01$. •: Codimension 2 points. The other markers show parameter values where simulations are performed.

corresponding 3-parameter bifurcation diagram is constructed in [5], and contains the basic structures that can be obtained in this problem.

5. Bifurcations of 3-Dimensional Patterns

The study of patterns performed here can clearly be extended to 3-dimensional flow patterns, and some results are obtained in [2, 9, 10, 17].

As an illustration, we consider the natural question whether the flow in the vortex breakdown really is axisymmetric. In fact, high-resolution vizualisation experiments [27] and recent 3-dimensional simulations [26] indicate that the flow is not perfectly symmetric, but the separatrices of the re-circulation zones exhibit a complex folded structure near the downstream stagnation point as shown in figure 8. To understand this phenomenon, it is natural to extend the analysis of § 3 to a full 3-dimensional flow, which is done in [17, 18]. The relevant situation to analyze is an unfolding of the velocity field close to a non-hyperbolic critical point with a Jacobian in the normal form

$$\begin{pmatrix} 0 & -\omega & 0 \\ \omega & 0 & 0 \\ 0 & 0 & 0 \end{pmatrix} \tag{37}$$

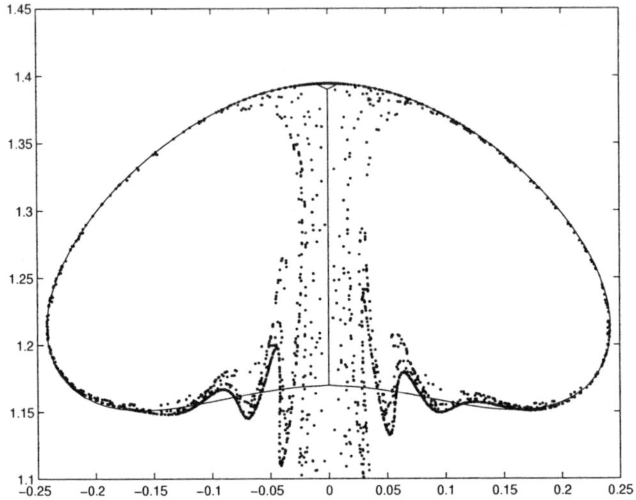

Figure 8. A small asymmetric perturbation of the axisymmetric flow in a vortex breakdown. The line shows the recirculating zone under perfect asymmetry. The markers are particles that are released in the flow on M_U^N very close to the top stagnation point N to mimic the visualisation experiment in [27] which it stronly resembles.

with eigenvalues 0 and $\pm i\omega$. The complex eigenvalues give a swirling motion, and the eigendirection of 0 is the axis of the cylinder. With standard normal form techniques, one can obtain a normal form which turns out to be axisymmetric to any finite order, even if no a priori assumptions about symmetry are made. That is, if a normal form is truncated as in § 3, no new information is obtained by considering the full 3-dimensional flow. Hence, there is no hope of obtaining a versal unfolding by this approach, and it is a signal that the 3-dimensional creation of a re-circulation zone has infinite codimension [3], even if there is only one parameter in the normal form.

There is a simple geometrical reason that this behaviour is to be expected. Consider a full 3-dimensional representation of an axisymmetric re-circulation zone as in figure 3(a) for $\mu > 0$. The top stagnation point N has a *stable manifold* consisting of the streamlines $\mathbf{x}(t)$ for which $\mathbf{x}(t) \to N$ for $t \to \infty$. This is one-dimensional, and consists of two streamlines on the axis. The *unstable manifold* M_U^N consists of the streamlines $\mathbf{x}(t)$ for which $\mathbf{x}(t) \to N$ for $t \to -\infty$. This is a two-dimensional surface which constitutes the surface of the re-circulation zone. For the bottom stagnation point S, the unstable manifold is the axis, and the stable manifold M_S^S is the bubble surface. In axisymmetric flow, M_U^N and M_S^S must be identical, but an arbitrarily small generic asymmetric perturbation of the velocity field will break this configuration. The two manifolds will then intersect transversally, and the bubble surface changes from being closed to open, and chaotic streamlines will occur. In other words, the closed axisymmetric bubble is structurally unstable with respect to asymmetric perturbations. In an experiment, where imperfections are unavoidable, and in a 3-dimensional simulation where asymmetric discretization errors are always present, one cannot expect to see an axisymmetric re-circulation zone. This does not mean that the mathematical boundary value problem which models the axisymmetric flow does not have an axisymmetric solution, but it implies that the Lagrangian properties of this solution cannot be realized. A numerical experiment with a small asymmetric perturbation in the velocity field shown in figure 8 displays the profound influence on the flow.

Thus, even in very simple 3-dimensional situations, one must expect a complicated bifurcation structure and chaotic streamlines. It is tempting to speculate that this is of importance for the transition from laminar to turbulent flow.

Acknowledgement

Jens Nørkær Sørensen has kindly provided the axisymmetric simulation of the vortex breakdown which forms the basis of figure 8.

References

1. Bakker, P. G. (1991) *Bifurcations in Flow Patterns.* Kluwer Academic Publishers, Dordrecht.
2. Boon, R. J. P. (1997) *Bifurcation in Fluid Flow near a Boundary Surface.* Ph.D. thesis, Delft University of Technology.
3. Broer, H. W. & Vegter, G. (1984) Subordinate Šil'nikov bifurcations near some singularities of vector fields having low codimension. *Ergodic Theo. Dynam. Sys.* **4**, 509–525.
4. Brøns, M. (1994) Topological fluid dynamics of interfacial flows. *Phys. Fluids* **6**(8), 2730–2737.
5. Brøns, M. (1999) Topological fluid mechanics of axisymmetric flows. In *Simulation and Identification of Organized Structures in Flows* (ed. J. N. Sørensen *et al.*), pp. 213-222. Kluwer Academic Publishers, Dordrecht.
6. Brøns, M. & Hartnack, J. N. (1999) Streamline topologies near simple degenerate critical points in two-dimensional flow away from boundaries. *Phys. Fluids* **11**(2), 314–324.
7. Brøns, M., Voigt, L. K., & Sørensen, J. N. (1999) Streamline topology of steady axisymmetric vortex breakdown in a cylinder with co- and counter-rotating end-covers. *J. Fluid Mech.* **401**, 275–292.
8. Brøns, M., Voigt, L. K., & Sørensen, J. N. (2001) Topology of vortex breakdown bubbles in a cylinder with rotating bottom and free surface. *J. Fluid Mech.* **428**, 133–148.
9. Chong, M. S., Perry, A. E., & Cantwell, B. J. (1990) A general classification of three-dimensional flow fields. *Phys. Fluids* **A2**(5), 765–777.
10. de Winkel, M. E. M. (1996) *Three-Dimensional Viscous Flow Patterns Near A Plane*

Wall. Ph.D. thesis, Delft University of Technology.

11. Dlery, J. M. (2001) Robert Legendre & Henri Werl: Toward the elucidation of three-dimensional separation. *Ann. Rev. Fluid Mech.* **33**, 129–154.
12. Escudier, M. P. (1984) Observations of the flow produced in a cylindrical container by a rotating endwall. *Exp. Fluids* **2**, 189–196.
13. Gelfgat, A. Y., Bar-Yoseph, P. Z., & Solan, A. (1996) Stability of confined swirling flow with and without vortex breakdown. *J. Fluid Mech.* **311**, 1–36.
14. Goldstein, H. (1950) *Classical Mechanics.* Addison-Wesley, Reading Mass.
15. Golubitsky, M. & Schaeffer, D. G. (1985) *Singularities and Groups in Bifurcation Theory I, Applied Mathematical Sciences* **51**. Springer Verlag, New York.
16. Hartnack, J. N. (1999) Streamline topologies near a fixed wall using normal forms. *Acta Mechanica* **136**(1-2), 55–75.
17. Hartnack, J. N. (1999) *Structural Changes in Incompressible Flow Patterns.* Ph.D. thesis, Department of Mathematics, Technical University of Denmark.
18. Hartnack, J. N., Brøns, M., & Spohn, A. (2000) The role of asymmetric perturbations in steady vortex breakdown bubbles. DCAMM Report 628, Technical University of Denmark.
19. Legendre, R. (1956) Séparation de l'écoulement laminaire tridimensionel. *La Rech. Aérospatiale* **54**, 3–8.
20. Leibovich, S. (1984) Vortex stability and breakdown: Survey and extension. *AIAA J.* **22**(9), 1192–1206.
21. Lugt, H. J. (1987) Local properties at a viscous free surface. *Phys. Fluids* **30**, 3647–3652.
22. Oswatitsch, K. (1958) Die Ablösungsbedingung von Grenzschichten. In *Boundary Layer Research*, pp. 357–364. IUTAM Symposium, Springer Verlag, Berlin.
23. Perry, A. E. & Chong, M. S. (1987) A description of eddying motions and flow patterns using critical-point concepts. *Ann. Rev. Fluid Mech.* **19**, 125–155.
24. Roesner, K. G. (1989) Recirculation zones in a cylinder with rotating lid. In *Topological Fluid Mechanics* (ed. A. Tsinober and H.K. Moffatt), pp. 699–708. IUTAM Symposium, Cambridge University Press.
25. Sørensen, J. N. & Loc, T. P. (1989) High-order Axisymmetric Navier-Stokes Code: Description and Evaluation Of Boundary Conditions. *Int. J. Num. Methods Fluids* **9**, 1517–1537.
26. Sotiropoulos, F. & Ventikos, Y. (2001) The three-dimensional structure of confined swirling flows with vortex breakdown. *J. Fluid Mech.* **426**, 155–175.
27. Spohn, A., Mory, M., & Hopfinger, E. J. (1998) Experiments on vortex breakdown in a confined flow generated by a rotating disc. *J. Fluid Mech.* **370**, 73–99.
28. Tobak, M. & Peake, D. J. (1982) Topology of three-dimensional separated flows. *Ann. Rev. Fluid Mech.* **14**, 61–85.
29. Wiggins, S. (1990) *Introduction to Applied Nonlinear Dynamical Systems and Chaos.* Springer Verlag, New York.

TOPOLOGICAL FEATURES OF INVISCID FLOWS

ROBERT GHRIST & RAFAL KOMENDARCZYK
School of Mathematics & CDSNS
Georgia Institute of Technology
Atlanta, GA 30332-0160, USA

E-mail: *ghrist@math.gatech.edu* ; *rako@math.gatech.edu*

Abstract. The Euler equations for an incompressible inviscid fluid in dimension three possess a wealth of topological phenomena woven into the dynamical and geometric properties of the fluid. Focusing first on steady Euler fields, we outline known results, giving special attention to the Beltrami fields and the contemporary topological techniques required to elucidate their dynamical features. We also propose a topological perspective for understanding the global dynamics of the Euler equations on the space of velocity fields.

1. Introduction to Geometric Euler Flows

The behavior of the velocity field, $\mathbf{u}$, of a perfect, inviscid, incompressible fluid is governed by the Euler equations:

$$\frac{\partial \mathbf{u}}{\partial t} + \nabla_{\mathbf{u}} \mathbf{u} = -\nabla p \quad ; \quad \operatorname{div}(\mathbf{u}) = 0. \tag{1}$$

Here, p is a real-valued (time-dependent) function measuring pressure. On Euclidean 3-space, the term $\nabla_{\mathbf{u}}\mathbf{u}$ is also written $(\mathbf{u} \cdot \nabla)\mathbf{u}$. These equations are, for the obvious reasons, usually considered only on a Euclidean domain. However, since we are interested in taking a topological perspective, there is no reason not to consider these equations in their most general setting.

Let M denote a manifold of dimension n (we will usually consider the case $n = 3$ for simplicity). For any time-dependent velocity field $\mathbf{u}$ on M, all of the terms in (1) save the first require some addition structure in order to compute. Thus, choose a Riemannian metric g on M: this g is a smoothly-varying inner-product on each tangent space to M. Via g, one

R.L. Ricca (ed.), An Introduction to the Geometry and Topology of Fluid Flows, 183–201.

may then compute all the standard quantities of vector calculus, including the gradient of $p : M \to \mathbb{R}$ and the covariant derivative $\nabla_{\mathbf{u}}\mathbf{u}$ of the vector field $\mathbf{u}$ in the direction $\mathbf{u}$. In addition, to talk of divergence-free fields, one needs a notion of volume on M: this too can be chosen as some density (a volume form μ to be precise) which may or may not be equal to that derived from the metric g. For the reader not familiar with the basics of differential geometry in the context of fluids, we would recommend selected passages of [4] or [1].

The wealth of topological phenomena that arise within the Euler equations cannot be under-estimated. We restrict to the goal of convincing the reader that cutting-edge topological techniques are not only efficacious but crucial for understanding even the most basic types of fluid flows. The reader who thirsts for more will find a nearly inexhaustible supply of interesting ideas and results in the recent excellent monograph of Arnold and Khesin [4].

The Euler equations are difficult enough on the tame Euclidean 3-space we so often work in: even the existence of solutions for all time is unknown and not at all clear. In order to get any concrete results for arbitrary geometric manifolds, we must begin modestly. For almost the entire remainder of this work, we restrict to the class of *steady* solutions to the Euler equation: hence,

$$\nabla_{\mathbf{u}}\mathbf{u} = -\nabla p \quad ; \quad \mathrm{div}(\mathbf{u}) = 0, \tag{2}$$

where neither $\mathbf{u}$ nor $p : M \to \mathbb{R}$ is time-dependent.

While the reader (and the authors!) may be most interested in proving results about the case where $M = \mathbb{R}^3$ and g is the Euclidean metric, it is nevertheless both interesting and instructive to see how steady Euler flows manifest themselves in other geometric contexts.

Example 1.1 [Hopf field] Consider the *3-sphere*, denoted S^3, which consists of all points in $\mathbb{R}^4$ a unit distance from the origin. Via stereographic projection, this 3-manifold is equivalent to $\mathbb{R}^3$ with an added "point at infinity." The *round* metric on S^3 is that inherited by it as a subset of Euclidean $\mathbb{R}^4$, much in the same manner as we would describe the geometry of the round 2-sphere $S^2 \subset \mathbb{R}^3$. The simplest example of a steady Euler field on the round S^3 is that given by the *Hopf field*, pictured in Figure 1. One may realize this field in Euclidean coordinates on $\mathbb{R}^4$ via

$$X_H := \left\{ \begin{array}{ccc} \dot{x} = -y & ; & \dot{z} = -w \\ \dot{y} = x & ; & \dot{w} = z \end{array} \right. \tag{3}$$

These equations correspond to the motion of two identical uncoupled simple harmonic oscillators at a fixed energy level. Thus, every orbit of the flow is periodic, or *closed.* Indeed, this is an extraordinary flow: every orbit is a geodesic on S^3, and the entire flow is symmetric.

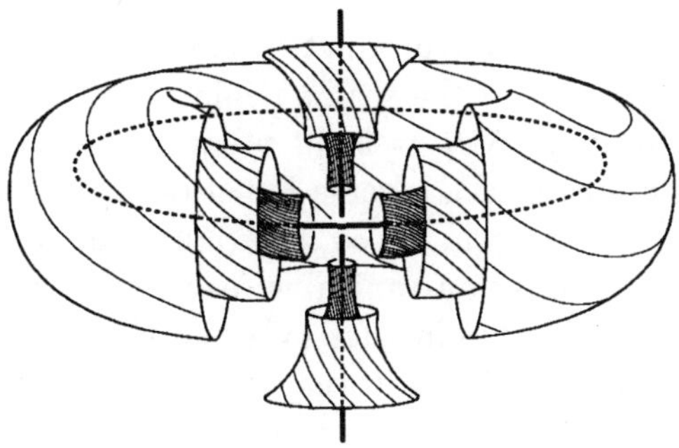

Figure 1. The Hopf flow on $S^3 = \mathbb{R}^3 + \{\infty\}$ is filled with closed unknotted flowlines which fill a 1-parameter family of invariant tori limiting onto a pair of circles.

Note that this fluid flow is filled with flowlines that are (1) closed, and (2) unknotted. It is a remarkable feature of inviscid fluids that both of these properties persist in some vestigial form no matter how one deforms the metric on S^3 or the conserved volume or the dynamics of the steady flow:

Theorem 1.2 ([11]) *Every non-vanishing solution to (2) on any Riemannian S^3 possesses a closed, unknotted flowline, assuming that the vector field and metric are sufficiently smooth.*[1]

Such is not the case for non-Euler vector fields [27, 28]: being a steady solution to the Euler equations is restrictive. This result hints that there are features of steady Euler flows which are entirely *topological*, independent of the *geometry* of the underlying manifold.

2. Integrable Flows

One's first thought is that the problem of analyzing steady Euler flows on a Riemannian manifold is trivial: consider the following adaptation of the Bernoulli theorem (cf. Arnold's theorem [2]).

'Not-quite-true' Theorem 2.1 *Given* **u** *a steady non-vanishing Euler flow on M a compact 3-manifold without boundary. If everything is of smoothness class C^ω (real-analytic), then M is filled almost everywhere by invariant tori.*

'Almost-correct' Proof. Transform Equation (2) via a standard vector-calculus identity into the Bernoulli form of the equation,

$$\mathbf{u} \times (\nabla \times \mathbf{u}) = -\nabla H, \tag{4}$$

[1] Real analytic, or C^ω, is the simplest set of sufficient conditions.

where $H : M \rightarrow \mathbb{R}$ is the Bernoulli function and $\nabla \times \mathbf{u}$ is the *vorticity* field of the fluid. Take the dot-product of both sides of the equation with $\mathbf{u}$. Since the left-hand side is the scalar triple product of $\mathbf{u}$ with itself and with $\nabla \times \mathbf{u}$, we obtain $\mathbf{u} \cdot \nabla H = 0$. In other words, H is constant along orbits of $\mathbf{u}$. Now, thanks to the Inverse Function Theorem, the level sets of H in M are generically smooth, closed, invariant submanifolds of dimension two. As the vector field $\mathbf{u}$ is non-vanishing, each level set $H^{-1}(c)$ must have Euler characteristic zero: this implies (via the Classification Theorem for Surfaces) that $H^{-1}(c)$ is either a torus or a Klein bottle. However, being oriented by the vector ∇H, the non-degenerate level sets of H are invariant tori. Since everything is C^ω, at most a finite number of level sets of H can be degenerate, and these must be of codimension strictly greater than zero.

QED

If this proof were correct, then it would imply that, under the stated hypotheses, non-vanishing steady Euler flows would be very rare beasts. They would simply not exist on most three-dimensional manifolds! Indeed, any closed manifold which can be outfitted with a hyperbolic geometry cannot be filled almost-everywhere with tori, no matter what the actual geometry of the manifold is. All such steady Euler flows would qualitatively look like the Hopf flow of Figure 1, and would not be terribly interesting.

The mistake, of course, lies in the assertion that the function $H : M \rightarrow \mathbb{R}$ is not a constant function. In that (and *only* in that) case,[2] the conclusions fail since the level set of H is *all* of M. Notice that if H is a constant on M, then $\mathbf{u} \times (\nabla \times \mathbf{u})$ is necessarily zero: $\mathbf{u}$ is parallel to its curl.

The true result that can be proved (the modern formulation of the Bernoulli Theorem, due to Arnold[3]) is the following:

Theorem 2.2 ([2]) *Given* $\mathbf{u}$ *a steady non-vanishing Euler flow on* M *a compact 3-manifold without boundary. If everything is of smoothness class* C^ω *(real-analytic), then either (1)* M *is filled almost everywhere by invariant tori; or (2)* $\mathbf{u}$ *is an eigenfield of the curl operator.*

This one case will occupy the remainder of our attention.

3. Beltrami Fields

Definition 3.1 *A volume-preserving vector field* $\mathbf{u}$ *is* **Beltrami** *if it is an eigenfield of the curl operator: i.e., if* $\nabla \times \mathbf{u} = \lambda \mathbf{u}$ *for some* $\lambda \in \mathbb{R}$.

[2]This is where the C^ω assumption is critical.

[3]Arnold actually proves quite a bit more than stated — his proof concerns arbitrary integrable Hamiltonian systems and consists in showing that the velocity and vorticity fields commute. It is curious that, for three-dimensional vector fields, one does not require that the flow be volume-preserving: only an integral of motion is needed.

Figure 2. The ABC fields represented in a fundamental domain of $\mathbb{R}^3$, projected onto a square. The dynamics display integrable behavior (left) when one of the parameters is zero; increasing the parameter (middle,right) reveals increasingly non-integrable dynamics.

Example 3.2 [Hopf field] The Hopf field on S^3 is an example of a Beltrami field on the round S^3 with eigenvalue 1. Notice that, although it is Beltrami, the dynamics of the field still satisfy the conclusions of Arnold's Theorem.

Example 3.3 [ABC fields] The most historically influential example of a Beltrami field is the following:

$$\dot{x} = A\sin z + C\cos y \quad ; \quad \dot{y} = B\sin x + A\cos z \quad ; \quad \dot{z} = C\sin y + B\cos x \tag{5}$$

In this equation, A, B, and C are parameters. For all such parameter values, one has a Beltrami field on the Euclidean $\mathbb{R}^3$, or, by quotienting out the periodic boundary conditions, on the three-dimensional torus T^3. For parameter values satisfying $A^2 > B^2 + C^2$, the field is non-vanishing. As displayed in Figure 2, the dynamics can be rather complicated — one appears to have regions of chaotic, non-integrable flow.

Despite the initial appearance of being the "exceptional" steady Euler fields, Beltrami fields turn out to be surprisingly important objects [4, 31]:

1. As already mentioned, Beltrami fields are the only steady solutions to the Euler equations which may exhibit "chaotic" Lagrangian dynamics.
2. The L^2-*energy* of a fluid velocity $\mathbf{u}$ is the integral of $\|\mathbf{u}\|^2$ over the manifold. If one considers this functional restricted to all vector fields $\phi_*\mathbf{u}$ obtained from $\mathbf{u}$ by deforming it with a volume-preserving diffeomorphism ϕ, then the Beltrami fields are critical points of this functional [3].
3. Every Beltrami field with non-zero eigenvalue has a non-vanishing *helicity*, a measure of average asymptotic linking of flowlines [3] which forms a lower bound on the L^2-energy of the fluid.

4. Given a perfectly-conducting fluid, the Beltrami fields yield steady solutions to the MHD equations linking the fluid flow and its induced magnetic field. This is an important class of steady plasmas. In ideal MHD, Beltrami fields are thus called *force-free fields* [8].

Despite their relative importance, there have been relatively few rigorous results about the dynamics of Beltrami fields. Nearly everything known about their dynamics comes from either numerical experiments [9] or from applying Melnikov analysis to near-integrable examples [7, 20], a very narrow band of fields.

There are numerous open problems concerning the dynamics and topology of Beltrami fields.

Question 3.4 Does every Riemannian 3-manifold admit a Beltrami field without fixed points?

Every three-manifold admits volume-preserving vector fields without fixed points: the question is whether any of these can be Beltrami. The spectral theorem applied to the curl operator (whose square is the Laplacian) quickly yields the existence of infinitely many Beltrami fields on every Riemannian 3-manifold. The analytic approach provides no control over the existence of fixed points, however. We believe that when the manifold is sufficiently negatively curved, Beltrami fields always possess fixed points.

Question 3.5 Given a Beltrami field on a closed 3-manifold, how many periodic orbits must exist?

This problem is a distant cousin of the Arnold Conjecture in the setting of fluids (see [30] for an introduction to the Arnold Conjecture in symplectic dynamics). One's first guess is that every Beltrami field must have infinitely many periodic orbits. However, this is not true: by deforming the round 3-sphere into something ellipsoidal, one can perturb the Hopf flow into a Beltrami field which has but two periodic orbits. It is an open conjecture that the minimum for the 3-sphere (in any metric) is two, and that any more than two implies infinitely many. A far-reaching tool for this conjecture is the *symplectic field theory* of [10].

Question 3.6 Are there any parameter values for which the ABC fields possess a closed orbit on $\mathbb{R}^3$?

Such orbits definitely do not exist when any of the parameters A, B, or C is equal to zero. In §5, we outline a program for finding such an orbit.

Question 3.7 Characterize the topology of the flowlines of Beltrami fields. What kinds of knots must/can be traced out by periodic flowlines?

Several partial results will be stated in §6.

Question 3.8 Under what circumstances is a Beltrami field forced to possess a closed flowline which is *hyperbolic*? That is, the eigenvalues of the linearization of the Poincaré return map are not equal to one.

This problem is, thanks to [19], a route to establishing hydrodynamic instability for steady Euler fields. We comment more on this problem in §7.

4. An Introduction to Contact Topology and Dynamics

Surprisingly, understanding the dynamics of Beltrami fields seems to require a foray into some rather esoteric-looking branches of three-dimensional topology. But a self-enforced exile to the furthest abstraction is rarely permanent and certainly not without reward. Most every question is §3 can be approached via techniques involving *contact structures.*

4.1. CONTACT STRUCTURES AND FORMS

For our purposes, a contact structure on a 3-manifold M is a certain type of smoothly-varying plane field ξ. Just as a vector field assigns a vector to each tangent space of M, a plane field assigns some 2-dimensional subspace to each tangent space. For example, given a vector field X on a Riemannian 3-manifold M, the 2-D subspace orthogonal to X in each tangent space define a plane field on M.

While any smooth vector field can always be "integrated" into a collection of smooth curves tangent to the field, such is not the case for a general plane field. Only in very rare circumstances can one stitch together the planes (like scales on a fish) to fill M with two-dimensional sheets tangent to the planes. Most often, the planes twist in such a way as to break the integrability. Occasionally, there can be found plane fields which twist so much as to be nowhere-integrable, not even at a single point.

Definition 4.1 *A* **contact structure** *on a 3-manifold M is a smoothly-varying plane field ξ which is completely non-integrable.*

Two more precise definitions can be formulated:

1. ξ is a contact structure iff for any pair of locally non-zero vector fields U, V tangent to ξ, their Lie bracket $[U, V]$ is nowhere tangent to ξ.
2. ξ is a contact structure iff it can be written [locally] as the kernel of a *contact form* — a differential 1-form α satisfying

$$\alpha \wedge d\alpha \neq 0. \tag{6}$$

That these definitions are all equivalent is really the Frobenius Theorem [1]. The two above definitions are most useful in computations, while the "twisted plane field" definition is perhaps the most readily grasped. The

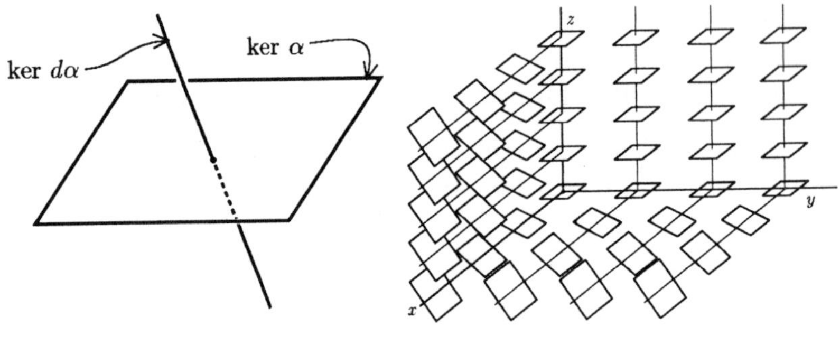

Figure 3. Visualizing a contact form at a tangent space (left); every contact structure is locally equivalent to the kernel of $dz + xdy$ (right).

best way to "visualize" a contact form is in terms of the kernels of α and $d\alpha$. On each three-dimensional tangent space, the kernel of α is a two-dimensional plane: the contact element. The 2-form $d\alpha$ has a 1-dimensional kernel consisting of those vectors which annihilate $d\alpha$ when plugged into the first slot. Equation (6) translates into saying that α is a contact form iff these subspaces are never tangent, see Figure 3(left).

Example 4.2 [Standard contact structure on $\mathbb{R}^3$] Consider the 1-form

$$\alpha_1 := dz + x\,dy. \tag{7}$$

where $\{dx, dy, dy\}$ are the duals (co-vectors) to the standard basis vectors in $\mathbb{R}^3$. Since $d\alpha_1 = dx \wedge dy$, we have that $\alpha_1 \wedge d\alpha_1 = dz \wedge dx \wedge dy$ which is nowhere zero (a volume form). Thus, the kernel of α_1 is a contact structure. This kernel is defined by solving $\alpha_1 = 0$ to obtain $dz/dy = -x$. That is, the slope of the plane field is given by $-x$: see Figure 3(right). If an observer walks along any x-axis, the plane field ξ twists counter-clockwise.

The Darboux theorem states that the standard contact form is a "building block" for any contact form on any 3-manifold, i.e., any contact form under appropriate change of coordinates can be locally expressed by (7).

Example 4.3 [Standard contact structure on S^3] Recall Example 1.1 of the Hopf field on the round S^3. The plane field orthogonal to this vector field is a contact structure. One may check this by recalling the vector field X_H of (3) and defining the 1-form α_2 via $\alpha_2(V) := X_H \cdot V$. One readily shows that $\alpha_2 = \frac{1}{2}(x\,dy - y\,dx + z\,dw - w\,dz)$, and that this is indeed a contact form satisfying (6) on S^3.

Example 4.3 gives several hints that contact structures are dynamically interesting objects. First, being related to the Hopf field, this contact structure is related to the global topological features of the three-sphere. Second, the Hopf field actually lies in the kernel of $d\alpha_2$, as a simple computation reveals. It thus follows (from the Cartan formula) that the flow of the Hopf field preserves the contact form.

Indeed, if we consider an arbitrary contact form α, the "cartoon" representation of Figure 3(left) indicates that the kernel of $d\alpha$ provides a "natural" or "canonical" direction on each tangent space. Choosing a non-zero vector in each such line yields a vector field on M whose dynamics ought to be related to the geometry of the contact structure. Indeed, this observation naturally leads one to the following definition:

Definition 4.4 *The* **Reeb field** *associated with a contact form α is defined as the unique vector field X satisfying*

$$i) \quad d\alpha(X, \cdot) = 0 \quad \text{and} \quad ii) \quad \alpha(X) = 1. \tag{8}$$

A **Reeb-like field** *is one for which we relax the second condition to $\alpha(X) > 0$.*

The flow of a Reeb field leaves the contact structure and the contact form invariant. They are always volume-preserving for the natural volume form $\alpha \wedge d\alpha$.

Example 4.5 [Reeb fields] The Reeb field associated with the local contact form α_1 of Example 4.2 is the vertical field $\frac{\partial}{\partial z}$. The Reeb field associated with the contact form α_2 orthogonal to the Hopf field *is* precisely the Hopf field X_H.

It is by no means a coincidence that the definition of a contact structure as a plane field which "everywhere twists" and the intuition of a rotational Beltrami field as a vector field which "everywhere rotates about itself" are so similar. The following correspondence asserts roughly that all Beltrami fields are Reeb fields, and vice versa (up to reparametrizations by time, which changes none of the topological features):

non-singular rotational Beltrami fields for all metrics, all non-zero scalings	$\Leftrightarrow$	*all Reeb fields for all contact forms, all non-zero scalings*

One direction is trivial: the orthogonal plane field associated with any Beltrami field is always a contact structure. To go in the other direction, one begins with a Reeb-like field and constructs a metric for which the field is Beltrami (see [12] for details and simple proof).

The category of Beltrami fields is geometric and rigid. But, after collecting all Beltrami fields for all metrics together, the resulting class is the set of Reeb fields — a decidedly topological category. This allows one to prove a number of non-trivial result about Beltrami fields by relaxing the metric constraints and treating them as topological objects.

One immediate application is to the problem of existence of *non-singular* Beltrami fields (Question 3.4). According to [29], every 3-manifold admits a contact structure. Any Reeb field associated with this yields a Beltrami field; thus,

Corollary 4.6 *Every 3-manifold admits a* non-singular *Beltrami flow for some Riemannian structure.*

Contact perspectives also allow us to construct examples of Beltrami fields on higher dimensional manifolds, e.g., analogues of the ABC fields (5) on the 5-torus T^5 [21].

4.3. HOFER'S THEOREM AND OVERTWISTED STRUCTURES

A compelling reason for understanding contact structures and their topological properties comes from recent work of Hofer and collaborators [25, 26, 10]. The programme outlined in [10] indicates that the topological properties of a contact structure are delicately entwined with the dynamical properties of the Reeb fields associated with that contact structure, much in the same way that the topology of a manifold can be recovered via the dynamics of a gradient field. The first hint of this relationship appears in the paper [25] which finds that certain types of contact structures are forced to have periodic orbits in their Reeb fields. Understanding and applying this result requires a deeper understanding of the topological classification of contact structures.

To summarize roughly the past twenty years of contact topology in one idea: the best way to analyze a contact structure ξ on M is to consider how ξ intersects various two-dimensional discs in M. Dynamical properties of these intersections correlate to dynamical properties of Reeb fields.

Definition 4.7 *Given a contact structure ξ and a disc D embedded in M, the* **characteristic foliation**, *D_ξ, is the set of curves carved out on D by intersecting the planes of ξ with the tangent planes of D in M.*

Definition 4.8 *The contact structure ξ is* **overtwisted** *if there exists an embedded disc D in M whose characteristic foliation D_ξ contains a limit cycle (an attracting periodic orbit): see Figure 4. If no such disc exists, then ξ is called* **tight**.

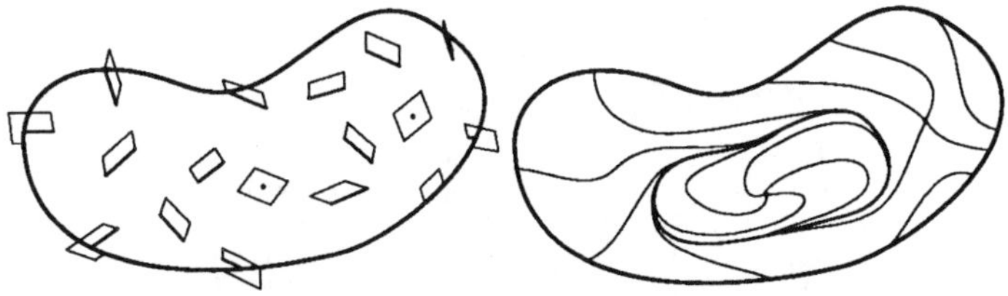

Figure 4. A contact structure slices a disc into curves and fixed points (of tangency). An overtwisted contact structure carves a limit cycle into some disc.

Example 4.9 [Tight vs. overtwisted] The standard contact structures for $\mathbb{R}^3$ and S^3 given in Examples 4.2 and 4.3 are tight. A concrete example of an overtwisted structure on $\mathbb{R}^3$ is that given by the form (in polar coordinates):

$$\alpha_3 = \cos(r)dz + r\sin(r)d\theta. \tag{9}$$

For this contact form, the slope of the contact form is given by $dz/d\theta = -r\tan r$; the planes "flip" over many times, in contrast to Example 4.2.

Hofer's theorem states the surprising result that a limit cycle in the dynamics of the contact planes (intersecting a disc) implies the existence of a periodic orbit in the transverse Reeb dynamics:

Theorem 4.10 ([25]) *Any Reeb field for an overtwisted contact structure on a closed 3-manifold M possesses a periodic orbit (some multiple of which can be shrunk to a point in M).*

This theorem, a tour-de-force of symplectic geometry and non-linear analysis, is essentially the *only* available global tool for analyzing the dynamics of arbitrary 3-D Beltrami fields (though see [10]).

5. Existence of Closed Orbits

5.1. CLOSED ORBITS

Hofer's techniques, combined with the correspondence between Reeb and Beltrami fields, quickly yields existence results for closed orbits of arbitrary Beltrami fields on certain 3-manifolds. We state a few examples.

Theorem 5.1 ([12, 14]) *Let* **u** *be a steady non-vanishing Euler flow of class* C^ω *(or, if restricting to Beltrami fields,* C^2 *suffices). Then* **u** *possesses a closed flowline in the following cases:*

On any Riemannian 3-sphere, for any such **u**.
On any Riemannian solid torus, for any **u** *tangent to the boundary.*

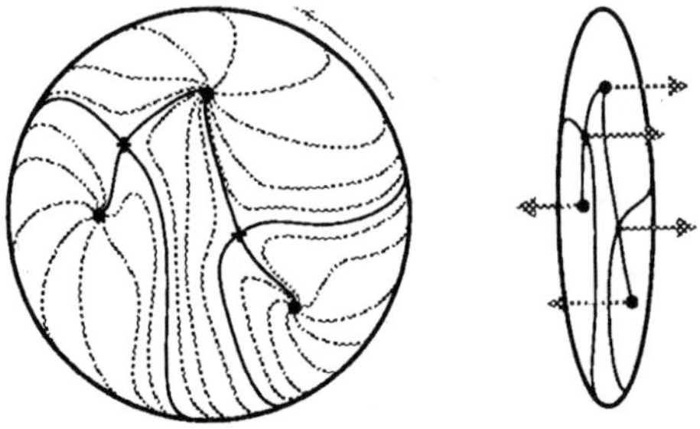

Figure 5. Projecting a Beltrami field to a cross-sectional disc yields a vector field with weighted index -2. The Beltrami field thus has a closed orbit in $D \times \mathbb{R}$.

On any Riemannian 3-torus, for any **u** *which is homotopically nontrivial.*[4]

The proofs of these results all have three parts: (1) proving in the case of an integrable Euler field (this requires the C^ω assumption); (2) determining which Beltrami fields are tight, and which are overtwisted; (3) using Hofer's Theorem in the overtwisted cases and some clever techniques for the tight cases, which depends on the manifold. The genuinely difficult case is that of the solid torus $D^2 \times \mathbb{R}$; this also is the case most closely tied to applications in plasma dynamics. We elaborate on the solid torus below.

5.2. A 3-D INDEX

Consider a non-vanishing Beltrami field **u** on a tube $D^2 \times \mathbb{R}$ which is periodic in the third variable and tangent to the boundary of the tube. Choose any convenient cross-sectional disc D, and project the vector field **u** orthogonally onto D. This generates a vector field $\mathbf{u}^\perp$ on D tangent to the boundary which generically has a finite number of fixed points of type source/sink or saddle. The standard Euler-Poincaré index theory says that the sum of the indices of the fixed points on D ($+1$ for sources/sinks and -1 for saddles) equals the Euler characteristic of D, which is 1. Such arguments are often used to find fixed points or periodic orbits in 2-D flows.

However, for D and $\mathbf{u}^\perp$ arising from a Beltrami field, there is a vestige of this 2-D index theory. Consider the fixed points of $\mathbf{u}^\perp$ on D: these are the points at which **u** is orthogonal to D. There is thus an additional sign depending on whether **u** points to the "right" or the "left" of the disc (see Figure 5).

[4]Homotopically trivial means that one can rotate the vectors of **u** so that they all point in the same direction.

Theorem 5.2 ([15]) *With D, $\mathbf{u}^{\perp}$ as above, compute the Euler-Poincaré index of the vector field $\mathbf{u}^{\perp}$ on D, weighting each term with a $(+)$ if the vector field* $\mathbf{u}$ *points to the right and $(-)$ if it points to the left of D. If this weighted index is not ± 1, then the Beltrami field* $\mathbf{u}$ *has a closed flowline in $D^2 \times \mathbb{R}$.*

The proof of this result is to use the stated information about $\mathbf{u}^{\perp}$ to reconstruct the contact planes orthogonal to $\mathbf{u}$ on D. Through topological techniques, the weighted index computation indicates that the contact structure on $D \times \mathbb{R}$ is overtwisted. One then applies Hofer's techniques.

This is really oblique: one knows *nothing* about the field $\mathbf{u}$ except that (1) it is Beltrami; and (2) it is orthogonal to some disc D at certain points with certain behavior. Nothing else is known about $\mathbf{u}$ away from the disc D. Genuinely two-dimensional data yields a genuinely three-dimensional implication. Of course, the result holds in an arbitrary metric, and the modified index is independent of the choice of disc D. It is remarkable that the chosen disc D *might not intersect the closed orbit it detects.* Even so, deforming the metric and the Beltrami field by an arbitrary amount off of the disc will not destroy the closed orbit.

This is a prime example of a topological result: it is stable with respect to arbitrarily large perturbations.

6. Knots

In dimension three, periodic flowlines are embedded closed curves: *knots*, which together form a *link* of closed curves. There is a long beautiful string of theory about the relationships between the dynamics of a three-dimensional flows and the topology of the link of closed orbits (see, e.g., [5, 23] and references therein).

By Theorem 2.2, it is necessary to partition our discussion of knot types into two classes: the integrable case and the Beltrami case.

6.1. INTEGRABLE FLOWS

As hinted at by Example 1.1, the knot types present in an integrable vector field are especially simple. Since almost all of the orbits are constrained to lie on a torus, one expects a certain rigidity.

Definition 6.1 *The* **connected sum** *of two knots K and K' is the knot obtained by splicing the two together as in Figure 6(left). Given a knot K, the* **cables** *of K are those knot types which can be drawn as simple closed curves on the surface of a torus having K at its core (Figure 6(right)). The* **torus knots** *are those knot types which are the cables of the unknot.*

Figure 6. Connected sums (left) and a cable of the figure-eight knot (right).

The following theorem classifying knots of integrable fields has a convoluted history. It was first stated (with a slightly stronger assumption) by Fomenko and Nguyen [16]. Casasayas et al. [6] observed that their result follows effortlessly from earlier work of Wada [33] and Morgan [32].

Theorem 6.2 *Every closed orbit in a non-vanishing C^ω integrable vector field on S^3 is a knot obtained from the unknot by repeated cabling and connected sum.*

Thus, the presence of a single figure-eight knot[5] in a non-vanishing flow is enough to conclude that the flow is not integrable.

6.2. BELTRAMI FIELDS

Beltrami flows, in contrast to their integrable cousins, display the broadest possible variety of behavior with respect to flowline topology. The trick is in the correspondence between Reeb fields and Beltrami fields. Reeb fields, since they are topological in nature, are subject to all of the manipulations a topologist can muster: surgery, cut-and-paste, and the like. After a customized contact form is built, one invokes the correspondence theorem to realize this as a genuine Beltrami field for the appropriate Riemannian metric. Using a very delicate sequence of topological constructions, one can construct a Beltrami field which exhibits the most complex type of field line topology imaginable:

[5]Or, indeed, any knot whose complement in S^3 admits a *hyperbolic* metric...

Theorem 6.3 ([13]) *There exists a non-vanishing Beltrami field on a Riemannian 3-sphere*[6] *which possesses closed flowlines of all possible knot and link types.*

The velocity field and the metric are all of class C^ω (real-analytic). These fields are by no means degenerate: any C^1-close vector field still possesses this set of knotted and linked flowlines. However, the construction most certainly does not yield a round (or Euclidean) metric.

Question 6.4 Is it possible for a smooth, steady Euler flow on Euclidean $\mathbb{R}^3$ to possess closed flowlines of all knot types?

The paper [22] gives sufficient conditions for this to occur in terms of existence of certain homoclinic or heteroclinic orbits. Whether such can be found in a steady Euler flow is not clear.

7. Moving Beyond Steady Flows

Given a dynamical system, the first task is to find and classify the fixed points. The next step of a qualitative analysis is to determine the existence and nature of any periodic orbits or connecting orbits in the global flow.

In the case of the Euler equations, one has a dynamical system on the space of all divergence-free vector fields. This survey has covered the basics of what can be said about the fixed points of this dynamical system: the steady Euler flows.[7] The next logical step after finding the fixed points is to determine the stability of these solutions: this in itself is a delicate problem.

7.1. INSTABILITY

The problem of classifying the local behavior of fixed points is precisely the classical problem of hydrodynamic stability. Though there are manifold notions of how hydrodynamical stability or instability should be defined (see [17]), the goal is to determine whether nearby velocity fields (perturbations of a steady Euler field) remain in a vicinity of the steady field, or whether perturbations grow large.

The answer seems quite plain, but there appears to be no way to prove it in general. The conjecture is this: *Almost every steady Euler field in dimension three is hydrodynamically unstable.* It is a genuine challenge to make this a mathematically well-defined question. The first point is to settle

[6] The proof also works and is indeed easier for a Riemannian $\mathbb{R}^3$.

[7] We have restricted attention to the case of dimension three, but the contact topology perspective is efficacious in all odd dimensions. For even dimensions, the appropriate framework is that of symplectic topology, as pointed out by Ginzburg and Khesin [24]. There are some beautiful ways in which steady fluids in high dimensions have radically different behaviors depending on the parity of the dimension.

on the type of instability one seeks: linear or nonlinear? energy or velocity norm? The paper [17] contains an excellent discussion of this issue.

The second order of business is to determine what possible sense one could ascribe to the phrase *"almost every"*? For the Beltrami fields, the idea of "almost every" is strained — for a generic Riemannian metric, the eigenvalues of the curl operator are simple, and there are thus a countable number of Beltrami fields, each of which is isolated in the space of volume-preserving vector fields. The only possible sense that "almost every" can take is "all but a finite number" or, equivalently, "for all sufficiently large eigenvalues." In this (most typical) setting, one most decidedly cannot perturb a Beltrami field in any manner which will yield a new Beltrami field (ignoring scalar multiples of the field).

In the particular case of the Euclidean metric, the symmetries provide for multiple eigenspaces of curl leading to parametrized families of Beltrami fields. The ABC fields are one such example (thus explaining the existence of three parameters) for the first eigenvalue of curl. In this context, one may restrict to such families and try to prove generic instability (as is done for most singular ABC fields in [18]).

The situation for steady integrable fields is much cloudier. Are such fields isolated in the space of volume-preserving vector fields? Can they appear in parametrized families? Clearly, in the presence of enough symmetry, whole families can arise: but it seems reasonable to expect that such families would have identical stability properties. Indeed, if a symmetric stable family could be found, this could provide a counter-example to the Instability Conjecture. We point out the following simple test case:

Question 7.1 Are the Hopf fields on the round 3-sphere hydrodynamically unstable?

As we see the situation, the best hope for establishing a universal result on hydrodynamic instability is to use the criteria of a *fluid Lyapunov number* developed by Friedlander and Vishik using techniques from geometric optics.

Theorem 7.2 ([19]) *Any steady Euler flow which possesses a hyperbolic periodic orbit is (linearly) hydrodynamically unstable (in L^2 norm).*

The contact-topological techniques provide a partial answer. In the case of an overtwisted Beltrami field, the periodic orbit forced is *not* elliptic: it is either hyperbolic or degenerate This orbit is thus hyperbolic generically (among Reeb fields). In like vein, the *symplectic field theory* of [10] indicates that, generically in the space of Reeb fields on T^3 (even in the tight case!), there exist hyperbolic periodic orbits. When one includes the metric as a variable (thus working in the space of all Reeb fields as opposed to the discrete space of Beltrami fields for a fixed metric), one can rigorously prove that hyperbolic orbits (and thus instability) are generic.

7.2. CONNECTIONS AND PERIODIC ORBITS

The next step in a global analysis concerns simple recurrence. Recall that a *homoclinic* orbit for a dynamical system is one which approaches a certain fixed point asymptotically both forwards and backwards in time. A *heteroclinic* orbit (or "connection") is an orbit which approaches two distinct fixed points as time goes to $\pm\infty$.

Question 7.3 Do the Euler equations admit any homoclinic or heteroclinic orbits?

There are no obvious topological obstructions to the existence of homoclinic orbits of the Euler equations. Heteroclinic orbits have what appear to be extraordinarily rigid obstructions. Assume that there is an orbit $\mathbf{u}(t)$ of Equation (1) which limits to Beltrami fields $\mathbf{u}^+$ and $\mathbf{u}^-$ as $t \to \pm\infty$ respectively. Since the vorticity field is "frozen in" by the Euler flow, the vorticity fields $\nabla \times \mathbf{u}(t)$ are all equivalent by volume-preserving diffeomorphisms, and, in the limits, $\mathbf{u}^+$ and $\mathbf{u}^-$ are themselves related: an unlikely situation apart from unusual symmetries.

Question 7.4 Classify the periodic orbits of the Euler equations in the space of volume-preserving vector fields.

Surely one could construct such a periodic orbit using a periodic pressure function $p_t : M \to \mathbb{R}$: take for example $\mathbb{R}^3$ with linear flow and use a family of pressure functions $p_t(x, y, z)$ to rotate the velocity vectors uniformly. Examples would be most helpful for obtaining a clearer picture of the global structure of the Euler equations as a dynamical system on the space of volume-preserving vector fields.

Acknowledgements

The first author acknowledges the hospitality of the Isaac Newton Institute, and the support of the National Science Foundation (DMS-9971629).

References

1. Abraham, R., Marsden, J. & Ratiu, T. (1988) *Manifolds, Tensor Analysis, and Applications.* Applied Math. Sci. **75**, Springer-Verlag, Berlin, Heidelberg, New York.
2. Arnold, V.I. (1966) Sur la géométrie differentielle des groupes de Lie de dimension infinie et ses applications à l'hydrodynamique des fluides parfaits. *Ann. Inst. Fourier* **16**, 316–361.
3. Arnold, V.I. (1986) The asymptotic Hopf invariant and its applications. *Sel. Math. Sov.* **5**, 327–345.

4. Arnold, V.I. & Khesin, B. (1998) *Topological Methods in Hydrodynamics.* Springer-Verlag, Berlin, Heidelberg, New York.
5. Birman, J. & Williams, R. (1983) Knotted periodic orbits in dynamical systems–I : Lorenz's equations. *Topology* **22**(1), 47–82.
6. Casasayas, J., Martinez Alfaro J. & Nunes, A. (1993) Knotted periodic orbits and integrability. In *Hamiltonian Systems and Celestial Mechanics* , pp. 35–44. Adv. Ser. Nonlinear Dynam. **4**, World Sci. Pub., River Edge, NJ.
7. Chicone, C. (1996) A geometric approach to regular perturbation theory with an application to hydrodynamics. *Trans. Amer. Math. Soc.* **347**(12), 4559–4598.
8. Childress, S. & Gilbert, A. (1995) *Stretch, Twist, Fold: The Fast Dynamo.* Springer-Verlag.
9. Dombre, T., Frisch, U., Greene, J., Hénon, M., Mehr, A. & Soward, A. (1986) Chaotic streamlines in the ABC flows. *J. Fluid Mech.* **167**, 353–391.
10. Eliashberg, Y., Givental, A. & Hofer, H. (2000) An introduction to symplectic field theory. *ArXiV Preprint* `math.SG/0010059`.
11. Etnyre, J. & Ghrist, R. (1999) Stratified integrals and unknots in inviscid flows. *Cont. Math.* **246**, 99–112.
12. Etnyre, J. & Ghrist, R. (2000) Contact topology and hydrodynamics I: Beltrami fields and the Seifert conjecture. *Nonlinearity* **13**, 441–458.
13. Etnyre, J. & Ghrist, R. (2000) Contact topology and hydrodynamics III: knotted flowlines. *Trans. Amer. Math. Soc.* **352**, 5781–5794.
14. Etnyre, J. & Ghrist, R. (2001) Contact topology and hydrodynamics II: solid tori. *Ergod. Theory & Dyn. Sys.*, to appear. [Also in *ArXiv Preprint* `math.SG/9907112`.]
15. Etnyre, J. & Ghrist, R. (2001) An index for closed orbits in Beltrami fields. *ArXiv Preprint* `math.DS/0101095`.
16. Fomenko, A.T. & Nguyen, T.Z. (1991) Topological classification of integrable nondegenerate Hamiltonians on isoenergy three-dimensional spheres. In *Advances in Soviet Mathematics* (ed. A.T. Fomenko), pp. 267–296. Amer. Math. Soc. **6**, Providence, RI.
17. Friedlander, S. & Yudovich, V. Instabilities in fluid motion. (1999) *Notices Amer. Math. Soc.* **46**(11), 1358–1367.
18. Friedlander, S., Gilbert, A. & Vishik, M. (1993) Hydrodynamic instability for certain ABC flows. *Geophys. Astrophys. Fluid Dynam.* **73**, 97–107.
19. Friedlander, S. & Vishik, M. (1992) Instability criteria for steady flows of a perfect fluid. *Chaos* **3**, 455–460.
20. Gautero, J.-L. (1985) Chaos lagrangien pour une classe d'écoulements de Beltrami. *C. R. Acad. Sci. Paris. S. II Méc. Phys. Chim. Sci. Univers. Sci. Terre* **301**(15), 1095–1098.
21. Ghrist, R. (2001) Steady nonintegrable high-dimensional fluids. *Lett. Math. Phys.*, to appear.
22. Ghrist, R. & Holmes, P. (1996) An ODE whose solutions contain all knots and links. *Intl. J. Bifurcation & Chaos* **6**(5), 779–800.
23. Ghrist, R., Holmes, P. & Sullivan, M. (1997) *Knots and Links in Three-Dimensional Flows.* Lecture Notes in Mathematics **1654**, Springer-Verlag, Berlin, Heidelberg, New York.
24. Ginzburg, V. & Khesin, B. (1994) Steady fluid flows and symplectic geometry. *J. Geom. and Phys.* **14**, 195–210.
25. Hofer, H. (1993) Pseudoholomorphic curves in symplectizations with applications to the Weinstein conjecture in dimension three. *Invent. Math.* **114**, 515–563.
26. Hofer, H., Wysocki, K. & Zehnder, E. (1998) The dynamics on three-dimensional strictly convex energy surfaces. *Ann. of Math.* **148**(1), 197–289.
27. Kuperberg, G. (1996) A volume-preserving counterexample to the Seifert conjecture. *Comment. Math. Helv.* **71**(1), 71–97.
28. Kuperberg, G. & Kuperberg, K. (1994) Generalized counterexamples to the Seifert conjecture. *Ann. of Math.* **144**, 239–268.

29. Martinet, J. (1971) Formes de contact sur les variétés de dimension 3. In *Proceedings of Liverpool Singularities - Symposium II*, (ed. C.T.C. Wall), pp. 142–163. Springer Lecture Notes in Math. **209**, Springer-Verlag.
30. McDuff, D. & Salamon, D. (1995) *Introduction to Symplectic Topology.* Oxford University Press, New York.
31. Moffatt, H.K., Zaslavsky, G.M., Comte, P. & Tabor, M. (Ed.) 1992 *Topological Aspects of the Dynamics of Fluids and Plasmas.* NATO ASI Series E: Applied Sciences **218**. Kluwer.
32. Morgan, J. (1978) Nonsingular Morse-Smale flows on 3-dimensional manifolds. *Topology* **18**, 41–54.
33. Wada, M. (1989) Closed orbits of nonsingular Morse-Smale flows on S^3. *J. Math. Soc. Japan* **41**(3), 405–413.

GEOMETRIC AND TOPOLOGICAL ASPECTS OF VORTEX MOTION

RENZO L. RICCA
Department of Mathematics
University College London
Gower Street, London WC1E 6BT, UK

E-mail: *ricca@math.ucl.ac.uk*

Abstract. In this paper we review some results on geometric and topological vortex dynamics. After some background on flow maps, topological equivalence of frozen fields and conservation laws, we discuss geometric aspects of vortex filament motion (intrinsic equations, connections with integrable dynamics and extension to higher dimensional manifolds) and the topological interpretation of kinetic helicity in terms of linking numbers. We recall basic results on evolution of vortex knots and links and outline possible applications of algebraic, geometric and topological measures to evaluate structural complexity of vortex flows.

1. Flow Maps and Topological Equivalence of Frozen Fields

Let $\mathcal{D}$ denote a fluid domain in $\mathbb{R}^3$, that is an open, unbounded, singly connected region of the Euclidean, three-dimensional space. The domain is filled by a homogeneous, inviscid fluid, which is treated as a mathematical continuum (i.e. neglecting microscopic and dissipative effects). Of course this is a mathematical simplification of reality, justified by assuming that the typical evolution time of the corresponding real process is much greater than the dissipative and diffusive characteristic time.

Fluid motion determines a continuous time-dependent re-arrangement of fluid regions in $\mathcal{D}$. We assume that fluid particles move with a velocity $\mathbf{u} = \mathbf{u}(\mathbf{x}, t)$, smooth function of the position vector $\mathbf{x}$ and time t, satisfying

$$\nabla \cdot \mathbf{u} = 0 \quad \text{in } \mathcal{D} \, ; \qquad \mathbf{u} = 0 \quad \text{as } \mathbf{x} \to \infty \, , \tag{1}$$

R.L. Ricca (ed.), An Introduction to the Geometry and Topology of Fluid Flows, 203–228.

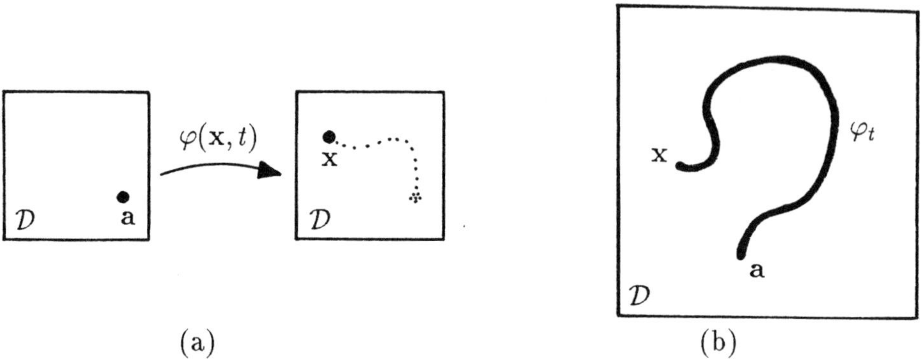

Figure 1. (a) A fluid particle moves in $\mathcal{D}$ by the action of a flow map φ associated with the fluid velocity $\mathbf{u}$; (b) as t varies, the collection of φ_t determines the trajectory followed by the fluid particle in time.

and that fluid motion is governed by the Euler equations

$$\frac{\partial \mathbf{u}}{\partial t} + (\mathbf{u} \cdot \nabla)\mathbf{u} = -\nabla p \, . \tag{2}$$

The velocity field $\mathbf{u}$ induces a map $\varphi = \varphi(\mathbf{x}, t)$ that at each time sends the fluid particle from the initial position $\mathbf{a}$ and time t_0 to the position $\mathbf{x}$ and time t: $\varphi_t : \mathbf{a} \to \mathbf{x}$, $\forall t \in I$ (I finite). Since this map advances each fluid particle from $\mathbf{a}$ to $\mathbf{x}$, as t varies the collection of φ_t determines the *particle path* followed by the fluid particle in time (see Figure 1).

We have:

Definition 1.1 *A* **fluid flow map** *is a functional element $\varphi \in \Phi$, where*

$$\Phi \equiv \left\{ C^\nu : \exists \, \varphi^{-1} \, | \, \mathbf{a} = \varphi^{-1}(\mathbf{x}, t) \, \forall t \in I; \; \varphi \textit{ volume-preserving} \right\} , \tag{3}$$

where Φ is a differential manifold of infinite dimension and C^ν denotes the class of functions continuously differentiable of order ν.

For obvious reasons we may want to take $\nu \geq 3$. Under the action of φ any portion of $\mathcal{D}$ will move in the fluid while changing shape, but keeping volume constant.

Remark 1.2 In this paper we concentrate on *non-degenerate* ('tame') flow maps, i.e. flow maps that are single-valued and continuous everywhere. Flow patterns exhibiting cusp formation, bifurcations and other types of physical singularities are actually ubiquitous in nature; however, we need to develop appropriate mathematical descriptions to be able to include these type of 'degeneracies' in a comprehensive mathematical theory of fluid flows.

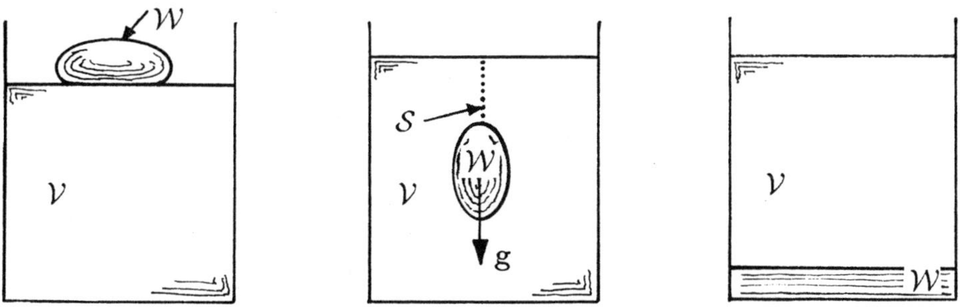

(a) (b) (c)

Figure 2. (a) A box is filled by some fluid in $\mathcal{V}$ with a blob of denser fluid in $\mathcal{W}$. Initially the two fluids are at rest; (b) under the action of the gravitational force **g** the fluid in $\mathcal{W}$ starts gradually to sink and penetrate $\mathcal{V}$, leaving behind a subset of 'virtual' separation $\mathcal{S}$ (dotted line). (c) As $t \to \infty$ the denser fluid precipitates to the bottom surface of the box.

Example 1.3 [Degenerate flow maps] A box is filled by two different fluids: one fluid occupies a region $\mathcal{V}$ and a blob of denser fluid is concentrated in $\mathcal{W}$, floating on top of $\mathcal{V}$ (see Figure 2a). Suppose that the two fluids have different, high surface tension, providing strong coherency of the bounding surfaces (so as to avoid mixing). The two fluid system is initially at rest, but is subject to a uniform gravitational field **g**. Since gravity pulls down the blob of denser fluid, $\mathcal{W}$ starts gradually to deform to penetrate the fluid in $\mathcal{V}$.

At two instances the flow map becomes degenerate.

i) As the blob sinks into $\mathcal{V}$, the boundary $\partial\mathcal{W}_t = \varphi_t(\partial\mathcal{W}_{t_0})$ becomes indefinitely stretched and since φ_t is assumed continuous on the entire fluid domain $\mathcal{D} = \bar{\mathcal{W}} \cup \bar{\mathcal{V}}$ (over-bar denotes closure), $\mathcal{V}$ cannot be severed by the motion of $\mathcal{W}$. A subset $\mathcal{S}$ gradually develops (dotted line in Figure 2b): this is where the inverse map φ_t^{-1} is multi-valued. The region $\mathcal{S}$, of 'virtual' separation, is mathematically interesting and physically important, but in most situations is ignored.

ii) As time passes the denser fluid gradually precipitates to the bottom surface of the box to reach equilibrium. As $t \to \infty$ the flow map φ_t becomes ill-defined for the points at the bottom surface of the box (Figure 2c).

Try this experiment with oil and vinegar.

Question 1.4 Does the topology of $\mathcal{V}$ change during the two stages of the process described in Example 1.3?

Problem 1.5 To classify degenerate flow patterns by using concepts from dynamical systems, graph theory and singularity theory.

For any fluid region $\mathcal{W}$ in $\mathcal{D}$ we have that

$$\partial\mathcal{W}_t = \varphi_t(\partial\mathcal{W}_{t_0}) \, , \tag{4}$$

which assures that any bounding fluid surface that is a material surface at time t_0, remains a material surface at all subsequent times t and moves with the flow (this follows from the definition of Φ). The *Jacobian* J of the transformation is defined by

$$J = \det\left(\frac{\partial x_i}{\partial \mathrm{a}_j}\right) \, , \tag{5}$$

where the tensor $\partial x_i/\partial \mathrm{a}_j$ represents the deformation of the infinitesimal volume as it is transported by the motion. The symmetric part of this tensor is associated with the distortion and change of the volume element, while the skew-symmetric part is associated with its rotation. From the assumption that φ possesses a differentiable inverse it follows that $0 < J < \infty$.

Vorticity is defined by $\boldsymbol{\omega} = \nabla \times \mathbf{u}$ and we assume that $\nabla \cdot \boldsymbol{\omega} = 0$ in $\mathcal{D}$. We have:

Definition 1.6 *The* **circulation** *(or* **flux of vorticity***) is defined by*

$$\kappa = \oint_{\mathcal{C}} \mathbf{u} \cdot d\mathbf{l} = \int_{\mathcal{S}} \boldsymbol{\omega} \cdot \hat{\boldsymbol{\nu}} \, d\sigma \, , \tag{5}$$

where $\mathcal{C}$ (of elementary directional length $d\mathbf{l}$) is a simple unknotted, closed circuit $\mathcal{C} \equiv \partial\mathcal{S}$, and $\mathcal{S}$ (of elementary area $d\sigma$) is a simply connected two-dimensional surface of unit normal $\hat{\boldsymbol{\nu}}$, pointing in the positive direction induced by the $\boldsymbol{\omega}$-field.

The two integrals are related by Stokes's theorem and in ideal conditions (Euler's equations) the common value $\kappa =$ constant is an *invariant* of fluid motion (Helmholtz's III law and Kelvin's theorem; see [30]). Moreover,

Definition 1.7 *The vorticity field $\boldsymbol{\omega}$ is said to be* **frozen** *in $\mathcal{D}$ if and only if it satisfies the transport (Helmholtz) equation*

$$\frac{\partial \boldsymbol{\omega}}{\partial t} = \nabla \times (\mathbf{u} \times \boldsymbol{\omega}) \, . \tag{6}$$

Exercise 1.8 Derive Helmholtz's equation from Euler's equations (2), assuming conservative body forces.

A formal solution to (6) is represented by the Cauchy equations

$$\omega_i(\mathbf{x}, t) = \omega_j(\mathbf{a}, t_0)\frac{\partial x_i}{\partial \mathrm{a}_j} \, , \tag{7}$$

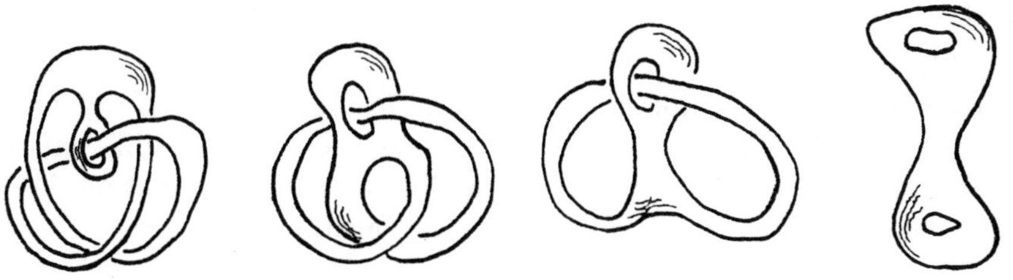

Figure 3. Topological equivalence between different configurations of the same fluid structure: (a) $\sim$ (b) $\sim$ (c) $\sim$ (d).

that encapsulate both convection of the $\boldsymbol{\omega}$-field from $\mathbf{a}$ to $\mathbf{x}$, and rotation and distortion of the fluid elements by the deformation tensor $\partial x_i/\partial \mathrm{a}_j$ (incompressibility is given by the condition $\det(\partial x_i/\partial \mathrm{a}_j) = 1$). Since the tensor is a time-dependent diffeomorphism of position, it maps continuously (i.e. without cuts or reconnections) the initial field $\boldsymbol{\omega}(\mathbf{a}, t_0)$ to $\boldsymbol{\omega}(\mathbf{x}, t)$, thus establishing a *topological equivalence* between initial and final configuration (Figure 3). We write

$$\boldsymbol{\omega}(\mathbf{a}, t_0) \sim \boldsymbol{\omega}(\mathbf{x}, t) , \tag{8}$$

and we regard equation (6) as a master equation for frozen fields and equation (7) as a topological equivalence statement for the initial and final configuration fields.

For more information on general aspects of fluid mechanics a standard reference is [6]; on specific aspects of vortex dynamics a standard reference is [30].

2. Conserved Quantities in Ideal Fluids

In absence of dissipative and diffusive effects the invariance of circulation is of course just one manifestation of the ideal conditions of fluid motion. In this context it is natural to expect the existence of families of such quantities (not all necessarily scalars). One possible classification of invariants is based on their nature:

(i)	*local*	*(metric)*	$\longleftrightarrow$	*pointwise*
(ii)	*global*	*(metric)*	$\longleftrightarrow$	*integral*
(iii)	*topological*	*(non-metric)*	$\longleftrightarrow$	*algebraic*

As will be shown in §5 there are deep connections between continuum fluid mechanics and topology. To this regard a fundamental question is:

Problem 2.1 To classify and relate ideal fluid invariants and topological invariants.

Problem 2.1 finds a more natural setting in the language of differential forms. We shall briefly illustrate this point by showing how different conservation laws can be cast into one single differential form equation. We do this by concentrating the discussion on local invariants.

2.1. LOCAL INVARIANTS AND DIFFERENTIAL FORM CONSERVATION LAWS

Local fluid invariants can be classified in four categories:

I type: *conserved quantity* ρ (e.g. mass per unit volume).
Governing equation for scalar quantities as a balance conservation law:

$$\frac{\partial \rho}{\partial t} + \nabla \cdot (\rho \mathbf{u}) = 0 \ .$$

II type: *Lagrangian invariant* $\Im$ (e.g. a passive scalar, like ink).
Governing equation for scalar quantities advected Lagrangian invariantly by the flow:

$$\frac{d\Im}{dt} = \frac{\partial \Im}{\partial t} + (\mathbf{u} \cdot \nabla)\Im = 0 \ .$$

III type: *frozen-in vector field* $\boldsymbol{\omega}$ (e.g. vorticity).
Governing equation for vector quantities advected along the flow streamlines:

$$\frac{\partial \boldsymbol{\omega}}{\partial t} = \nabla \times (\mathbf{u} \times \boldsymbol{\omega}) \ .$$

IV type: *Frobenius invariant* $\mathbf{S}$ (e.g. momentum of a vortex ring).
Governing equation for vector quantities advected by Frobenius-type surfaces frozen in the flow (see Figure 4):

$$\frac{d\mathbf{S}}{dt} = (\mathbf{S} \times \nabla) \times \mathbf{u} \ .$$

All local fluid invariants can be classified in these four categories. These four types of invariants can be expressed in terms of differential forms, each one corresponding to an invariant ω^p-form ($p = 0, 1, 2, 3$) obeying the conservation law

$$\mathcal{L}_{t,\mathbf{u}}\omega^p \equiv \frac{\partial \omega^p}{\partial t} + \mathcal{L}_{\mathbf{u}}\omega^p = 0 \ . \tag{9}$$

Note that since $d\mathcal{L}_{\mathbf{u}} = \mathcal{L}_{\mathbf{u}} d$, equation (9) holds true also for $d\omega^p$, where $d\omega^p = \omega^{p+1}$; the four types of conservation laws are immediately recovered by setting $\omega^0 = \Im$, $\omega^1 = \mathbf{S}$, $\omega^2 = \boldsymbol{\omega}$ and $\omega^3 = \rho$.

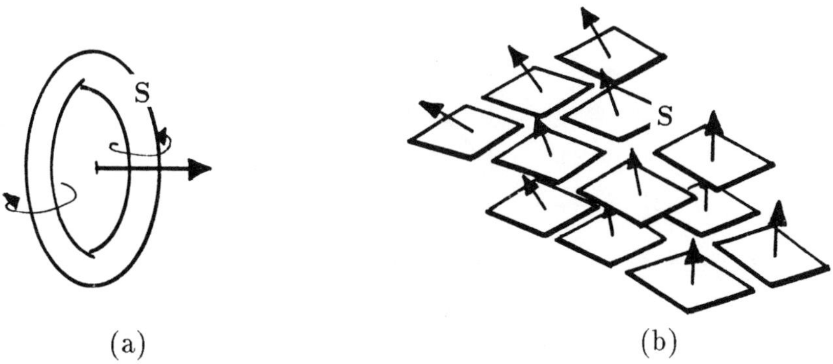

Figure 4. (a) The toroidal surface of a vortex ring translating in an ideal fluid provides an example of a Frobenius surface for the linear momentum; (b) Frobenius surfaces frozen in the flow.

Exercise 2.2 Prove that the only non-vanishing differential forms in $\mathbb{R}^3$ are the ω^p-forms with $p = 0, 1, 2, 3$.

General relations between invariant p-forms and conservation laws can be derived from properties of differential forms [1, 10]. Similar considerations can be applied to global types of invariants. For an extensive study of these relations and possible connections with topological invariants see [34].

3. Geometric Dynamics of Vortex Filaments

Let us consider an isolated vortex filament (closed on itself or extending to infinity) in $\mathcal{D}$. We want to study the motion of the filament in terms of geometric quantities such as curvature and torsion and highlight aspects of motion that have interesting connections with integrable dynamics in soliton theory and problems in applied differential geometry of minimal surfaces. The analysis presented below has a long and fascinating history that goes back at least a hundred years (for more information see the review paper [26]).

3.1. INTRINSIC EQUATIONS OF MOTION IN EUCLIDEAN SPACE

To study the evolution of the filament in space we simply identify the filament with its vortex centreline Γ, smooth and free from inflections and self-intersections. The filament has, however, finite thickness and for simplicity we assume that its cross-section is circular, of radius $a \ll R$ (where R is the radius of curvature of Γ) and circulation κ. At an arbitrary point on Γ we position the origin of the intrinsic Frenet frame $(\hat{\mathbf{t}}, \hat{\mathbf{n}}, \hat{\mathbf{b}})$, unit tan-

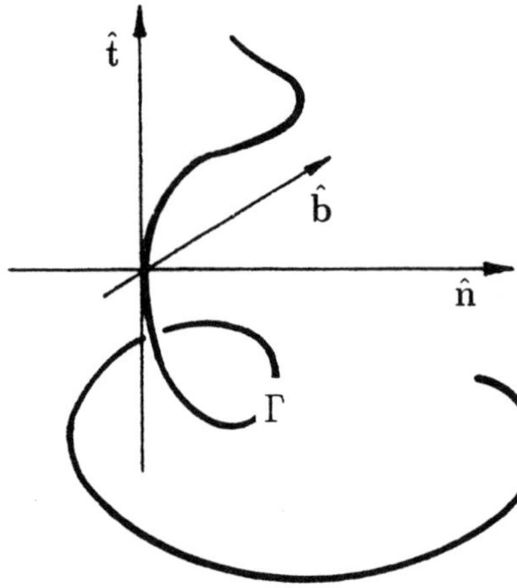

Figure 5. Intrinsic Frenet frame on a curve Γ.

gent, normal and binormal vectors to Γ (see Figure 5). Vorticity is simply given by $\boldsymbol{\omega} = \bar{\omega}\hat{\mathbf{t}}$, where $\bar{\omega}$ is a constant.

Remark 3.1 Given the distribution of vorticity $\boldsymbol{\omega}$, the standard problem is to find the induced velocity field $\mathbf{u}$ given by the Biot-Savart integral [30].

Let $\mathbf{X} = \mathbf{X}(s,t)$ (s arc-length) denote the vortex line Γ and $\hat{\mathbf{t}} \equiv \mathbf{X}'$ (prime denotes arc-length derivative). Tangent, normal and binormal are well-defined everywhere on the curve and satisfy the *Serret-Frenet equations*

$$\hat{\mathbf{t}}' = c\hat{\mathbf{n}} \ , \quad \hat{\mathbf{n}}' = -c\hat{\mathbf{t}} + \tau\hat{\mathbf{b}} \ , \quad \hat{\mathbf{b}}' = -\tau\hat{\mathbf{n}} \ , \tag{10}$$

where τ is torsion. Let $\mathbf{v} \equiv \dot{\mathbf{X}}$ be the velocity induced by vorticity; in intrinsic components we have $\mathbf{v} = (v_t, v_n, v_b)$, where everything is a smooth function of s and t (over-dot denotes time derivative). Since $\dot{\hat{\mathbf{t}}} = \dot{\mathbf{X}}' = \mathbf{v}'$, after some algebra we can derive the intrinsic equations of motion of the vortex filament in $\mathcal{D}$. Assuming (for simplicity) inextensibility of Γ we have *two intrinsic equations*

$$\dot{c} = (cv_t + v_n' - \tau v_b)' - (\tau v_n + v_b')\tau \ , \tag{11}$$

$$\dot{\tau} = \left[\frac{(cv_t + v_n' - \tau v_b)\tau + (\tau v_n + v_b')'}{c}\right]' + (\tau v_n + v_b')c \ . \tag{12}$$

that give the time-evolution of curvature and torsion of the filament in terms of the initial geometry and induced velocity. The condition of inextensibility is satisfied when the curve is parametrized by arc-length; this condition is expressed by the third equation (*congruence condition*)

$$v_t' = cv_n \ . \tag{13}$$

A full derivation of equations (11)–(13) can be found in [26].

Exercise 3.2 (i) Prove the validity of condition (13) for arc-parametrized curves. (ii) Using eqs. (11)–(12) study the case of a circle that shrinks with a velocity proportional to the curvature.

Exercise 3.3 Suppose $\mathbf{v} = v_b(c,\tau)\hat{\mathbf{b}}$ where v_b is a smooth function of c and τ. Prove that in the stationary case (i.e. with $\dot{c} = 0$ and $\dot{\tau} = 0$) eqs. (11-12) admit two integrals of motions given by:

$$W_1 = v_b^2 \tau = \text{constant} , \tag{14}$$

and

$$W_2 = \frac{v_b'' - v_b \tau^2}{c} + v_b c - \int_\Gamma v_b c' \, ds = \text{constant} . \tag{15}$$

3.2. INTEGRABILITY AND GLOBAL GEOMETRIC INVARIANTS

In the previous section we didn't specify the induced velocity $\mathbf{v}$ of the filament. In actual fact this velocity is determined by the prescribed vorticity by 'un-curling' (through the Biot-Savart integral) the equation $\boldsymbol{\omega} = \nabla \times \mathbf{v}$. Given the vorticity (i.e. its distribution over the cross-section) we thus obtain specific functional relationships for v_t, v_n and v_b. We have further (analytical) difficulties when we consider more realistic vortex motions, for instance in presence of a varying cross-section and vorticity distribution). The simplest case of a vortex line with $\boldsymbol{\omega} = \bar{\omega}\hat{\mathbf{t}}$ (see §3.1) reveals interesting connections with integrable systems. The asymptotic theory developed for this case [26] shows that the motion, goverened by the so-called *localized induction approximation* (LIA), is given (after appropriate re-scaling) by:

$$\text{LIA :} \qquad v_t = 0 \ , \quad v_n = 0 \ , \quad v_b = c \ . \tag{16}$$

Under LIA the intrinsic equations (11)–(12) take a much simpler form known as *Da Rios–Betchov equations.*

Exercise 3.4 Under LIA reduce eqs. (11)–(12) to the Da Rios–Betchov equations and directly from these latter find the two integrals of motion, particular cases of W_1 and W_2 of Exercise 3.3.

By applying a Madelung transformation in the form

$$\psi(s,t) = c(s,t)\, e^{i \int_\Gamma \tau(\bar{s},t)\, d\bar{s}} \ , \qquad \psi(s,t) \in \mathbb{C}^1 \ , \tag{17}$$

we can map points from the (c, τ)-space to the complex plane in $\mathbb{C}^1$, thus reducing the two intrinsic equations (11)–(12) in c and τ to one single equation in ψ. By using eq. (17) Hasimoto [12] finds a remarkable relationship between LIA and the *non-linear Schrödinger equation* (NLSE)

$$\frac{1}{i}\frac{\partial \psi}{\partial t} = \frac{\partial^2 \psi}{\partial s^2} + \frac{1}{2}|\psi|^2 \psi \ , \tag{18}$$

that in one dimension admits solutions in terms of solitary waves (*solitons*). From soliton theory we know that NLSE is indeed completely integrable and has an infinite (countable) number of *conservation laws* in involution. Under LIA vortex filament motion and integrable dynamics of soliton theory are thus related so that we can interpret mathematical aspects of soliton systems in terms of fluid dynamical properties. One interesting aspect regards the existence of soliton invariants. Since ψ is a well-defined one-to-one mapping from the solution space in (c, τ) to the complex plane, its inverse is also well-defined. By applying the inverse map ψ^{-1} we can express the family of conservation laws, obtained directly from (18), in terms of global geometric functionals; the first three of them (neglecting the constant quantities) are

$$\int_\Gamma c^2 \, ds \ , \quad \int_\Gamma c^2 \tau \, ds \ , \quad \int_\Gamma \left(\frac{c^4}{4} - c'^2 - c^2\tau^2 \right) ds \ , \quad \ldots \ , \tag{19}$$

where these integrals are all constants of the motion (invariant in time). The set, however, is not complete and other integral quantities, not captured by the NLSE theory, are also conserved. Few of these quantities admit physical interpretation in terms of kinetic energy, helicity, linear and angular momenta of the vortex filament (all conserved quantities under LIA as well as under Euler's equations), whereas the remaining provide a purely geometric information (including total torsion, writhing, etc.) [24].

In recent years more work has been dedicated to extend these connections between vortex motion and integrability. As Fukumoto & Miyazaki show [11], in presence of axial flow vortex filament motion can be reduced to a Korteweg-de Vries type of equation, another soliton equation with infinite number of conservation laws in involution. Further generalizations (to include Hirota type equations and sine-Gordon equations) are also possible. How general can we go to preserve integrability on one hand and vortex motion on the other? And, is this a route towards integrability of Euler's equations? Asymptotic analyses based on Biot-Savart integral reveal an increasing analytical complexity and it is doubtful that additional information (given, for example, by tube thickness and distribution of vorticity) can preserve integrability. Work in this direction is carried out, among others, by Fukumoto and collaborators. An interesting open question seems to be the following:

Problem 3.5 To determine under which general geometric conditions the motion of a vortex line is completely integrable.

Langer & Perline [17] discovered interesting geometric relationship (not necessarily satisfying the laws of vortex motion) between LIA-type of equations and the hierarchy of integrable one-dimensional systems. Their approach, based on the study of the Hamiltonian structure associated with a class of soliton equations and their Poisson geometry, shows that it is possible to define a recursive operator $\Re$, that, starting from LIA, generates integrable dynamics in cascade. The operator can be explicited as

$$\Re : \quad \begin{cases} \dot{\mathbf{X}}^{(0)} = c\hat{\mathbf{b}} \ , \\ \dot{\mathbf{X}}^{(j)} = \dot{\hat{\mathbf{t}}}^{(j-1)} \times \hat{\mathbf{t}} + \left[\int c(\bar{s},t)(v_b'^{(j-1)} + \tau v_n^{(j-1)}) d\bar{s} \right] \hat{\mathbf{t}} \ , \end{cases} \tag{20}$$

where $j = 1, 2, \ldots$. Note that the recursive operator (20) preserves arc-length parametrization. General conditions for integrability of space curves and relations to Bäcklund transformations and Lax pairs can be found in [22].

Finally, since the motion of Γ is governed by the principle of least action, the surface (in s and t) swept out by Γ during the motion is a minimal surface for kinetic energy. In the case of integrable dynamics, we may have a natural connection between soliton theory and Plateau problems, and more precisely between characteristic properties of soliton surfaces and geometric aspects of minimal surfaces in the metric given by the kinetic energy of the system. The situation is illustrated by the graph below:

$$\begin{array}{ccc} \textit{Intrinsic kinematics} & \overset{\psi(s,t)}{\longleftrightarrow} & \textit{Integrable dynamics} \\ \varphi_t(s) \updownarrow & & \updownarrow \Re \\ \textit{Minimal surfaces} & \overset{?}{\longleftarrow \cdots \longrightarrow} & \textit{Poisson geometry} \end{array}$$

3.3. EXTENSION TO HIGHER DIMENSIONAL MANIFOLDS

The intrinsic equations (11)–(13) can be extended to $2n+1$-dimensional manifolds $\mathcal{M} \cong \mathbb{R}^{2n+1}$, where Γ is now an arc-parametrized curve in $\mathcal{M}$. The $(0,2)$-type *metric tensor* g is a smooth and positive-definite section of the bundle of the symmetric bi-linear two-forms on $\mathcal{M}$, given by $g =$

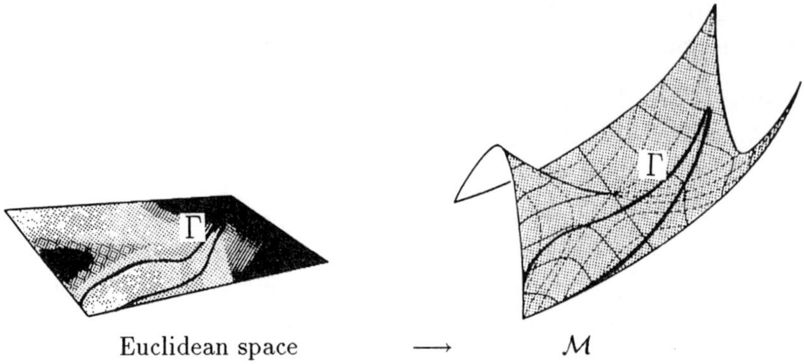

Figure 6. Geometrization of compressibility in the Euclidean space by the curvature tensor of $\mathcal{M}$.

$g_{ij}\, dx^i \otimes dx^j$. Given a C^r linear connection ∇ on $\mathcal{M}$, one can define a C^{r-1} *torsion tensor* T of type $(1,2)$ by

$$\mathsf{T}(\mathsf{Y},\mathsf{Z}) = \frac{1}{2}\left(\nabla_{\mathsf{Y}}\mathsf{Z} - \nabla_{\mathsf{Z}}\mathsf{Y} - [\mathsf{Y},\mathsf{Z}]\right) ,$$

and a C^{r-1} *Riemann (curvature) tensor* R of type $(3,1)$ by

$$\mathsf{R}(\mathsf{Y},\mathsf{Z})(\mathsf{V},\mathsf{W}) = \left[\nabla_{\mathsf{Y}}\nabla_{\mathsf{Z}}g(\mathsf{W}) - \nabla_{\mathsf{Z}}\nabla_{\mathsf{Y}}g(\mathsf{W}) - \nabla_{[\mathsf{Y},\mathsf{Z}]}g(\mathsf{W})\right](\mathsf{V}) ,$$

where $[\mathsf{Y},\mathsf{Z}]$ is the Lie derivative of Z with respect to Y and $\mathsf{V},\mathsf{W},\mathsf{Y},\mathsf{Z}$ are arbitrary C^{r+1} fields.

The Serret-Frenet equations (10) take the general form given by the $SO(2n+1)$-structure for the generalized $2n$ curvatures $\Omega_i^j = -\Omega_j^i$ $(i,j = 1,\ldots,2n+1)$, where Ω_i^j are sufficiently smooth functions of s and t. The generalized induced velocity has components v^j in the basis $\{e_j\}$ given by the metric g. By applying Ricci formula, and after some tedious but straightforward algebra, we obtain the *intrinsic equations in generalized form* [23]:

$$v^1_{\ ,s} = 2\mathsf{T}_{st}^{\ t} v^1 - \Omega_2^1 v^2 + 2\mathsf{T}_{st}^{\ s} \qquad \text{for} \quad k = 1 , \tag{21}$$

$$\dot{\Omega}_{k-1}^k = A_k^k - \Omega_{k-1}^k \left(\sum_{i=1}^{k-2} \frac{\dot{\Omega}_i^{i+1}}{\Omega_i^{i+1}} - 2k\mathsf{T}_{st}^{\ s} \right) , \tag{22}$$

where

$$A_1^k = v^k_{\ ,s} + v^{\bar{k}}\Omega_{\bar{k}}^k - 2\mathsf{T}_{st}^{\ t} v^k ,$$

$$A_k^k = A_{(k-1),s}^k + A_{k-1}^{\bar{k}}\Omega_{\bar{k}}^k - 2\mathsf{T}_{st}^{\ t} A_{k-1}^k - \Omega_{k-1}^{k-2} A_{k-2}^k ,$$

for $k = 2, \ldots, 2n+1$, and $\bar{k} = k-1, k+1$. Equation (21) represents the congruence condition for the arc-parametrized curve Γ and (22) gives $2n$ intrinsic equations for the time evolution of the curvatures Ω^i_j in $\mathcal{M}$.

Remark 3.6 By taking g, T and R pointwise functions of density (via the Jacobian J) we can geometrize the compressibility of the ambient space by prescribing appropriate functional relationships between the connection coefficients (hence the components of the torsion and curvature tensors) and J (see Figure 6). In this way extension to higher dimensional manifolds provides a route to a geometrization of fluid mechanical properties.

More information on geometric properties of fluid flows can be found in the seminal papers by Arnold [3] and Ebin & Marsden [9], and in more recent works by Holm, Marsden & Ratiu [13] and Shkoller [32].

4. Vortex Knots and Links and Reidemeister's Moves

Vortex flux tubes are coherent bundles of vortex lines embedded in a tube-like region. If the tube axis (which is a vortex line) is in the shape of a knot or a link, the corresponding vortex flux tube, formed by the tubular neighbourhood of vortex lines, is also knotted or linked. Formally, knotted and linked flux tubes can be constructed in a standard way as indicated, for example, by Moffatt [19] (see also [20]), that is via standard embedding, Dehn's surgery and appropriate cross-switchings. A formal definition of knotted or linked vortex tubes is given here. Let us consider first the following:

Definition 4.1 *A surface $\mathcal{S}$ is a* **vortex surface** *in $\mathcal{D}$ if it is made of vortex lines that are everywhere tangent to this surface.*

As we remarked in §1 an ideal vortex surface $\mathcal{S}$ at time $t = t_0$ remains a vortex surface $\mathcal{S}_t$ at every subsequent time, since $\mathcal{S}_t = \varphi_t(\mathcal{S}_{t_0})$.

Consider the standard solid torus T in $\mathbb{R}^3$ given by

$$((2+\epsilon\cos\beta)\cos\alpha,\ (2+\epsilon\cos\beta)\sin\alpha,\ \epsilon\sin\beta)\ , \tag{23}$$

where $\alpha \in [0, 2\pi]$, $\beta \in [0, 2\pi]$ and $\epsilon \in [0,1]$. Let $\mathcal{F}_{p,q}$ ($p > q > 1$ co-prime integers) denote the foliation of T by the curves $\Gamma_{\epsilon,\beta}$ given by

$$\Gamma_{\epsilon,\beta}(s) = \left\{ \begin{array}{c} [2+\epsilon\cos(\beta+qs)]\cos(ps) \\ [2+\epsilon\cos(\beta+qs)]\sin(ps) \\ \epsilon\sin(\beta+qs) \end{array} \right\}, \tag{24}$$

where $s \in [0, 2\pi]$. We have

Figure 7. Stretching of a vortex filament is produced by tangential actions on the tube strands. This deformation leaves the filament in the same topological class.

Definition 4.2 *A collection of* **linked (knotted) vortex tubes** $\mathcal{L}_v$ *is defined as a smooth embedding in* $\mathcal{D}$ *of finitely many disjoint standard solid tori* $\sqcup_i T_i$, *and smooth vorticity* $\boldsymbol{\omega}$ *such that:*

(i) $\mathcal{L}_v$ *is an embedding when restricted to the interior of* $\sqcup_i T_i$;

(ii) the bounding surface $\mathcal{S} \equiv \sqcup_i \mathcal{L}_v(\partial T_i)$ *is a vortex surface;*

(iii) for each component $\mathcal{L}_v(T_i)$ *there exists a pair* $\{p_i, q_i\}$ ($p_i > q_i > 1$ *co-prime integers) such that* $\mathcal{L}_v$ *maps the foliation* $\mathcal{F}_{p_i,q_i}$ *of* T_i *onto the integral curves of* $\boldsymbol{\omega}$ *in* $\mathcal{L}_v(T_i)$.

If p and q are real numbers, p/q is irrational and we can extend the definition to vortex tubes formed by non-closed vortex lines (of support $\Gamma_{\epsilon,\beta}$) space-filling the tori $\sqcup_i T_i$.

Since vorticity is frozen in $\mathcal{D}$, linked (knotted) vortex filaments are also frozen in their topological equivalence class. Topological fluid mechanics deals essentially with the study of fluid structures (thought of as embeddings) that can be continuously deformed one into another by ambient isotopies. For example, a natural isotopic deformation is represented by stretching of vortex tubes (geometrically equivalent to a time-dependent re-parametrization of the tube centreline, see Figure 7).

From knot theory we know that topology is preserved under the action of Reidemeister's moves. In the context of the Euler equations these moves are performed quite naturally by the action of local flows on the strands of fluid structures. If the background fluid in the complement $(\mathcal{D} - \mathcal{L}_v)$ is irrotational and at rest, then these flows must satisfy the Dirichlet problem for the Laplacian of the stream function ψ, given by

$$\begin{cases} \mathbf{u} = \nabla\psi \\ \nabla^2\psi = 0 \end{cases} \quad \text{in} \quad (\mathcal{D} - \mathcal{L}_v)\,, \tag{25}$$

with boundary conditions

$$\begin{aligned} \mathbf{u}\cdot\hat{\boldsymbol{\nu}} &= u_\perp && \text{on} \quad \mathcal{S} \equiv \sqcup_i \mathcal{L}_v(\partial T_i) \ , \\ \mathbf{u} &= 0 && \text{as} \quad \mathbf{x} \to \infty \ , \end{aligned} \tag{26}$$

where $u_\perp = u_\perp(\mathbf{x}, t)$ is the normal component of the velocity on the vortex boundary surface, with $\hat{\boldsymbol{\nu}}$ denoting the normal vector to the surface. Equations (25) and (26) have a unique solution in terms of local flows [6]. These flows act by performing a sequence of Reidemeister's moves on the tube strands through continuous deformations. Note that the boundary condition (26) does not prescribe the tangential component of the velocity on the bounding vortex surface. As we pointed out earlier, tangential effects (present for instance if the tube gets stretched) preserve vortex topology. This means that Reidemeister's moves are performed by local flows that are solutions to (25–26), up to arbitrary tangential actions.

For some historical information and a simple introduction to topological fluid mechanics see the review article by Ricca & Berger [28].

5. Helicity and Linking Numbers

A fundamental question in topological fluid mechanics is to understand whether and how fluid and topological invariants relate one another. A fundamental result regards kinetic helicity and its topological interpretation.

Definition 5.1 *The* **kinetic helicity** *of a linked (knotted) vortex system $\mathcal{L}_v$ in $\mathcal{D}$ is defined by*

$$H(\mathcal{L}_v) = \int_{\sqcup_i \mathcal{L}_v(T_i)} \mathbf{u}\cdot\boldsymbol{\omega}\, d^3\mathbf{x} \ . \tag{27}$$

Helicity is the fluid dynamical version of the Hopf integral, i.e. the integral of the inner product of a solenoidal vector field and its curl; isotopy invariance of this quantity was discovered by Whitehead in 1947. However, in the context of ideal fluid mechanics conservation of helicity (kinetic and magnetic) was shown by the works of Woltjer (1958), Moreau (1961), Moffatt (1969) and Arnold (1974) (see [18, 4]).

A fundamental result that establishes a bridge between topology and fluid mechanics regards the topological interpretation of helicity in terms of linking numbers. We have:

Theorem 5.2 ([18, 8, 20]) *Let $\mathcal{L}_v$ be a collection of vortex links (knots). Then*

$$H(\mathcal{L}_v) = \sum_i Lk_i\, \kappa_i^2 + 2\sum_{i\neq j} Lk_{ij}\, \kappa_i\kappa_j \ , \tag{28}$$

where Lk_i denotes the (Călugăreanu-White) linking number of the tube axis of $\mathcal{L}_v(T_i)$ with respect to the framing induced by the embedding of the $\boldsymbol{\omega}$-field, and Lk_{ij} denotes the (Gauss) linking number of $\mathcal{L}_v(T_i)$ with $\mathcal{L}_v(T_j)$.

Let $\mathcal{L}_v(T) \equiv \mathcal{K}_v(T)$ be a single vortex knot. Then, equation (28) reduces to

$$H(\mathcal{K}_v) = Lk\,\kappa^2 = (Wr + Tw)\,\kappa^2 \ , \tag{29}$$

where we write the (Călugăreanu-White) linking number Lk in terms of its decomposition given by the writhing number Wr of the tube axis and total twist Tw of the vortex tube (for a precise definition of these quantities see the article of Langevin, this volume, and Moffatt & Ricca [20]). Note that writhe and twist are purely geometric quantities and their values change with a change of shape. This means that Wr and Tw change continuously under continuous deformation, their sum remaining constant in time.

The writhing number Wr is characterized by the following properties:

i) Wr depends only on the geometry of the tube axis;

ii) Wr is invariant under rigid motions or dilations of the ambient space (conformal invariant), but its sign changes under reflection;

iii) in passing from an under-crossing to an over-crossing of the tube strands (in a given projection plane), its value jumps by +2.

Exercise 5.3 Show that the writhing number of a curve Γ (as defined by the integral formula given in Langevin's article, this volume) admits physical interpretation in terms of the sum of the signed crossings of the diagram of Γ in a given projection plane, averaged over all projections, that is

$$Wr = < n_+(\hat{\boldsymbol{\nu}}) - n_-(\hat{\boldsymbol{\nu}}) > \ , \tag{30}$$

where the angular brackets denote averaging over all directions $\hat{\boldsymbol{\nu}}$ of projection, and $n_\pm$ denotes the number of apparent $\pm$ crossings, from the direction of projection $\hat{\boldsymbol{\nu}}$. Derive eq. (30) directly from the integral formula.

Remark 5.4 For a nearly plane curve (except small indentations to allow crossings) the writhe of the curve can be estimated by counting the sum of the signed crossings that are apparent from that plane projection (see Figure 8).

The normalized total twist Tw is given by the sum of the total torsion of the tube axis and the intrinsic twist of the $\boldsymbol{\omega}$-lines in the tube T, divided by 2π, and has the following properties:

i) Tw is a continuous function of the tube axis;

ii) Tw is invariant under rigid motions or dilations of the ambient space (conformal invariant), but changes sign under reflection;

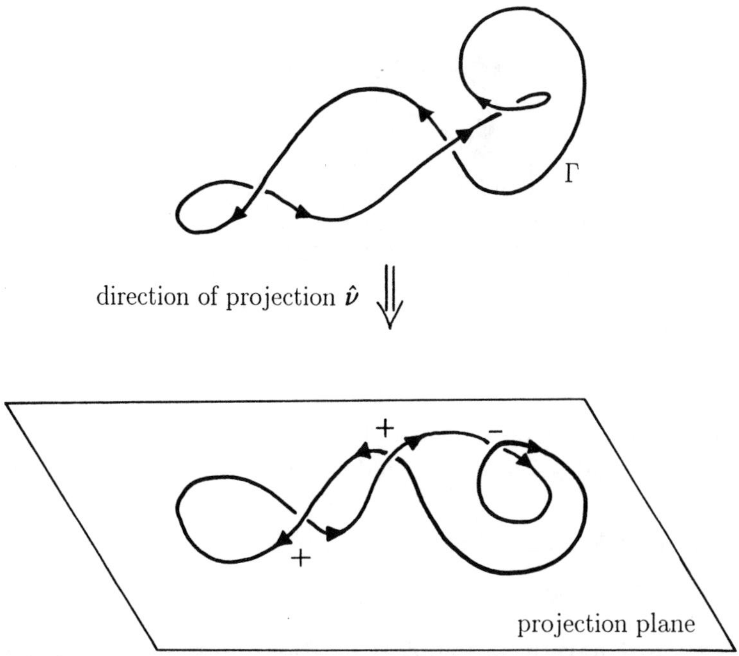

Figure 8. The writhing number of an oriented curve can be estimated by counting the sum of the signed crossings of the diagram of the curve projected on the plane along the direction $\hat{\boldsymbol{\nu}}$, then averaging over all directions.

iii) Tw is additive for contiguous tubular segments of T.

Part of the twist contribution to helicity is therefore associated with torsion of the tube axis and part with what may be described as 'intrinsic twist' of the field lines in the vortex flux tube.

If the embedding of the $\boldsymbol{\omega}$-lines corresponds to a zero-framing of each component $\mathcal{L}_v(T_i)$ (i.e. $Lk_i = 0$ for each i-th component), then

$$H(\mathcal{L}_v) = 2\sum_{i,j} Lk_{ij}\,\kappa_i\kappa_j \ . \tag{22}$$

Note that as shown in Figure 9 there are cases of non-trivial zero-framed vortex links with zero (Gauss) linking number Lk_{ij} and thus zero total helicity.

Higher-order linking numbers able to classify topologies otherwise not captured by the standard Gauss linking number (as in the case of the Borromean rings) have been studied by Berger [7]. Since links are close relatives of braids, a hierarchy of linking integrals generates a family of winding numbers for braids. Higher-order helicity integrals for braided fluid structures are based on these invariants.

Figure 9. The Borromean configuration for three zero-framed linked vortex tubes provides an example of zero (Gauss) linking number: the total helicity of this system is zero!

For more information on aspects of helicity and fluid flows see the review article by Moffatt and Tsinober [21] and the book by Arnold and Khesin [4].

6. Evolution of Vortex Knots and Links

6.1. THIN CORED VORTEX KNOTS

Thin vortex knots have been found as solutions to the localized induction approximation (LIA; see §3.2). Remember that this is an approximation of the Biot-Savart law of Euler's equations. Existence and steadiness of knotted solutions to LIA have been studied by Kida [15] and Keener [14]. Kida's solutions are torus knots in the physical space. We have:

Theorem 6.1 ([15]) *Let $\mathcal{K}_v$ denote the embedding of a knotted vortex filament in an ideal fluid in $\mathcal{D}$. If $\mathcal{K}_v$ evolves under LIA, then there exists a class of steady solutions in the shape of torus knots $\mathcal{K}_v \equiv \mathcal{T}_{p,q}$.*

In geometric terms Kida's solutions are closed curves embedded on a mathematical torus Π, wrapping the torus $p > 1$ times in the longitudinal direction and $q > 1$ times in the meridian direction (p, q co-prime integers). The winding number is given by $w = q/p$, and self-linking given by $Lk = pq$, two topological invariants of the knot type. Kida finds torus knot solutions in terms of fully non-linear relationships that involve elliptic functions of traveling waves.

Torus knots have two isotopes $\mathcal{T}_{p,q}$ and $\mathcal{T}_{q,p}$ (for given p and q), that are topologically equivalent but geometrically different. Since vortex filament motion is influenced by the curve geometry (and in particular by curvature), the question of evolution and stability of the two isotopes is particularly interesting. By using linear perturbation techniques and cylindrical polar coordinates (r, α, z) it is possible to obtain 'small-amplitude' torus knot solutions that are asymptotically equivalent to those of Kida.

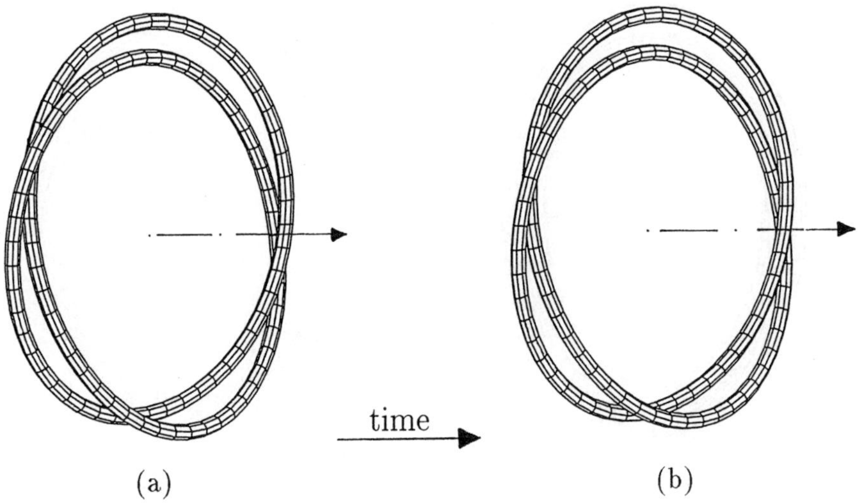

Figure 10. Evolution of torus knot $\mathcal{T}_{2,3}$ under LIA. (a) Initial configuration; (b) configuration at a later time. The knot evolution is stable as predicted by the LIA analysis. The tube shown is merely an aid to visualize vortex evolution [28].

The advantage of this approach is that we can write the solutions in terms of simple trigonometric functions amenable to further analysis. A linear stability analysis based on this approach gives the following result:

Theorem 6.2 ([25]) *Let $\mathcal{T}_{p,q}$ denote the embedding of a 'small-amplitude' vortex torus knot $\mathcal{K}_v$ evolving under LIA. $\mathcal{T}_{p,q}$ is steady and stable under linear perturbations iff $q > p$ ($w > 1$).*

Numerical simulations have been carried out to check and investigate properties of torus knot evolution based on the result of Theorem 6.2 [29]. Under LIA torus knots with winding number $w > 1$ translate and rotate uniformly and steadily in space as rigid bodies. In the case of $w < 1$, however, instabilities develop almost immediately and the knot unfolds towards reconnection events.

Figure 10 shows two snapshots of the stable knot $\mathcal{T}_{2,3}$ and Figure 11 shows the knot $\mathcal{T}_{3,2}$ when it becomes unstable and unfolds. These simulations led to the discovery of a strong stabilizing effect present when the full Biot-Savart law governs the evolution of LIA unstable knots. This is a rather intriguing effect that merits further investigation. A beautiful applied mathematics problem is:

Problem 6.3 Given a vortex filament in the shape of a torus knot $\mathcal{T}_{p,q}$, with circular cross-section and uniform distribution of vorticity, to find the induced velocity in terms of analytical solution of the Biot-Savart integral.

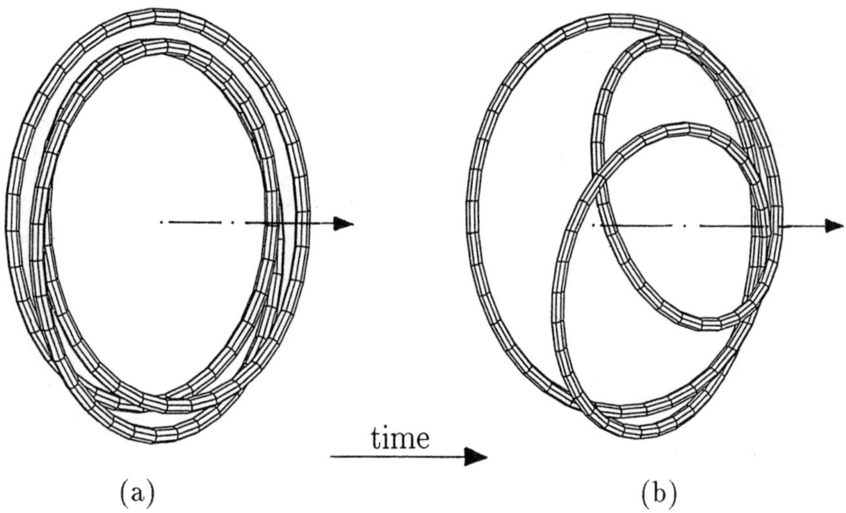

Figure 11. Evolution of torus knot $\mathcal{T}_{3,2}$ under LIA. (a) Initial configuration; (b) configuration at a later time. The knot evolution is unstable according to the LIA analysis and unfolds immediately [28].

Another interesting aspect of current research is represented by possible links between soliton invariants and torus knot solutions. Since LIA and NLSE are related via the Hasimoto map (17), the infinite number of conserved quantities in involution expressed in terms of global geometric functionals (19) are also constants of motion of Kida's torus knot solutions. Finding connections between this family of invariants and the polynomial invariants of knots could represent an important step towards closer links between differential geometry, geometric topology and integrability theory.

6.2. THIN CORED VORTEX LINKS

From a mathematical viewpoint little work is done on vortex links, the only known results being those of J.J. Thomson [33] more than a century ago. Thomson considers particularly symmetric systems of links that travel in steady motion as rigid bodies in the fluid. The simplest case is represented by the Hopf link obtained by embedding two vortex rings Γ_1 and Γ_2 equally spaced on the surface of a mathematical torus Π of radius R and small diameter d. This link system can be realized by the following 'thought construction': choose a meridian plane of Π and place two point vortices (representing the cross-sections of the vortex filaments) on the circumference in diametrically opposite position. Consider now the simultaneous uniform rotation of the two point vortices around the common

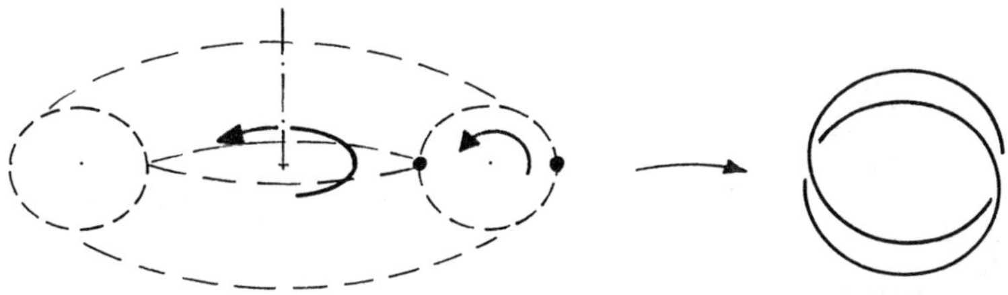

Figure 12. Interpretation of J.J. Thomson's 'thought construction' of a Hopf vortex link system ($n = 2$, $Lk = 1$).

center of mass (i.e. the center of the meridional circumference) *and* around the principal axis of the torus, along the great circle of radius R, in the longitudinal direction (see Figure 12). The vortex link system results from the collection of the two point vortex positions occupied after their full (double) revolution.

Let $\lambda = \max|\mathbf{X}_i - \mathbf{X}_i^*|$, for points $\{\mathbf{X}_i, \mathbf{X}_i^*\} \in \Gamma_i$, $i = 1, 2$, and $\delta = \min|\mathbf{X}_1 - \mathbf{X}_2|$ for points $\mathbf{X}_1 \in \Gamma_1$ and $\mathbf{X}_2 \in \Gamma_2$. Assuming that $\lambda \gg \delta$, where $\lambda = O(R)$ and $\delta = O(d)$, we have:

Theorem 6.4 ([33]) *Consider the Hopf link given by two vortex rings of equal circulation κ and relative linking number Lk, embedded and equally spaced on a torus Π in $\mathcal{D}$. The vortex system is steady and stable iff*

$$\frac{M(2\pi\rho\kappa)^{1/2}}{LkP^{3/2}} < 1 \;, \tag{31}$$

where ρ is the fluid density (constant) and $M = |\mathbf{M}|$ and $P = |\mathbf{P}|$ are the intensities of the angular momentum $\mathbf{M}$, and the linear momentum $\mathbf{P}$ of the system.

This is a remarkable result that combines geometry, topology and fluid dynamics. The simplest link system (with $Lk = 1$) rotates and translates with angular velocity Ω and translational velocity V given by

$$\Omega = \frac{\kappa}{\pi d^2} \;, \quad V = \frac{\kappa}{4\pi R} \log \frac{64R^2}{a^2} \;. \tag{32}$$

Higher-order two-component link systems (i.e. with $Lk > 1$) are obtained by increasing the number of full revolutions of the point vortex system around the great circle (see Figure 13a).

Consider now n vortex rings linked together. As above, the system can be thought of as generated by full revolutions of n point vortices (equally spaced on the small circumference) around the two principal axes of Π (in the longitudinal and meridian direction) to obtain n-vortex components linked in space (see Figure 13b). We have

Figure 13. (a) A higher-order Hopf link with $n = 2$, $Lk = 2$; (b) a link of $n = 3$ components with relative linking $Lk = 1$.

Theorem 6.5 ([33]) *Consider the link of n vortex rings of equal circulation κ and relative linking number Lk, embedded and equally spaced on a torus* Π *in $\mathcal{D}$. The vortex system is steady and stable iff $n \leq 6$.*

This result, applied to a system of n point vortices in the plane, has been refined by other authors (for more information see [30]). For recent numerical works on simulations of vortex link production see [2] and for reconnection and dissipation of a trefoil vortex knot see [16].

7. Algebraic, Geometric and Topological Measures of Flow Complexity

Measuring structural complexity in vortex flows is becoming an important aspect of fluid mechanics research. On one hand progress in geometric and topological fluid mechanics and dynamical system theory give new tools to explore dynamical aspects of fluid flows, including knotting and linking of flow patterns, on the other hand numerical and computational progress make now possible analyses and visualizations (visiometrics) of mechanisms (such as vortex reconnection) with unprecedented details. Many mathematical concepts already developed are available for numerical implementation and new ones are being put forward for application on test cases.

Based on the idea of a *tropicity domain*, given by the 'numerical' domain determined by a computed vortex tangle (see, for example, Figure 14), we can determine principal directional axes (*tropicity axes*) to measure degrees of *tubeness*, *sheetness* and *bulkiness* of the vortex system and relative spread of flow lines (particle paths, streamlines, vortex lines) [27].

For discrete vortex tangles (produced numerically by a simulation or naturally in superfluids) *directional writhing* $Wr(\hat{\boldsymbol{\nu}})$ provides an estimate of the geometric average coiling of projected vortex lines. This measure is

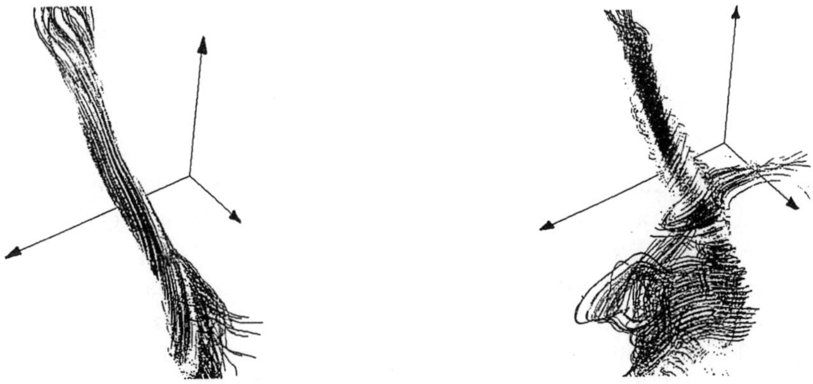

(a) (b)

Figure 14. (a) Computed vortex lines and (b) streamlines of a vortex tube produced by direct numerical simulation of homogeneous, isotropic turbulence [30].

simply given by the writhing number Wr of the oriented graph resulting from the projection of the tangle onto a projection plane of normal $\hat{\boldsymbol{\nu}}$. By keeping track of the orientation of the curve (induced by vorticity) and by assigning the value $\epsilon_r = \pm 1$ to each projected crossing r (according to the standard convention on crossing signs; cf. eq. 30 of §5), we can calculate the directional writhing by

$$Wr(\hat{\boldsymbol{\nu}}) = \sum_r \epsilon_r \ , \qquad Wr = <\sum_r \epsilon_r >_{\mathcal{D}} \ , \tag{33}$$

whereas the writhing number Wr is given by averaging the directional writhe over the whole domain. We can show [5] that this quantity is well aproximated by the *estimated writhing*

$$Wr_\perp = <\sum_r \epsilon_r >_\perp \approx Wr \ , \tag{34}$$

obtained by taking the algebraic mean over the three principal orthogonal planes ($x = y = z = 0$), that gives a simple, direct measure of the average coiling of the tangle in space.

Structural complexity is also measured by counting the total number of apparent crossings present at a given time. This quantity, which is associated with un-oriented tangle diagrams, is defined by the algebraic measure

Definition 7.1 *The* **average crossing number** $\bar{C}$ *is given by the total number of apparent un-signed crossings of the tangle, averaged over the whole domain $\mathcal{D}$. We have*

$$\bar{C} = <\sum_r |\epsilon_r| >_{\mathcal{D}} \ . \tag{35}$$

Once again, it is computationally convenient to approximate this measure by the algebraic mean taken over the three principal orthogonal planes, hence

$$\bar{C}_{\perp} = <\sum_{r} |\epsilon_r|>_{\perp} \approx \bar{C} \ . \tag{36}$$

Recent work done by Barenghi, Ricca & Samuels [5] shows that $\bar{C}_{\perp}$ provides indeed a good approximation to $\bar{C}$ and it seems very effective to detect structural complexity.

Finally, it is of fundamental importance to relate measures of algebraic, geometric and topological complexity to physical properties of the system, such as kinetic helicity and energy. To this end it may be convenient to re-write eq. (28) in compact form. Consider the linking numbers Lk_{ij} ($i,j \in [1,\ldots,n]$; $Lk_{ii} \equiv Lk_i$) as elements of a square matrix ($n \times n$); since $Lk_{ij} = Lk_{ji}$, we can reduce the linking matrix to diagonal form

$$\begin{pmatrix} Lk_{11} & Lk_{12} & \ldots & Lk_{1n} \\ Lk_{21} & Lk_{22} & \ldots & Lk_{2n} \\ \ldots & \ldots & \ldots & \ldots \\ Lk_{n1} & Lk_{n2} & \ldots & Lk_{nn} \end{pmatrix} \rightarrow \begin{pmatrix} M_{11} & 0 & \ldots & 0 \\ 0 & M_{22} & \ldots & 0 \\ \ldots & \ldots & \ldots & \ldots \\ 0 & 0 & \ldots & M_{nn} \end{pmatrix}$$

where each element M_{ii} takes into account self- and mutual linking of the vortex lines. We can therefore re-cast eq. (28) in the form

$$H(\mathcal{L}_v) = \sum_{i=1,\ldots,n} M_{ii} \, f(\kappa) \ , \tag{37}$$

where $f(\kappa)$ is a linear function of quadratic terms of the vortex circulations.

If the tangle is made of vortex filaments of roughly same length L (obtained by an average measure over the tropicity domain $\mathcal{D}$), we can show that on dimensional grounds the enstrophy Θ of the system is given by a relationship of the form

$$\Theta = \int_{\mathcal{D}} |\boldsymbol{\omega}|^2 \, d^3\mathbf{x} = \frac{1}{L} \sum_{i=1,\ldots,n} M_{ii} \, f(\kappa) \ , \tag{38}$$

that provides an interesting connection with helicity. In steady state conditions we can expect to find bounds for minimum enstrophy levels or for other types of 'ground state energies' in relation to the complexity of the physical system.

A combination of algebraic, geometric and topological measures together with kinetic helicity and energy measures provide indeed useful tools to explore complexity and relate flow complexity to energy levels. Work in this direction is in progress.

Acknowledgements

I would like to acknowledge the hospitality of the Isaac Newton Institute for Mathematical Sciences (Cambridge, UK) during the period August–December, 2000. Financial support from UK EPSRC (Grant GR/K99015) is also kindly acknowledged.

References

1. Abraham, R., Marsden, J. & Ratiu, T. (1988) *Manifolds, Tensor Analysis, and Applications.* Springer-Verlag, Berlin.
2. Aref, H. & Zawadzki, I. (1991) Linking of vortex rings. *Nature* **354**, 50–53.
3. Arnold, V.I. (1966) Sur la géométrie différentielle des groupes de Lie de dimension infinie et ses applications à l'hydrodynamique des fluides parfais. *Ann. Inst. Fourier* (Grenoble) **16**, 319–361.
4. Arnold, V.I. & Khesin, B.A. (1998) *Topological Methods in Hydrodynamics.* Applied Mathematical Sciences **125**, Springer-Verlag, New York.
5. Barenghi, C.F., Ricca, R.L. & Samuels, D.C. (2001) How tangled is a tangle? *Physica* D **157**, 197–206.
6. Batchelor, G.K. (1967) *An Introduction to Fluid Mechanics.* Cambridge University Press.
7. Berger, M.A. (1990) Third order link invariants. *J. Phys. A: Math. & Gen.* **23**, 2787–2793.
8. Berger, M.A. & Field, G.B. (1984) The topological properties of magnetic helicity. *J. Fluid Mech.* **147**, 133–148.
9. Ebin, D.G. & Marsden, J.E. (1970) Groups of diffeomorphisms and the motion of an incompressible fluid. *Ann. Math.* **92**, 102–163.
10. Flanders, H. (1963) *Differential Forms with Applications to the Physical Sciences.* Academic Press.
11. Fukumoto Y. & Miyazaki, T. (1991) Three-dimensional distortions of a vortex filament with axial velocity. *J. Fluid Mech.* **222**, 369–416.
12. Hasimoto, H. (1972) A soliton on a vortex filament. *J. Fluid Mech.* **51**, 477–485.
13. Holm, D.D., Marsden, J.E. & Ratiu, T.S. (1998) Euler-Poincaré models of ideal fluids with nonlinear dispersion. *Phys. Rev. Lett.* **80**, 4173–4177.
14. Keener, J.P. (1990) Knotted vortex filaments in an ideal fluid. *J. Fluid Mech.* **211**, 629–651.
15. Kida, S. (1981) A vortex filament moving without change of form. *J. Fluid Mech.* **112**, 397–409.
16. Kida, S. & Takaoka, M. (1988) Reconnection of vortex tubes. *Fluid Dyn. Res.* **3**, 257–261.
17. Langer, J. & Perline, R. (1991) Poisson geometry of the filament equation. *J. Nonlin. Sci.* **1**, 71–93.
18. Moffatt, H.K. (1969) The degree of knottedness of tangled vortex lines. *J. Fluid Mech.* **35**, 117–129.
19. Moffatt, H.K. (1990) The energy spectrum of knots and links. *Nature* **347**, 367–369.
20. Moffatt, H.K. & Ricca, R.L. (1992) Helicity and the Călugăreanu invariant. *Proc. R. Soc. Lond.* A **439**, 411–429.
21. Moffatt, H.K. & Tsinober, A. (1992) Helicity in laminar and turbulent flow. *Annu. Rev. Fluid Mech.* **24**, 281–312.
22. Nakayama, K., Segur, H. & Wadati, M. (1992) Integrability and the motion of curves. *Phys. Rev. Lett.* **69**, 2603–2606.
23. Ricca, R.L. (1991) Intrinsic equations for the kinematics of a classical vortex string in higher dimensions. *Phys. Rev.* A **43**, 4281–4288.

24. Ricca, R.L. (1992) Physical interpretation of certain invariants for vortex filament motion under LIA. *Phys. Fluids.* A **4**, 938–944.
25. Ricca, R.L. (1993) Torus knots and polynomial invariants for a class of soliton equations. *Chaos* **3**, 83–91 [(1995) Erratum. *Chaos* **5**, 346].
26. Ricca, R.L. (1996) The contributions of Da Rios and Levi-Civita to asymptotic potential theory and vortex filament dynamics. *Fluid Dyn. Res.* **18**, 245–268.
27. Ricca, R.L. (2000) Towards a complexity measure theory for vortex tangles. In *Knots in Hellas '98* (ed. McA. Gordon *et al.*), pp. 361–379. Series on Knots and Everything **24**, World Scientific, Singapore.
28. Ricca, R.L. & Berger, M.A. (1996) Topological ideas and fluid mechanics. *Phys. Today* **49** (12), 24–30.
29. Ricca, R.L., Samuels, D.C. & Barenghi, C.F. (1999) Evolution of vortex knots. *J. Fluid Mech.* **391**, 29–44.
30. Saffman, P.G. (1992) *Vortex Dynamics.* Cambridge University Press.
31. She, Z.-S., Jackson, E. & Orszag, S.A. (1990) Intermittent vortex structures in homogeneous isotropic turbulence. *Nature* **344**, 226–228.
32. Shkoller, S. (1998) Geometry and curvature of diffeomorphism groups with H_1 metric and mean hydrodynamics. *J. Funct. Anal.* **160**, 337–365.
33. Thomson, J.J. (1883) *A Treatise on the Motion of Vortex Rings.* Macmillan & Co., London.
34. Tur, A. & Yanovsky, V. (1993) Invariants in dissipationless hydrodynamic media. *J. Fluid Mech.* **248**, 67–106.

TOPOLOGY BOUNDS THE ENERGY

BORIS A. KHESIN
Department of Mathematics
University of Toronto
Toronto, ON M5S 3G3, Canada
E-mail: *khesin@math.toronto.edu*

Abstract. We survey several results related to topological obstructions to energy relaxation of a magnetic field in a perfectly conducting medium.

1. Introduction

In this expository note we survey several results related to topological obstructions to energy relaxation of a magnetic field in a perfectly conducting medium. Many more details and results, as well as a guide to the literature on this topic can be found in [2] or [12].

A motivation for this problem is the following model of a star. The magnetic field is supposed to be *frozen in* the perfectly conducting medium (plasma) filling the star, i.e. the topology of the field's trajectories does not change under the fluid flow. On the other hand, the magnetic energy can and does change, and the conducting fluid keeps moving (due to Maxwell's equations) until the excess of magnetic energy over its possible minimum is fully dissipated (this process is called "energy relaxation"). It turns out that mutual linking of magnetic lines may prevent complete dissipation of the magnetic energy. The problem is to describe the energy lower bounds of the magnetic field in terms of topological characteristics of its trajectories.

Consider a divergence-free (magnetic) vector field ξ in a (simply connected) bounded domain $M \subset \mathbb{R}^3$. The *energy* of the field ξ is the square of its L_2-norm, i.e., the integral $E(\xi) = \int_M ||\xi||^2 \, d^3x$. (This definition makes sense for any compact Riemannian manifold M, possibly with boundary. In this case, $||\xi||^2 = (\xi, \xi)$ is the square of the Riemannian length of a vector,

R.L. Ricca (ed.), An Introduction to the Geometry and Topology of Fluid Flows, 229–238.

d^3x is replaced by the Riemannian volume form, and the field ξ is supposed to be tangent to the boundary of M.)

Let us act on the field ξ by a volume-preserving diffeomorphism h: $\xi_h := h_*\xi$. The main problem is:

Problem 1.1 Given a divergence-free field ξ, to find a good lower bound for the energy $\inf_h E(\xi_h)$ in terms of the topology of the field ξ.

(Here we minimize the energy over the action of all volume-preserving diffeomorphisms h of M.)

Example 1.2 A topological obstruction to the energy relaxation can be seen in the example of a magnetic field confined to two linked solitori. Assume that the field vanishes outside those tubes and the field trajectories are all closed and oriented along the tube axes inside.

To minimize the energy of a vector field with closed orbits by acting on the field by a volume-preserving diffeomorphism, one has to shorten the length of most trajectories. (Indeed, the orbit periods are preserved under the diffeomorphism action; therefore, a reduction of the orbits' lengths shrinks the velocity vectors along the orbits.) In turn, the shortening of the trajectories implies a fattening of the solitori (since the acting diffeomorphisms are volume-preserving).

For a linked configuration, as in Figure 1, the solitori prevent each other from endless fattening and therefore from further shrinking of the orbits. Therefore, heuristically, in the volume-preserving relaxation process the magnetic energy of the field supported on a pair of linked tubes is bounded from below and cannot attain too small values.

2. Helicity Bounds the Energy

To describe the first obstruction to the energy minimization we need the following notion.

Definition 2.1 *The* **helicity** *of the field ξ in the domain $M \subset \mathbb{R}^3$ is*

$$\mathcal{H}(\xi) = \int_M (\xi, \operatorname{curl}^{-1}\xi)\, d^3x,$$

where $(\ ,\ )$ is the Euclidean inner product, and $A = \operatorname{curl}^{-1}\xi$ is a divergence-free vector potential of the field ξ, i.e., $\nabla \times A = \xi$, $\operatorname{div} A = 0$.

One can easily see that the integral is independent of the particular choice of A (which is defined up to addition of the gradient ∇f of a harmonic function, since M is simply connected).

The word "helicity" was coined by K. Moffatt in [8] and it reveals the topological meaning of this characteristic of a vector field (see [9], [12] for the relevant historical surveys).

Theorem 2.2 **([1])** *For a divergence-free vector field ξ,*

$$E(\xi) \geq C \cdot |\mathcal{H}(\xi)|,$$

where C is a positive constant dependent of the shape and size of the compact domain M.

The proof is a composition of the Schwarz inequality and the Poincaré inequality, applied to the potential vector field $A = \mathrm{curl}^{-1}\xi$.

Note that the inverse (nonlocal) operator curl^{-1} sends the space of divergence-free vector fields (tangent to the boundary) on a simply connected manifold onto itself. This operator is symmetric, and its spectrum accumulates at zero on both sides. Thus the constant C can be taken equal to the largest absolute value of the operator curl^{-1} eigenvalues. The corresponding eigenfield has the minimal energy within the class of divergence-free fields obtained from this eigenfield by the action of volume-preserving diffeomorphisms.

Example 2.3 The Hopf field has minimal energy on S^3.

The above statement also holds for an arbitrary closed three-dimensional Riemannian manifold if one confines oneself to divergence-free fields that are *null-homologous*, i.e., have a single-valued divergence-free potential.

Remark 2.4 Note that one can give a metric-free definition of helicity as follows, see [1]. Let M be a simply connected manifold with a volume form μ, and ξ a divergence-free vector field on M. The latter means that the Lie derivative of μ along ξ vanishes: $L_\xi \mu = 0$, or, which is the same, the substitution $i_\xi \mu =: \omega_\xi$ of the field ξ to the volume form μ is a closed 2-form: $d\omega_\xi = 0$. On a simply connected manifold M the latter means that ω_ξ is exact: $\omega_\xi = d\alpha$ for some 1-form α (called a potential). (If M is not simply connected, then we have to require that the field ξ is null-homologous, i.e., that the 2-form ω_ξ is exact. If M is with boundary, we require that ξ is tangent to it.)

Definition 2.5 (2.1 bis) *The* **helicity** *$\mathcal{H}(\xi)$ of a null-homologous field ξ on a three-dimensional manifold M (possibly with boundary) equipped with a volume element μ is the integral of the wedge product of the form $\omega_\xi := i_\xi \mu$ and its potential:*

$$\mathcal{H}(\xi) = \int_M \alpha \wedge d\alpha = \int_M d\alpha \wedge \alpha, \text{ where } d\alpha = \omega_\xi.$$

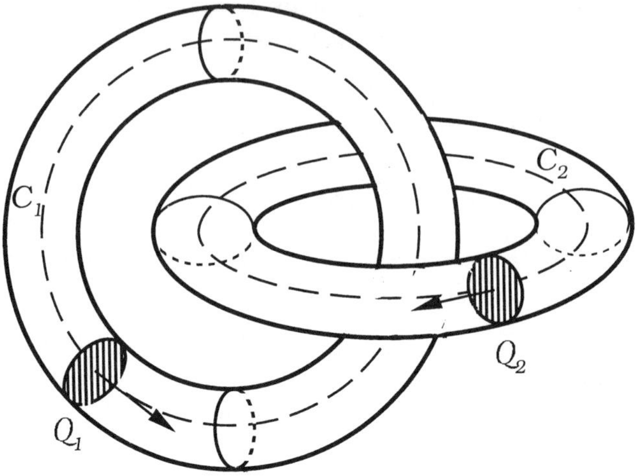

Figure 1. C_1, C_2 are axes of the tubes; Q_1, Q_2 are the corresponding fluxes.

Exercise 2.6 Prove the equivalence of Definitions 2.1 and 2.5 if M is a domain in $\mathbb{R}^3$.

An immediate consequence of this pure topological (metric-free) definition is the following

Theorem 2.7 **([1])** *The helicity $\mathcal{H}(\xi)$ is preserved under the action on ξ of a volume-preserving diffeomorphism of M.*

In this sense $\mathcal{H}(\xi)$ is a topological invariant: it was defined without coordinates or a choice of metric, and hence every volume-preserving diffeomorphism carries a field ξ into a field with the same helicity.

3. What is Helicity?

Example 3.1 (1.2 bis) To get familiar with helicity we first consider a magnetic (divergence-free) field ξ which is identically zero except in two narrow linked flux tubes whose axes are closed curves C_1 and C_2. The magnetic fluxes of the field in the tubes are Q_1 and Q_2 (Figure 1).

Suppose further that there is no net twist within each tube or, more precisely, that the field trajectories foliate each of the tubes into pairwise unlinked circles.

Exercise 3.2 The helicity invariant of such a field is given by

$$\mathcal{H}(\xi) = 2\, lk(C_1, C_2) \cdot Q_1 \cdot Q_2,$$

where $lk(C_1, C_2)$ is the linking number of C_1 and C_2.

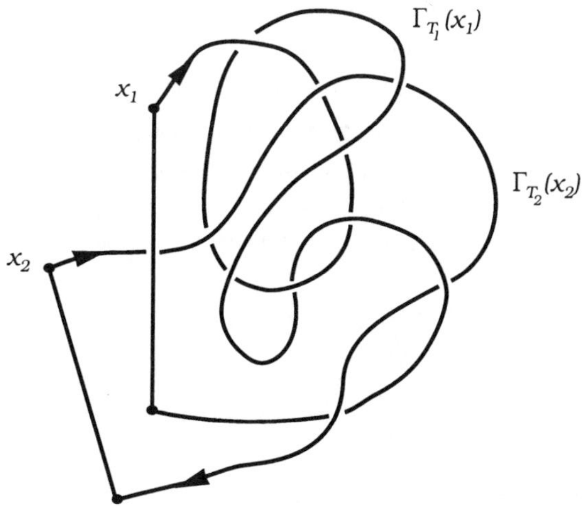

Figure 2. The long segments of the trajectories are closed by the "short paths."

Recall, that the (Gauss) *linking number* $lk(C_1, C_2)$ of two oriented closed curves C_1, C_2 in $\mathbb{R}^3$ is the signed number of the intersection points of one curve with an arbitrary (oriented) surface spanning the other curve.

V. Arnold proposed the following ergodic interpretation of helicity in the general case of any divergence-free field (when the trajectories are not necessarily closed or confined to invariant tori) as the average linking number of the field's trajectories.

Let ξ be a divergence-free field on M and $\{g^t : M \to M\}$ its phase flow. We will associate to each pair of points in M a number that characterizes the "asymptotic linking" of the trajectories of the flow $\{g^t\}$ passing through these points. Given any two points x_1, x_2 in M and two large numbers T_1 and T_2, we consider "long segments" $g^t x_1 (0 \le t \le T_1)$ and $g^t x_2 (0 \le t \le T_2)$ of the trajectories of ξ issuing from x_1 and x_2. Close these long pieces by the shortest geodesics between $g^{T_k} x_k$ and x_k. We obtain two closed curves, $\Gamma_1 = \Gamma_{T_1}(x_1)$ and $\Gamma_2 = \Gamma_{T_2}(x_2)$ (see Figure 2). Assume that these curves do not intersect (which is true for almost all pairs x_1, x_2 and for almost all T_1, T_2). Then the linking number $lk_\xi(x_1, x_2; T_1, T_2) := lk(\Gamma_1, \Gamma_2)$ of the curves Γ_1 and Γ_2 is well-defined.

Definition 3.3 *The* **asymptotic linking number** *[1] of the pair of trajectories $g^t x_1$ and $g^t x_2$ ($x_1, x_2 \in M$) of the field ξ is defined as the limit*

$$\lambda_\xi(x_1, x_2) = \lim_{T_1, T_2 \to \infty} \frac{lk_\xi(x_1, x_2; T_1, T_2)}{T_1 \cdot T_2},$$

where T_1 and T_2 are to vary so that Γ_1 and Γ_2 do not intersect.

It turns out that this limit exists (as an element of the space $L_1(M \times M)$ of the Lebesgue-integrable functions on $M \times M$) and is independent of the system of geodesics (i.e., of the Riemannian metric), see Remark 3.5 below.

Theorem 3.4 ([1]) *The helicity of a divergence-free vector field ξ on a simply connected manifold M with a volume element μ is equal to the average self-linking of trajectories of this field, i.e., to the asymptotic linking number $\lambda_\xi(x_1, x_2)$ of trajectory pairs integrated over $M \times M$:*

$$\mathcal{H}(\xi) = \int_M \int_M \lambda_\xi(x_1, x_2)\ \mu_1 \mu_2.$$

In the case of a manifold M with boundary, all the vector fields involved are supposed to be tangent to the boundary.

Remark 3.5 In the original paper [1], instead of segments of shortest geodesics, one considered systems of "short paths" between every two points of the manifold, which satisfy some conditions to provide the existence of $\lambda_\xi(x_1, x_2)$ almost everywhere as a pointwise limit as $T_1,\ T_2 \to \infty$. Such a system of "short paths" would, generally speaking, depend on a vector field. In [16] T. Vogel suggested to use the L_1-convergence, rather than the pointwise one, and showed that in the L_1-case it is sufficient to use the system of shortest geodesics for any vector field. His approach settled in a universal way the existence question for "short paths" systems. It might also shed some light on the following long-standing problem.

Problem 3.6 [1] Is the average self-linking number of a divergence-free vector field invariant under the action of homeomorphisms preserving the measure on the manifold?

Here, a measure-preserving homeomorphism is supposed to transform the flow of one smooth divergence-free vector field into the flow of the other, both fields having well-defined average self-linking numbers.

The above problem is a counterpart of the homotopy invariance of the classical Hopf invariant for maps $\pi : S^3 \to S^2$. The classical Hopf invariant is always an integer, and it is equal to the linking of the preimages of two generic points in S^2. One can also give an integral definition of the Hopf invariant for such a map. This invariant is equal to the helicity of a vector field tangent to the levels of π (so that all orbits of this field are closed). In the above problem one wonders whether the homotopy invariance can be extended to the fields with not necessarily closed trajectories.

4. Energy Estimates

As we have seen above, a nonzero helicity (or average linking of the trajectories) of a field ξ provides a lower bound for the energy. Note that the heuristic argument mentioned in the Introduction is more general in the following sense. It demonstrates that there exists a lower bound for the energy for a field which has at least one linked pair of solitori as in the example above. However, the helicity of such a field might turn out to be zero, if, e.g., it has another ("mirror") pair of solitori linked in the opposite direction which makes vanish the total averaged self-linking of trajectories of the vector field. This shows that one needs more subtle energy estimates, where, in particular, the contribution of any non-trivially linked "tube of trajectories" into the energy bound could not be canceled out.

Apparently, one of the the best results in this direction is the following.

Theorem 4.1 **([5])** *Suppose a vector field ξ in $I\!R^3$ has an invariant torus T forming a non-trivial knot of type K. Then*

$$E(\xi) \geq \left(\frac{16}{\pi \cdot \mathrm{Vol}(T)}\right)^{1/3} \cdot Q^2 \cdot (2 \cdot \mathrm{genus}(K) - 1),$$

where $Q = \mathrm{Flux}(\xi)$ is the flux of ξ through a cross-section of T, $\mathrm{Vol}(T)$ is the volume of the solid torus, and $\mathrm{genus}(K)$ is the genus of the knot K.

Recall, that for any knot its *genus* is the minimal number of handles of a spanning (oriented) surface for this knot. For an unknot the genus is 0, since one can take a disk as a spanning surface. For a nontrivial knot one has $\mathrm{genus}(K) \geq 1$ and, therefore, the above energy is bounded away from zero: $E(\xi) > 0$.

Remark 4.2 Note that there are no restrictions on the behavior of the divergence-free field inside this invariant torus, and hence this result has a wide range of applicability. In particular, it is sufficient for the field to have at least one closed linked trajectory of the *elliptic* type. The latter means that its Poincaré map has two eigenvalues of modulus 1. Then the KAM theory implies that a generic elliptic orbit is confined to a set of nested invariant tori. Hence any such orbit ensures that the energy of the corresponding field has a non-zero lower bound [7].

Problem 4.3 The question remains whether the presence of any non-trivially linked closed trajectory (of any type: hyperbolic, non-generic, etc.) or the presence of chaotic behavior of trajectories for a field could provide a positive lower bound for the energy (even if the averaged linking of all trajectories totals zero) and therefore could prevent a relaxation of the field to arbitrarily small energies.

Remark 4.4 In [5] one can find an extension of Theorem 4.1 from knots to links, i.e., explicit energy estimates for the fields having nontrivially linked invariant tubes. The paper also contains many sharper results in terms of asymptotic crossing numbers for knots, links, and field trajectories, as well as in terms of conformal modulus of solid tori. (We refer to [5] or [2] for necessary definitions.)

One should mention, that in certain cases one can give somewhat better estimates by assuming more from the vector field. In particular, following [6] we suppose that a vector field ξ not only has an invariant torus confining a knot (or a link) K but it is *strongly modeled* on K, cf. Example 3.1. The latter means that a tubular neighborhood T of K can be identified in a volume-preserving way with (possibly, several copies of) the product $D^2 \times S^1$ of the disk and circle, and this identification can be carried in such a way that the field becomes $\partial/\partial\phi$, i.e., constant and oriented along the circles S^1 in the product.

Theorem 4.5 ([6]) *The energy of a vector field ξ strongly modeled on a nontrivial knot (or on a nontrivial indecomposable link) K in $I\!R^3$ satisfies the inequality*

$$E(\xi) > \left(\sqrt{6/125}/\pi^2\right)^{4/3} \cdot \mathrm{Vol}(T)^{5/3} \approx 0.00624 \cdot \mathrm{Vol}(T)^{5/3}.$$

where $\mathrm{Vol}(T)$ *is the volume of the neighborhood, where we modeled the field.*

Note that for a field, strongly modeled in a tube T, the flux Q is proportional to the area of the crossection of T, and hence to its volume $\mathrm{Vol}(T)$, since the length of S^1 is fixed. This shows consistency of Theorems 4.1 and 4.5: both right-hand-sides have the order of $\mathrm{Vol}(T)^2/\mathrm{Vol}(T)^{1/3} = \mathrm{Vol}(T)^{5/3}$.

Remark 4.6 The rotation field in the three-dimensional ball is an example of an opposite type: all its trajectories are *pairwise unlinked.* It was suggested by A. Sakharov and Ya. Zeldovich, and proved by M. Freedman (see [4]), that this field can be transformed by a volume-preserving diffeomorphism to a field whose energy is less than any given ϵ.

The idea of Freedman's proof is to split the ball into two parts. By stretching the main part of the ball into a long snake and putting it back, one shrinks most of the field's orbits, and hence makes the corresponding energy arbitrarily small. After that one controls the energy gain on the small part, needed to make the first operation smooth. A very similar idea works in Shnirelman's proof of the finiteness of the diameter of the group of volume-preserving diffeomorphisms of a three-dimensional cube (see [13], [15]). It also consists of two steps: the first one is a combinatorial problem for small cubes (quite different from the above), while the second step is a (very similar to Freedman's) control of the stretching in the layer between the small cubes.

A recent very powerful techniques of *generalized flows* and their approximations by the regular flows provided a natural framework for existence theorems in ideal hydrodynamics, as well as they allowed one to give a much simpler estimates for the diameter and other geometric characteristics of the group of volume-preserving diffeomorphisms [3], [14], [15]. The notion of a generalized flow (or, a polymorphism) means that every fluid particle does not necessarily have a unique position at any moment, but moves simultaneously to every other point of the media with certain probability.

Problem 4.7 It would be very interesting to apply the techniques of generalized flows to energy estimates discussed above.

A dream would be to define an action of a "volume-preserving polymorphism" on a divergence-free vector field and to estimate the minimal energy of its image under the action of all such polymorphisms. Then one could hope to use an approximation-type theorem for polymorphisms to show that the energy infimum of a vector field acted upon by diffeomorphisms coincides with the energy minimum for the action of polymorphisms.

Acknowledgements

This paper is an expanded version of the talk given by the author at the Pedagogical Workshop on Topological Fluid Dynamics in Cambridge, UK, in September 2000. I am grateful to the organizers and the Isaac Newton Institute for Mathematical Sciences in Cambridge for hospitality and stimulating atmosphere.

The present work was partially sponsored by PREA and McLean awards, by an Alfred P. Sloan Research Fellowship, and by the NSF and NSERC research grants.

References

1. Arnold, V.I. (1973) The asymptotic Hopf invariant and its applications. *Proc. Summer School in Diff. Equations at Dilizhan,* Erevan (in Russian); English transl.: *Sel. Math. Sov.* **5** (1986), 327–345.
2. Arnold, V.I. & Khesin, B.A. (1998) *Topological Methods in Hydrodynamics.* Applied Mathematical Sciences **125**, Springer-Verlag, New York.
3. Brenier, Y. (1989) The least action principle and the related concept of generalized flows for incompressible perfect fluids. *J. Amer. Math. Soc.* **2**(2), 225–255.
4. Freedman, M.H. (1999) Zeldovich's neutron star and the prediction of magnetic froth. In *Proceedings of the Arnoldfest* (ed. E.Bierstone, *et al.*), pp. 165–172. Fields Institute Communications **24**, Toronto.
5. Freedman, M.H. & He, Z.-X. (1991) Divergence-free fields: energy and asymptotic crossing number. *Annals of Math.* **134**(1), 189–229.
6. Freedman, M.H. & He, Z.-X. (1991) Links of tori and the energy of incompressible flows. *Topology* **30**(2), 283–287.
7. Kozlovsky, O. (2001) Personal communication.
8. Moffatt, H.K. (1969) The degree of knottedness of tangled vortex lines. *J. Fluid. Mech.* **35**, 117–129.

9. Moffatt, H.K. (1981) Some developments in the theory of turbulence. *J. Fluid Mech.* **106**, 27–47.
10. Moffatt, H.K. (1985) Magnetostatic equilibria and analogous Euler flows of arbitrarily complex topology. *Part 1: J. Fluid Mech.* **159**, 359–378;
11. Moffatt, H.K. (1986) Magnetostatic equilibria and analogous Euler flows of arbitrarily complex topology. *Part 2: J. Fluid Mech.* **166**, 359–378.
12. Moffatt, H.K. and Tsinober, A. (1992) Helicity in laminar and turbulent flow. *Annual Review of Fluid Mechanics* **24**, 281–312.
13. Shnirelman, A. (1985) The geometry of the group of diffeomorphisms and the dynamics of an ideal incompressible fluid. *Matem. Sbornik* **128**(1), 82–109. [Also in English translation: (1987) *Math. USSR, Sbornik* **56**(1), 79–105.]
14. Shnirelman, A. (1994) Generalized fluid flows, their approximation and applications. *Geom. and Funct. Analysis* **4:5**, 586–620.
15. Shnirelman, A. (2001) Diffeomorphisms, braids and flows. In *An Introduction to the Geometry and Topology of Fluid Flows* (ed. R.L. Ricca), this volume. NATO-ASI Series: Mathematics, Kluwer, Dordrecht, The Netherlands.
16. Vogel, T. (2000) On the asymptotic linking number. *Preprint.* München University.

MEASURES OF TOPOLOGICAL STRUCTURE IN MAGNETIC FIELDS

MITCHELL A. BERGER
Department of Mathematics
University College London
Gower Street, London WC1E 6BT, UK

E-mail: *mberger@ucl.ac.uk*

Abstract. Observers and theorists describing magnetic field structures often use terms such as twisted, kinked, sheared and braided. All these aspects of field structure can be quantified using topological invariants. While topological quantities obey conservation laws in systems with no resistivity and simple boundary conditions, in more general circumstances they can change in time as the physical system evolves. Topological structure is often thought of as a global property of a magnetic system. However, some aspects of structure can be examined in sub-volumes of space. This allows us to examine transport of structure from one region of space to another. We illustrate by examining the transport of magnetic helicity through parts of the sun and heliosphere during the solar cycle.

1. Introduction

Many magnetic structures seen in nature are quiescent or slowly moving. For example, the solar atmosphere contains clouds of hot plasma called prominences which can live for months and arches of very hot plasma called X-ray loops which live for days or weeks (see Figure 1). In contrast, a dynamical timescale for such structures may be found from the time it takes for a wave to go from one end to another. For an X-ray loop of length 10^5 Km an Alfvén wave traveling at 3×10^3 Km/s will traverse the length in less than a minute. In human technology, we often like to have quiescent magnetic structures as well. For example, fusion devices require a strong magnetic field to contain hot plasma. This field should be as stable as

R.L. Ricca (ed.), An Introduction to the Geometry and Topology of Fluid Flows, 239–252.

Figure 1. An arcade of hot loops (at some 10^6 °K) in the solar corona, as seen by the TRACE mission. The loops are thought to trace out lines of magnetic force.

possible in order to prevent uncontrolled motion of the plasma.

Long-lived structures seen in nature or the laboratory must exist close to a stable equilibrium state. Now equilibria are (local or global) minima of the magnetic energy for given constraints. Boundary conditions supply some of these constraints. The remaining constraints are generally topological in nature.

The boundary constraints on a volume $\mathcal{V}$ tell us how field lines pass through the boundary S. For example, most magnetic fusion devices have toroidal metal walls providing a boundary. This boundary is meant to impose the condition $\mathbf{B} \cdot \hat{\mathbf{n}}|_S = 0$ where $\mathbf{B}$ is the magnetic field and $\hat{\mathbf{n}}$ is the normal to S. This boundary condition say that no field lines cross S. In the solar atmosphere (corona), on the other hand, field lines enter through the solar surface (photosphere) and may exit into interplanetary space. The plasma at the photosphere is far denser than in the atmosphere. Magnetic forces dominate the atmosphere while fluid forces dominate below. The influence of the coronal field upon the sub-photospheric field is small. Thus the distribution of magnetic flux B_n at the photosphere is determined much more by conditions below than conditions above. Here $B_n = B_n(\theta, \phi, t)$ is the normal component at position (θ, ϕ) and is a slowly varying function of time, with time-scales of minutes for small concentrations of flux to hours or longer for larger concentrations. Closed field structures form loops, with the field returning to the photosphere, while open field lines extend into the solar wind. The open field lines must eventually return, of course, as the net flux through the photosphere is zero in the absence of monopoles;

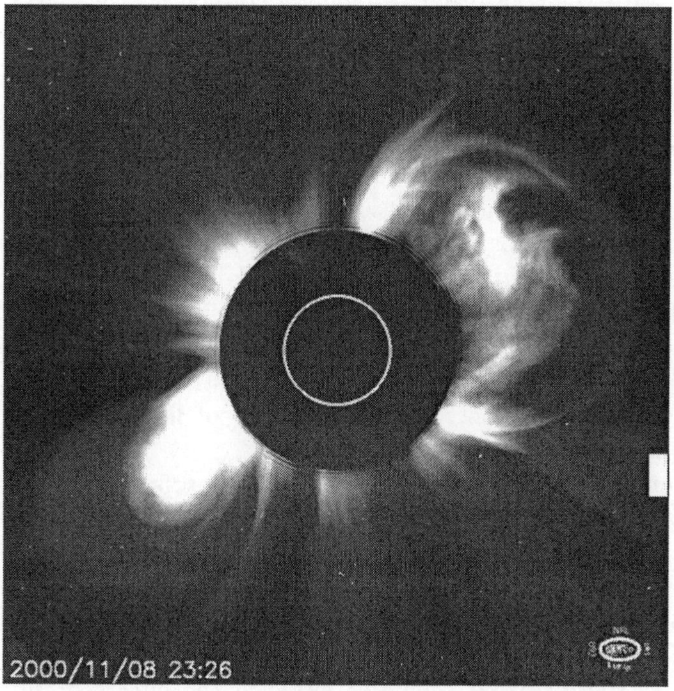

Figure 2. A coronal mass ejection. Image courtesy of SOHO/LASCO consortium.

but not after first visiting distant stars and galaxies.

There are also outer boundary conditions on solar magnetic structures. Magnetic pressure from the open fields provide constraints on the closed field structures in the form of a free boundary surface. Magnetic fields reduce their energy upon expanding, so without this confinement many coronal structures would simply blow out into space. And in fact they do. Once or twice a day on average vast coronal structures overcome outer constraints and eject plasma into the solar wind [11], [16] (see Figure 2).

After specifying boundary conditions, we need an equilibrium equation. For the situation where magnetic forces dominate, the Lorentz force must vanish, i.e. $\mathbf{J} \times \mathbf{B} = \mathbf{0}$ where $\mathbf{J}$ is the electric current density, or more simply $(\nabla \times \mathbf{B}) \times \mathbf{B} = 0$. If we include gas pressure then $(\nabla \times \mathbf{B}) \times \mathbf{B} = \nabla p$. However, the equation and boundary conditions still do not determine the equilibrium field. To pin down which field is correct we also need to specify the topology of the field lines [1].

By topology we mean the details of how positive flux ($B_n > 0$) at the boundary connects to negative flux, as well as how all the field lines wrap about each other. Each field line topology determines a different minimum energy state. Thus in order to model a coronal field we need to have some idea about how to specify the field topology. A simple way of doing this is to find topological quantities that describe characteristic features of a field, and measure those quantities for typical coronal structures. Models can then be made with similar amounts of topological structure. Section 2 reviews aspects of structure such as twist, shear, and kinking which can be measured by the magnetic helicity integral. Section 3 will then discuss measures such as crossing number appropriate to more chaotic fields.

The field topology is itself determined by how the equilibrium field is created and maintained. Solar fields are generated deep within the convection zone before rising to the surface. The emerging field may already contain helicity or some more complicated structural property. Additional structure may be injected by the motion of field line end-points at the photosphere.

As emphasized by Parker [14], for most topologies the minimum energy state is not actually a smooth field configuration. Such states instead possess (in the limit of zero resistivity) sheets of infinite electric current density. These current sheets are sites where the magnetic field direction changes discontinuously. As there always is some small electric resistivity present, the current sheets in practice will be current layers of finite thickness. High current densities enhance the effects of resistivity. Thus the presence of current layers can sometimes lead to violent events (flares), where the magnetic field lines reconnect with each other [15]. This converts magnetic energy directly to heat and particle acceleration. In addition, the field line topology changes once the field lines have been cut and reconnected. With a new topology comes a new minimum energy state. If the new minimum is significantly lower than the old minimum, then the field and plasma may thrash around until the new minimum is reached. In this case more energy will be released due to viscosity dissipating the kinetic energy of the motion.

2. Magnetic Helicity

2.1. CLOSED VOLUMES

The magnetic helicity integral is based on the famous linking number introduced by Gauss in the early 19th century. Consider two curves, 1 and 2, where σ parametrises curve 1, τ parametrises curve 2, $\mathbf{x}(\sigma)$ and $\mathbf{y}(\tau)$ are points on curves 1 and 2, and $\mathbf{r} = \mathbf{x} - \mathbf{y}$ is the relative position vector.

Gauss showed that the number of times two curves link can be measured by the integral

$$L_{12} = -\frac{1}{4\pi} \oint_1 \oint_2 \frac{d\mathbf{x}}{d\sigma} \cdot \frac{\mathbf{r}}{\mathbf{r}^3} \times \frac{d\mathbf{y}}{d\tau}\, d\sigma\, d\tau \,. \tag{1}$$

Figure 3. Two tubes in the shape of the Whitehead link. The Gauss linking number is actually 0, as there are two positive and two negative crossings between the two tubes. The central crossing contributes to the writhe of the figure 8 tube, but does not add to the linking number with the other tube.

In practice this integral is time-consuming to evaluate (although, as it always gives an integer a numerical evaluation does not need to be very precise). Fortunately one can evaluate the integral simply by projecting the curves onto a plane and counting the number of crossings (see Figure 3). One can prove that the integral equals twice the number of positive crossings minus the number of negative crossings. The Gauss linking integral is invariant to deformations of the two curves, as long as the two curves do not cross each other.

Similar integrals can be devised for continuous fields. Inside a volume $\mathcal{V}$ bounded by a magnetic surface ($\mathbf{B} \cdot \hat{\mathbf{n}}|_S = 0$) let [13]

$$H = -\frac{1}{4\pi} \int_{\mathcal{V}} \int_{\mathcal{V}} \mathbf{B}(\mathbf{x}) \cdot \frac{\mathbf{r}}{\mathbf{r}^3} \times \mathbf{B}(\mathbf{y}) \, d^3\mathbf{x} \, d^3\mathbf{y} \ . \tag{2}$$

The helicity H can be interpreted as the sum of linking numbers of all pairs of magnetic field lines, weighted by magnetic flux (compare the structure of Equation (1) and Equation (2)). Of course, field lines may not all close upon themselves; however, any field can be closely approximated as a collection of small closed flux tubes for which this interpretation makes sense [2].

While H is defined as a double integral, it can be simplified by employing the Coulomb gauge (Biot-Savart) vector potential

$$\mathbf{A}(\mathbf{x}) = -\frac{1}{4\pi} \int_{\mathcal{V}} \frac{\mathbf{r}}{\mathbf{r}^3} \times \mathbf{B}(\mathbf{y}) \, d^3\mathbf{y} \ ; \tag{3}$$

$$H_{\mathcal{V}} = \int_{\mathcal{V}} \mathbf{A} \cdot \mathbf{B} \, d^3\mathbf{x} \ . \tag{4}$$

Gauge transformations of the form $\mathbf{A} \rightarrow \mathbf{A} + \nabla\phi$ will leave H fixed, as long as $\mathbf{B} \cdot \hat{\mathbf{n}}|_S = 0$.

2.2. OPEN VOLUMES

Many interesting topological objects involve curves with fixed end-points on a boundary, for example tangles and braids. The field line structure within a volume not bounded by a magnetic surface can be describable in terms of such objects. We will need to define the helicity carefully, however, both to retain topological meaning and to avoid gauge ambiguity. We can do this within a volume $\mathcal{V}$ by measuring the helicity relative to a base configuration. In particular, given a field $\mathbf{B}_{\text{total}}$ defined in all space, write $\mathbf{B}_{\text{total}} = (\mathbf{B}, \mathbf{B}')$ where $\mathbf{B}$ is the field inside $\mathcal{V}$ and $\mathbf{B}'$ is the field outside. Unfortunately, $H(\mathbf{B}_{\text{total}}) = H(\mathbf{B}, \mathbf{B}')$ includes information about all the helical structure in $\mathbf{B}'$. We need to subtract this extra information. To do this, we suppose there exists some simple vector field $\mathbf{P}$ inside $\mathcal{V}$ for which we can calculate the helicity $H(\mathbf{P}, \mathbf{B}')$. Once we subtract this reference helicity, the dependence on the external field will vanish [3].

Now, if we are sitting inside $\mathcal{V}$, the only thing we know about the external field $\mathbf{B}'$ is the boundary information $\mathbf{B} \cdot \hat{\mathbf{n}}|_S$. This tells us the distribution of flux crossing the boundary S. The boundary information determines a unique vector field, the vacuum (or potential) field $\mathbf{P}$ satisfying

$$\mathbf{P} \cdot \hat{\mathbf{n}}|_S = \mathbf{B} \cdot \hat{\mathbf{n}}|_S \; ; \tag{5}$$

$$\nabla \times \mathbf{P} = 0 \, . \tag{6}$$

Thus $\mathbf{P}$ shares the same boundary information as $\mathbf{B}$. (If $\mathcal{V}$ is multiply connected the net flux through any non-contractible closed curve on S should also be the same for $\mathbf{B}$ and $\mathbf{P}$.) A simple variational calculation will show that the vacuum field is the minimum energy state consistent with the boundary data.

Thus the vacuum field has zero current, minimum helical structure (recall that magnetic field lines spiral about current lines), and minimum energy. It also requires a minimum of information for its specification. These qualities make it the ideal choice for the base configuration. We thus define

$$H_{\mathcal{V}} = H(\mathbf{B}, \mathbf{B}') - H(\mathbf{P}, \mathbf{B}') \, . \tag{7}$$

Because each H is integrated over all space, $H_{\mathcal{V}}$ is gauge-invariant. This definition can be expressed as an integral of $\mathcal{V}$ alone [10]:

$$H_{\mathcal{V}} = \int_{\mathcal{V}} (\mathbf{A} + \mathbf{A_P}) \cdot (\mathbf{B} - \mathbf{P}) \, d^3\mathbf{x} \, . \tag{8}$$

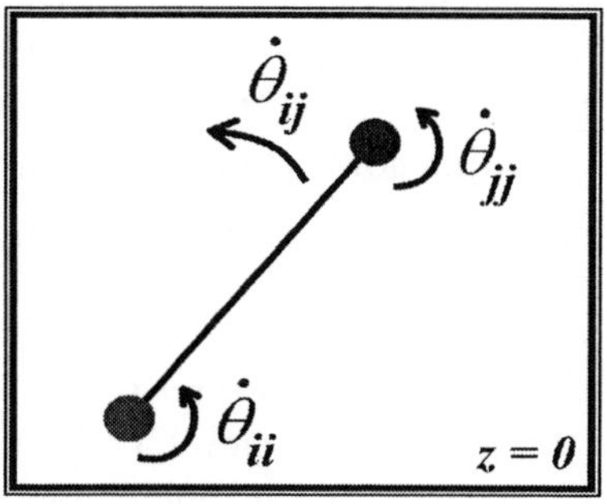

Figure 4. Two flux elements at the boundary $z = 0$. Fluid motions cause the elements to rotate and move around each other, causing the flux tubes above the boundary to twist and braid.

3. Time Evolution of Magnetic Helicity

The interpretation of $H_{\mathcal{V}}$ in terms of field morphology will be discussed in Section 4. First we describe the time evolution of $H_{\mathcal{V}}$. Suppose the plasma inside $\mathcal{V}$ has magnetic diffusivity η and a velocity field $\mathbf{V}$. Then one can derive a Poynting theorem [3], [4]

$$\frac{dH_{\mathcal{V}}}{dt} = -2\int_{\mathcal{V}} \eta \mathbf{J}\cdot\mathbf{B}\, d^3\mathbf{x} + 2\oint_{\partial\mathcal{V}} [(\mathbf{A_P}\cdot\mathbf{V})\mathbf{B} - (\mathbf{A_P}\cdot\mathbf{B})\mathbf{V}]\cdot\hat{\mathbf{n}}\, d^2\mathbf{x}\ , \quad (9)$$

where $\mathbf{A_P}$ (uniquely) satisfies

$$\hat{\mathbf{n}}\cdot\nabla\times\mathbf{A_P} = B_n\ , \quad (10)$$
$$\nabla\cdot\mathbf{A_P} = 0\ , \quad (11)$$
$$\mathbf{A_P}\cdot\hat{\mathbf{n}} = 0\ . \quad (12)$$

The volume integral in Equation (9) represents helicity dissipation. Helicity dissipation obeys strict inequalities [4], [5], and can be considered negligible in a high magnetic Reynolds number plasma. In particular, consider an arbitrary reconnection or dissipation process occuring over a time Δt which dissipates an amount of helicity H. Let E be the initial energy and

$$\tau_d = \frac{H^2}{\eta E^2} \quad (13)$$

be a resistive diffusion time. Then one can show that

$$\left|\frac{\Delta H}{H}\right| \le \sqrt{\frac{\Delta t}{\tau_d}}\ . \quad (14)$$

For any fast reconnection event, where $\Delta t \ll \tau_d$ the inequality shows that ΔH will be negligible . For a solar flare with $\Delta t \sim 1000$ s, $H/E \sim 10^3$ Km, and $\eta \sim 10^{-6}$ Km^{2}s^{-1}, we obtain $\tau_d \sim 10^{12}$ s, and $|\Delta H/H| < 3 \times 10^{-5}$.

The surface integral in Equation (9) measures the twisting of field lines in $\mathcal{V}$ due to motions on the boundary (first term), and bulk transport of helical structure through the boundary (second term). Note in particular that the second term involves $-(\mathbf{A_P} \cdot \mathbf{B})V_n$, i.e. it looks like the normal velocity V_n carrying with it a quantity $(\mathbf{A_P} \cdot \mathbf{B})$. In fact, this term can be interpreted another way. Consider the point of intersection of a field line with the boundary. If V_n transports the field line through the boundary, then that intersection point will move (unless the field line is itself normal to the boundary). The intersection point moves with an effective velocity

$$\mathbf{V}_{\text{eff}} = -\frac{V_n}{B_n}\mathbf{B}_{||} \tag{15}$$

where $\mathbf{B}_{||}$ is the part of $\mathbf{B}$ parallel to the boundary. Thus $(-\mathbf{A_P} \cdot \mathbf{B})V_n$ can be re-written $(\mathbf{A_P} \cdot \mathbf{V}_{\text{eff}})B_n$ just like the first term in the boundary integral. In other words, motions through the boundary generate effective field line motions parallel to the boundary which have the same effect on the helicity as true parallel motions.

To illustrate, let $\mathcal{V}$ be the upper half space $z > 0$. Suppose the magnetic field at $z = 0$ consists of n small discrete flux elements, with flux Φ_i, $i = 1, \ldots, n$. Consider a horizontal velocity field at $z = 0$ which moves the flux elements about each other. Let $\dot{\theta}_{ij} = d\theta_{ij}/dt$ be the angular velocity of element i about element j. Also let $\dot{\theta}_{ii} = d\theta_{ii}/dt$ be the rotational velocity of element i (see Figure 4). From Equation (9), ignoring dissipation [4]

$$\frac{dH_{\mathcal{V}}}{dt} = -\frac{1}{2\pi}\sum_{i=1}^{n}\sum_{j=1}^{n}\dot{\theta}_{ij}\Phi_i\Phi_j \; . \tag{16}$$

The $\dot{\theta}_{ii}$ terms correspond to internal twisting of field lines in the flux tubes above the plane, while the $\dot{\theta}_{ij}$ terms measure braiding of different tubes.

Similar expressions can be obtained for spherical boundaries. In the above equation, θ_{ij} gives the angle of the position vector from i to j. If i and j now refer to the positions of flux elements on a sphere, then θ_{ij} becomes spherical angle between i and j. (For example if i were placed at the North pole, then θ_{ij} would be the azimuthal or longitudinal angle at the position of j.)

4. Field Morphology

If we know how magnetic helicity relates to the field morphology we can avoid long computations involving vector potentials. This helps not only for analysing theoretical models, but also for observations, where magnetic fields and potentials are not directly obtained. What we do get from the telescope images is the shapes of filaments and loops.

Suppose we have an image of a collection of closed loops in the solar corona. We first need to deduce the positions of end-points in the photosphere, and which positive end-points ($\mathbf{B}\cdot\hat{\mathbf{n}}|_S > 0$) connect to which negative end-points. For a collection of flux elements that are not entangled (i.e. they do not cross each other as seen from above), this information will be sufficient to determine the helicity. If any two flux elements do cross each other, we will need to determine or guess which loop is on top.

In analogy with the terminology for inductance, we can call the internal twist of the field lines in a loop 'self helicity' [18]. The contribution to the helicity from the linking and shear between loops will be called 'mutual helicity'. In practice, mutual helicity is much more easy to observe than self helicity.

First consider two coronal loops 1 and 2 with fluxes Φ_1, Φ_2. Assume that the loops are uniformly twisted through T_1 and T_2 turns (e.g. a field line rotates about the axis of loop 1 T_1 times). The self-helicities due to their twist are

$$H_{11} = T_1\Phi_1^2; \qquad H_{22} = T_2\Phi_2^2 \,. \tag{17}$$

However, the two loops will also share a mutual helicity if they cross over each other or are mis-aligned. Summing the self and mutual helicities gives the total helicity

$$H = H_{11} + H_{22} + 2H_{12} \,. \tag{18}$$

For simplicity, let the corona be the upper half space $\{z > 0\}$. The photospheric ($z = 0$) ends of the loops are assumed to be small and located at the points 1^+, 1^-, 2^+, and 2^-. Here $B_z > 0$ at 1^+ and 2^+. As seen from above, either the two loops cross each other or do not cross (see figure 5). If they do cross, we assume that loop 1 is the upper loop.

We can use Equation (16) to determine the mutual helicity of loops 1 and 2. First, note that the magnetic helicity of a structure does not depend on the history of how that structure was formed. Thus if we imagine a simple history of formation for which integrating Equation (16) is easy, then we can find expressions for H_V. One uses a similar procedure for calculating the electric potential of a pair of point charges by bringing in one of the point charges from infinity.

Consider the quadrilateral $1_+2_+1_-2_-$ between the endpoints of the loops (at their final positions). Suppose the upper loop 1 emerges first from beneath the photosphere. At this stage (as there is no loop 2) $H_{12} = 0$. Next loop 2 emerges, at first appearing like a dipole, with endpoints 2^+

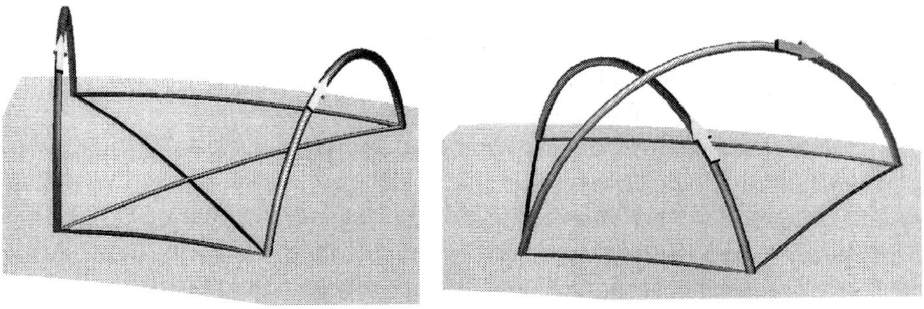

Figure 5. Two un-crossed loops and two crossed loops.

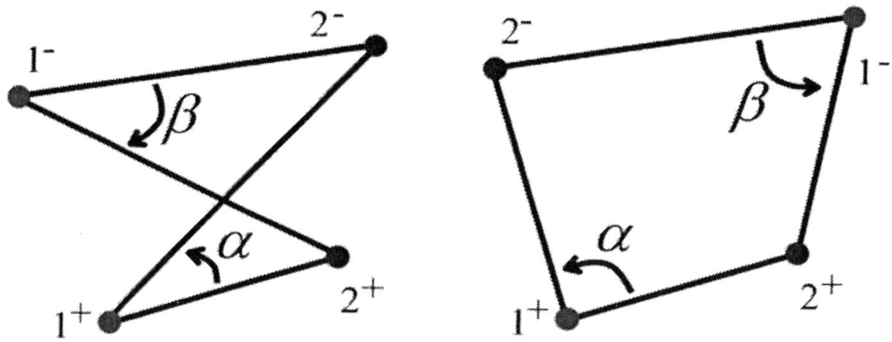

Figure 6. Definition of the angles in Equation (19).

and 2^- next to each other. These two endpoints then separate and move to their final positions. During this separation the line segments between endpoints (e.g. between 1_+ and 2_-) rotate. Let α and β be the (final) angles at vertices 1_+ and 1_- respectively (see Figure 6). Integrating Equation (16) over the emergence of loop 2 from initial dipole to final positions gives

$$H_{12} = H_{21} = \frac{\Phi_1 \Phi_2}{2\pi}(\alpha + \beta) \ . \tag{19}$$

Next consider N loops. In this case there are N self helicities and $N(N-1)$ mutual helicities:

$$H = \sum_{i=1}^{N} H_{ii} + \sum_{i=1}^{N} \sum_{\substack{j=1 \\ j \neq i}}^{N} H_{ij} \ . \tag{20}$$

Thus the total helicity H equals the sum of the entries in a matrix H_{ij}. If N is large then there will be many more mutual helicity terms. In this case ignoring the self helicities (if they are difficult to observe) may only give a small error.

Suppose our N loops can be divided into, say, three families – for example family A might be large loops overlying a filament, family B smaller loops with positive foot points near the neutral line, and family C smaller loops with negative foot points near the neutral line. The matrix H_{ij} can then be partitioned:

$$H_{ij} = \begin{pmatrix} H_{AA} & H_{AB} & H_{AC} \\ H_{BA} & H_{BB} & H_{BC} \\ H_{CA} & H_{CB} & H_{CC} \end{pmatrix} . \tag{21}$$

Here H_{AA} is the self-helicity of family A; for example, if loops 1–4 belong to family A then

$$H_{AA} = \sum_{i=1}^{4}\sum_{j=1}^{4} H_{ij} . \tag{22}$$

Similarly, H_{AB} is the helicity matrix containing all the mutual helicities between A loops and B loops.Thus if family B contains loops 5–10 then

$$H_{AB} = \sum_{i=1}^{4}\sum_{j=5}^{10} H_{ij} . \tag{23}$$

5. Applications to Solar Physics

We can apply magnetic helicity to solar physics in several ways (see [8] for extended discussions). First, one can attempt to determine the magnetic helicity of coronal structures such as prominences. Understanding the helicity of prominences and their overlying fields may provide clues as to how the prominence magnetic field formed, what the lifetime of the prominence might be, and how the prominence filament might behave during a coronal mass ejection [12]. Furthermore, conservation of magnetic helicity during reconnection can constrain the possible final states of the reconnection process. For example, mutual helicity might be converted to self-helicity, adding to the twist of the flux tubes [3] and perhaps initiating a kink instability.

We can also consider the magnetic helicity of the Sun on large scales. The rotation of the Sun sends out waves of twist into the solar wind. The corresponding helicity flow is negative in the North and positive in the South. The magnitude of the helicity flow can be estimated and compared to observations (Figure 7) [6], [7].

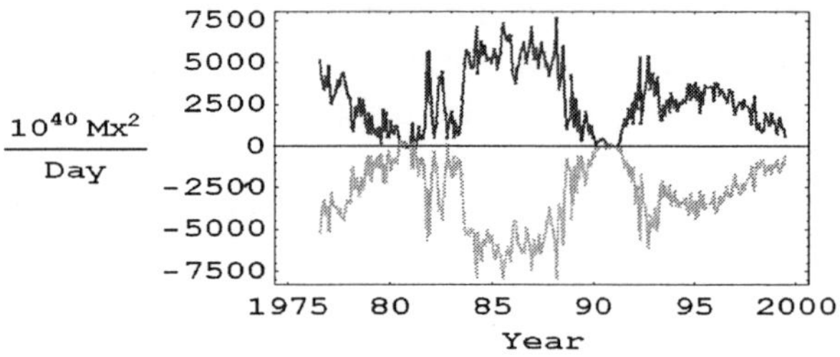

Figure 7. Helicity transfer into the Southern solar wind (predominantly positive curve) and into the Northern solar wind (predominantly negative curve) over a solar cycle.

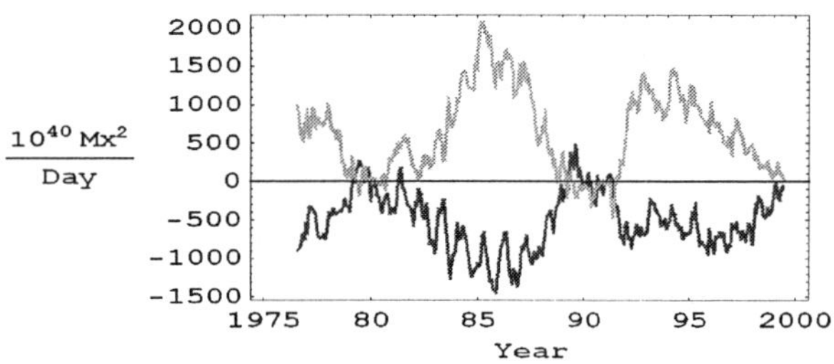

Figure 8. Helicity transfer dH_S/dt into the Southern interior of the Sun (predominantly positive curve) and dH_N/dt into the Northern interior (predominantly negative curve).

On somewhat smaller scales, numerous observations (see [8] for details) suggest that the helicity of coronal structures as well as magnetic clouds in the solar wind [16] also tend to have negative helicity in the North, and positive helicity in the South. We need to discover how each hemisphere obtains its characteristic sign of helicity. Also, if we can go further than the sign, and determine the magnitude of helicity storage and flow through each hemisphere, then we will have a valuable tool for studying the solar magnetic cycle and its underlying dynamo.

Let $\mathcal{V}_N$ be the Northern half of the solar interior. Its helicity $H_{\mathcal{V}N}$ is presumably negative, as the structures that form in its atmosphere have negative helicity. One source of negative helicity lies in differential rotation [17], [7]. The equator of the Sun rotates faster than the poles. As a result, field lines entering the Northern hemisphere near the poles and exiting through the equator (e.g. the field lines of a magnetic dipole) will become more and more twisted. Field lines in the Southern hemisphere become twisted in the opposite sense. The rotation rate as a function of longitude is known by tracking photospheric features. Recently, helioseismology measurements also enable us to determine the rotation rate as a function of depth [9]. In fact, the equator rotates at almost a constant rate as a function of depth, with a slight drop near the base of the convection zone. We will also need B_n to evaluate Equation (9). Fortunately, magnetogram observations of B_n as a function of position and time have been archived over the last (22 year) solar cycle by the Wilcox observatory [19]. The resulting helicity flow into the Northern and Southern hemispheres is plotted in Figure 8. The net inflow to a hemisphere is on average, $-5 \times 10^{42}\,\mathrm{Mx}^2/d$, or $-2 \times 10^{46}\,\mathrm{Mx}^2$ for an 11 year cycle to the Northern hemisphere ($6.5 \times 10^{42}\,\mathrm{Mx}^2/d$, or $2.5 \times 10^{46}\,\mathrm{Mx}^2$ for the South). Estimates [17] suggest only some 10% of this magnetic helicity input is needed to account for the helicity shed in mass ejections.

Acknowledgements

I would like to acknowledge the hospitality of the Isaac Newton Institute, and the support of PPARC grant GR/L63143. The Transition Region and Coronal Explorer, TRACE, is a mission of the Stanford-Lockheed Institute for Space Research (a joint program of the Lockheed-Martin Advanced Technology Center's Solar and Astrophysics Laboratory and Stanford's Solar Observatories Group), and part of the NASA Small Explorer program. SOHO is a project of international cooperation between ESA and NASA.

References

1. Aly, J.J. (1984) On some properties of force-free magnetic fields in infinite regions of space. *Astrophys. J.* **283**, 349–362.
2. Arnold, V.I. & Khesin, B.A. (1998) *Topological Methods in Hydrodynamics.* Springer, New York.
3. Berger M.A. & Field G.B. (1984) The topological properties of magnetic helicity. *J. Fluid Mech.* **147**, 133–148.
4. Berger, M.A. (1984) Rigorous new limits on magnetic helicity dissipation in the solar corona. *Geophys. & Astrophys. Fluid Dyn.* **30**, 79–104.
5. Berger, M.A. (1999) Magnetic Helicity in Space Physics. In *Magnetic Helicity in Space and Laboratory Plasmas* (ed. M.R. Brown, R.C. Canfield & A.A. Pevtsov), pp. 1–11. AGU Geophysical Monograph Series **111**, Washington D.C..
6. Bieber, J.W., Evenson, P.A. & Matthaeus, W.H. (1987) Magnetic helicity of the Parker field. *Astrophys. J.* **315**, 700–705.
7. Berger, M.A. & Ruzmaikin, A. (2000) Helicity production by differential rotation. *J. Geophys. Res.* **105**, 10481–10490.
8. Brown, M.R., Canfield R.C., & Pevtsov A.A. (editors) (1999) *Magnetic Helicity in Space and Laboratory Plasmas.* AGU Geophysical Monograph Series **111**, Washington D.C..
9. Charbonneau, P., Dikpati, M. & Gilman, P.A. (1999) Stability of the solar latitudinal differential rotation inferred from helioseismic data. *Astrophys. J* **526**, 523–537.
10. Finn, J. & Antonsen, T.M. (1985) Magnetic helicity: what is it, and what is it good for? *Comments on Plasma Phys. and Contr. Fusion.* **9**, 111–123.

11. Low, B.C. (1994) Magnetohydrodynamic processes in the solar corona - flares, coronal mass ejections, and magnetic helicity. *Physics Plasmas* **1**, 1684–1690.
12. Martin, S.F. & McAllister A. (1997) In *Coronal Mass Ejections* (ed. N. Crooker, J.-A. Josselyn & J. Feynman), pp. 127–138. Geophys. Monog. **99**, Amer. Geophys. Union.
13. Moffatt, H.K. (1969) The degree of knottedness of tangled vortex lines. *J. Fluid Mech.* **35**, 117–129.
14. Parker, E.N. (1979) *Cosmical Magnetic Fields.* Oxford University Press, Oxford.
15. Priest, E.R. & Forbes, T.R. (1999) *Magnetic Reconnection: MHD theory and applications.* Cambridge University Press, Cambridge.
16. Rust, D. (1994) Spawning and shedding helical magnetic fields in the solar atmosphere. *Geophys. Res. Lett.* **21**, 241–244.
17. Rust, D. (1997) Helicity conservation. In *Coronal Mass Ejections,* p. 127. (ed. N. Crooker, J.-A. Josselyn & J. Feynman), pp. 119–125. Geophys. Monog. **99**, Amer. Geophys. Union.
18. Wright, A. & Berger, M.A. (1989) The effect of reconnection upon the linkage and interior structure of magnetic flux tubes. *J. Geophys. Res.* **94**, 1295–1302.
19. Zhao, X.P. & Hoeksema, J.T. (1993) Unique determination of model coronal magnetic fields using photospheric observations. *Solar Phys.* **143**, 41–48.

DIFFEOMORPHISMS, BRAIDS AND FLOWS

ALEXANDER SHNIRELMAN
School of Mathematical Sciences
Tel Aviv University
69978 Ramat Aviv, Israel

E-mail: *shnirelm@math.tau.ac.il*

Abstract. Flows of ideal incompressible fluids may be regarded as geodesics on the group of volume preserving diffeomorphisms. The natural method of constructing geodesics is the variational method: we are looking for the shortest path connecting two fluid configurations. In the three dimensional case this approach meets serious difficulties, but in two dimensions it brings a partial success: there exists a reasonable generalized solution of variational problem. This success is possible because in the 2-dimensional case there is an additional structure: every flow is at the same time a continual braid. The same braid structure may be found in the problem of equilibrium configuration of a perfectly conducting fluid with a frozen-in magnetic field, where weak solution with possible singularities always exists.

1. Introduction. What is a Fluid?

1. Let M be a bounded domain in $\mathbb{R}^n$, or a compact Riemannian manifold. We consider the motion of an ideal incompressible fluid in M.

A fluid consists of a continuum of fluid particles, filling M with a constant density. So, fluid positions (or configurations) are defined by the positions of all fluid particles. If we arbitrarily choose some reference position, then any other position can be obtained from it by some permutation of fluid particles. In "classical" fluid dynamics these permutations are supposed to be smooth, volume-preserving diffeomorphisms of M. We shall assume this class of permutations, with some refinements, discussed below. Volume preserving diffeomorphisms form a group $D(M) = D$; we restrict ourselves to the component of unity of this group.

R.L. Ricca (ed.), An Introduction to the Geometry and Topology of Fluid Flows, 253–270.

Any element $g \in D$ is a mapping of the domain M into itself; but $M \subset \mathbb{R}^n$, and therefore g may be regarded as a mapping from M into $\mathbb{R}^n$. Let Y denote the Hilbert space $L^2(M, \mathbb{R}^n)$; then D is in a natural way a subset of Y. The Hilbert space Y may be regarded as a space of non-constrained configurations, while the condition that fluid configuration remains all the time in D is a constraint.

The set $D \subset Y$ is, formally speaking, an infinite-dimensional and infinite-codimensional manifold. It becomes a true manifold, if we consider it as a subset in the space of sufficiently smooth maps (say, the space X_s of Sobolev class maps $M \to \mathbb{R}^n$, for sufficiently high s); see [3]. These smooth spaces have however little or no physical meaning, while, as we shall see, in the L^2-metric the set D is quite bad. But we have no choice: only this case is relevant for fluid dynamics, and either we regard it as it is, or we don't tackle the subject at all. This is a God given system, and we have to study it as it is.

Nonetheless, on a formal level, D is regarded as a submanifold of X, and this approach is used to derive the equation of motion. The motion of a material point in the Euclidean space Y, constrained by the condition that at all time the point is tied to a manifold $D \subset Y$, is described by the Hamiltonian principle, which states that if g_t is the trajectory, then for any t_1, t_2, $t_1 < t_2$, the variation $\delta J\{g_t\}_{t_1}^{t_2} = 0$, provided $\delta g_{t_1} = 0$, $\delta g_{t_2} = 0$, where

$$J\{g_t\}_{t_1}^{t_2} = \int_{t_1}^{t_2} \int_M \frac{1}{2} |\dot{g}_t(x)|^2 \, dx dt \tag{1}$$

is the action of the fluid during the time interval $t_1 \le t \le t_2$. In other words, the fluid trajectory g_t in the configuration space D is a geodesic in the metric induced by the inclusion $D \subset Y$ (this fundamental fact and its deep implications were discovered by V. Arnold [1]). Proceeding formally, we get a necessary condition: the fluid acceleration $\ddot{g}_t$ is orthogonal in Y to the tangent space to D (this statement is called the D'Alembert principle). The tangent space to D consists of incompressible vector fields in M, tangent to ∂M; its orthogonal complement consists of potential vector fields. Thus, we obtain the Euler-Lagrange equations

$$\ddot{g}_t(x) + \nabla p(g_t(x)) = 0 \; ; \tag{2}$$

$$g_t \in D \; . \tag{3}$$

If $u(x,t) = \dot{g}_t(g_t^{-1}(x))$ is the Eulerian velocity field, then the Lagrange equations (2), (3) may be written as the usual Euler equations:

$$\frac{\partial u}{\partial t} + (u, \nabla) u + \nabla p = 0 \; , \tag{4}$$

$$\nabla \cdot u = 0 \; . \tag{5}$$

This derivation of Euler equations looks a little shaky, because the space D is not a smooth manifold; its local structure is extremely complicated. We describe below the irregularity of D (Theorem 2.6); this irregularity is the root of difficulties in the theory of an ideal incompressible fluid.

2. The main feature of incompressible fluid dynamics is the lack of nontrivial solutions of the Euler (and Navier-Stokes) equations. Even if we know (in the 2-dimensional case) that the solution exists and is unique for infinite time, and is as regular as the initial velocity, we usually have little idea how does this solution look like. All we have is a small number of nearly trivial solutions. For example, the solid rotation of the fluid is a solution; but there is little "fluid" about it. The same may be said about shear flows, potential flows, point vortices, and small perturbations of these flows. (Here I do not speak about numerical solutions, since from a mathematical viewpoint these should be regarded as results of experiments, and the aim of a theory is to explain them.)

Note that in gas dynamics the situation is different: there exists an excellent class of examples, namely shock waves, which are quite representative and capture the essence of compressible fluid flows. Just imagine, how this science would be, if we had a general theory of existence, but we did not know anything about these examples; the situation in compressible fluid dynamics would be immediately felt regrettable. Incompressible fluid dynamics is exactly in this position: we do not know what is the nature of most typical flows, which are far from the above 'degeneracies'. So, we desperately need a method for constructing truly nontrivial solutions of the Euler equations.

A general approach to this problem is based on the Hamiltonian principle. The idea is that fluid flow may be regarded as a motion along geodesics in an infinite-dimensional Riemannian manifold D. Any geodesic (i.e. the "minimal" trajectory) is, at least locally, the shortest one; so, we may attempt to use variational methods to construct geodesics. This approach has a great advantage: from the beginning we have a lot of information about the global behavior of trajectories. There is also a great disadvantage: solutions of variational problems sometimes do not exist; a variational problem in D is much more difficult than in a finite-dimensional Riemannian manifold.

This paper is devoted to some aspects of variational solution of Euler equations. Before studying geodesics on D we have to learn something about the geometry of this space. Then we describe some negative results concerning non-existence of minimal geodesics in the 3-dimensional case, and some positive results on existence in two dimensions. In the last part I explain the related problem of magnetic equilibrium.

2. Basic Facts on the Geometry of the Space D

1. For any curve $g_t \in D$, $t_1 \leq t \leq t_2$ we define two quantities, the action $J\{g_t\}_{t_1}^{t_2}$ and the length $L\{g_t\}_{t_1}^{t_2}$:

$$J\{g_t\}_{t_1}^{t_2} = \int_{t_1}^{t_2} \int_M \frac{1}{2} |\dot{g}_t(x)|^2 \, dx dt \ , \tag{6}$$

$$L\{g_t\}_{t_1}^{t_2} = \int_{t_1}^{t_2} \Big(\int_M |\dot{g}_t(x)|^2 dx \Big)^{1/2} dt \ . \tag{7}$$

Let us define the distance between two elements g and h of D as follows:

$$\mathrm{dist}(g,h) = \inf L\{g_t\}_0^1 \ , \tag{8}$$

where infimum is taken over all trajectories $g_t \in D$, $0 \leq t \leq 1$, such that $g_0 = g, g_1 = h$.

From now on, we consider only *attainable* diffeomorphisms g, i.e. such that $\mathrm{dist}(Id, g) < \infty$, where Id is the identical diffeomorphism (the toploogy in the group D is defined by C^N-distances on compact sets, $N = 0, 1, \ldots$). Generally speaking, not every diffeomorphism from the component of unity of the group D is attainable (see below); with a little abuse of notation, we shall denote by D the set of attainable diffeomorphisms (it is easy to see that they form a group). Thus, the group D is a metric space. Let us define the *diameter* of D:

$$\mathrm{diam}(D) = \sup_{g,h \in D} \mathrm{dist}(g,h) \ . \tag{9}$$

The first question is, whether the diameter of D is finite or infinite. The answer is given by the following

Theorem 2.1 *Suppose $M \subset I\!R^n$ is a unit cube: $M = (0,1)^n$. Then*

$$(i) \quad if \ \ n = 2, \quad then \ \ \mathrm{diam}(D) = \infty \ ; \tag{10}$$

$$(ii) \quad if \ \ n > 2, \quad then \ \ \mathrm{diam}(D) < \infty \ . \tag{11}$$

Proof. Part (i) has been proven by Eliashberg and Ratiu [5] as a special case of a general theorem on the group of symplectomorphisms. The main idea is the following: let M be a 2-dimensional disk, rather than a square (this doesn't make any difference for the problem); consider the self-map of a disk, which is defined in the polar coordinates as

$$(r, \varphi) \mapsto (r, \varphi + N\psi(r)) = h_N(r, \varphi) \ , \tag{12}$$

where $\psi(r)$ is an arbitrary non-constant function.

It turns out that dist $(Id, h_N) \geq cN$, that is proven with the help of the Calabi invariant of a symplectic flow. N may be arbitrarily large, and therefore diam $D(M^2) = \infty$. Part (ii) has been proven in [8]; the proof consists of a direct construction, for any given $g \in D$, of a "short" path g_t, connecting Id and g, so that its length is uniformly bounded for all $g \in D$.

QED

The main tool used in proving part (ii) is the discrete approximation of diffeomorphisms and flows. Let us divide the cube $M = (0,1)^n$ into N^n equal small cubes $\kappa_1, \dots, \kappa_{N^n}$ of size $1/N$. We shall approximate volume-preserving diffeomorphisms of M and flows in it by permutations of small cubes and, respectively, sequences of such permutations. Let M_N denote the set of all small cubes κ, and D_N the set of all their permutations. We call a permutation $\sigma \in D_N$ *elementary*, if for every small cube $\kappa \in M_N$ either $\sigma(\kappa) = \kappa$, or $\sigma(\kappa) = \kappa'$, where κ' is a neighbor of κ, i.e. they have a common $(n-1)$-dimensional face. A sequence $\sigma_1, \dots, \sigma_k$ of elementary permutations is called a *discrete flow*. We say that this discrete flow connects Id and $\sigma \in D$, if $\sigma = \sigma_1 \circ \dots \circ \sigma_k$. Any permutation of small cubes may be regarded as a discontinuous, volume preserving transformation of M (which is smooth in every small cube κ). So, our first result is the following:

Lemma 2.2 *Any diffeomorphism $g \in D$ can be approximated in L^2 by permutations of small cubes. Moreover, for any $\varepsilon > 0$ there exist N, a permutation $\sigma \in D_N$, and a piecewise-continuous incompressible flow g_t, which is smooth in every small cube κ, such that $g_0 = g$, $g_1 = \sigma$, and $L\{g_t\}_0^1 < \varepsilon$.*

The next basic fact is the following combinatorial

Lemma 2.3 *Any permutation $\sigma \in D_N$ of small cubes is a product of no more than CN elementary permutations, where C depends only on the space dimension n.*

So, any permutation of small cubes may be connected with Id by a discrete flow $\sigma_1, \dots, \sigma_k$, $k < CN$.

For a discrete flow we can define its length

$$L\{\sigma_1, \dots, \sigma_k\} = \sum_{i=1}^{k} \frac{1}{N} \left(\frac{m_i}{N^n} \right)^{1/2}, \tag{13}$$

where m_i denotes the number of small cubes moved by σ_i into their neighbors.

The next lemma brings us one step closer to the construction of the short path connecting Id and g.

Lemma 2.4 *For any discrete flow $\sigma_1, \ldots, \sigma_k$ connecting Id and some permutation σ, there exists a piecewise-smooth flow h_t (smooth in every small cube κ), such that $h_0 = Id$, $h_1 = \sigma$, and $L\{h_t\}0^1 \leq CL\{\sigma_1, \ldots, \sigma_k\}$.*

Concatenating the flow h_t with the flow g_t, connecting the permutation σ with g and having small length, as described in Lemma 1.1, we obtain a piecewise-smooth flow G_t, $0 \leq t \leq 2$, connecting Id and g.

The last step is to smooth this flow. Only at this step condition $\dim M \geq 3$ plays a crucial role. Namely, we use the fact that if Z is the space of all tuples $(x_1, \ldots, x_{N^n}) \in M^{N^n}$, such that $x_i \neq x_j$ for $i \neq j$, then the fundamental group $\pi_1(Z)$ is trivial.

Lemma 2.5 *If $n = \dim M \geq 3$, $g \in D$, and G_t is a piecewise-smooth flow, smooth in every small cube κ, such that $G_0 = Id$, $G_1 = g$, then for any $\varepsilon > 0$ there exists a smooth volume-preserving flow $\tilde{G}_t$, connecting Id and g, such that $L\{\tilde{G}_t\}_0^2 \leq L\{G_t\}_0^2 + \varepsilon$.*

2. If the map $g \in D$ is L^2-close to Id, then the approximating permutation of small cubes $\sigma \in D_N$ is also close to Id, in the sense that for most of the cubes κ the distance $|\sigma(\kappa) - \kappa|$ is small. We are then able to construct a short discrete flow $\sigma_1, \ldots, \sigma_k$ connecting Id and σ. This gives the following:

Theorem 2.6 *Let $M \subset \mathbb{R}^n$ be a unit cube. If $n \geq 3$, then there exist $C > 0$ and $\alpha > 0$, such that for any $g, h \in D(M)$*

$$\operatorname{dist}(g, h) \leq C||g - h||_{L^2}^{\alpha} \ . \tag{14}$$

The estimates of C and α depend on our success to construct a short discrete flow, connecting Id and σ. Our first estimates for $n = 3$ were $\alpha \geq \frac{1}{64}$, $C \leq 100$ [8]. Later, using more advanced methods, described below, these estimates were improved to $\alpha \geq \frac{2}{7}$, $C \leq 2$ [9]. This estimate is likely to be sharp.

This result shows the degree of regularity of embedding of the space D into the Hilbert space $X = L^2(M, \mathbb{R}^n)$. For a smooth submanifold in the Euclidean space, the exponent α is 1. So, D is very far from smoothness. On the other hand, situation for $n = 2$ is much worse: there exist $g, h \in D$, such that $||g - h||_{L^2}$ is arbitrarily small, while $\operatorname{dist}(g, h)$ is arbitrarily large. For example, we can take $g = Id$, and h a twist map, described above, which is confined to a disk of sufficiently small radius.

3. The space D with its internal metric is not complete, and it is interesting to describe its completion $\tilde{D}$. The answer is given by

Theorem 2.7 *If $n \geq 3$, then the the space $\tilde{D}$ coincides with the space $\bar{D}$, the closure of D in X. The last space consists of all Lebesgue measurable, measure-preserving maps $f : M \to M$.*

Recall that a Lebesgue measurable map $f : M \to M$ (which is generally not invertible) is called measure-preserving, if for every Lebesgue measurable set $A \subset M$, $\mathrm{mes}(f^{-1}(A)) = \mathrm{mes}(A)$.

To prove this result, it is sufficient to construct a sequence of smooth maps $f_k \in D$, converging to f in L^2; then Theorem 2.6 shows that this sequence is a Cauchy sequence in D. The sequence f_k is constructed by approximating the map f in L^2 by permutations of small cubes and then smoothing this piecewise-continuous map.

For $n = 2$ Theorem 2.7 is not true; completion of D has little to do with its closure in X, and the structure of $\tilde{D}$ appears to be transcendentally complicated.

A different, but related question, is whether any volume-preserving diffeomorphism g of a cube $M \subset \mathbb{R}^n$ is an element of D, i.e. may be connected with Id by a path of finite length (a problem may arise because of the arbitrarily wild behavior of g near ∂M). The previous theorem does not prove this, because the Cauchy sequence g_n does not, in general, converge to g in a compact-open topology, but only in L^2. Using some refinement of the proof of Theorem 2.1, though, it is possible to prove the following result [10]:

Theorem 2.8 *(i) If $n \geq 3$, then any volume preserving diffeomorphism of M can be connected with Id by a continuous path $g_t \in D$, such that $L\{g_t\}_0^1 < \infty$; thus, $g \in D$. (ii) If $n = 2$, there exists a volume preserving diffeomorphism g of M, that cannot be connected with Id by a path of finite length. This diffeomorphism may be chosen continuous in $\bar{M}$, and moreover, $g(x) = x$ for any point $x \in \partial M$. Moreover, there exists an uncountable family $g^{(s)}$ of such diffeomorphisms, $0 \leq s \leq 1$, so that if $s_1 \neq s_2$, then $g^{(s_1)}$ and $g^{(s_2)}$ cannot be connected by a path having finite length.*

So, the group (topologically connected) of area-preserving diffeomorphisms of the square, contains a continuum of components that are metrically connected.

3. Minimal Geodesics on D

1. Let us fix two fluid configurations $g_0, g_1 \in D$; because of the homogeneity of the space D we can always take $g_0 = Id, g_1 = g \in D$. Let us look for a path $g_t \in D$, $0 \leq t \leq 1$, such that $g_0 = Id$, $g_1 = g$, and such that the action of this path $J\{g_t\}_0^1$ is minimal among all paths connecting Id and g. Such a path, if exists, is a geodesic and satisfies the Lagrange-Euler equations (in our case (2), (3)). This idea, going back to Hilbert and Hadamard, is very successful in the finite-dimensional Riemannian geometry; but for our case it requires much more care, as we shall see below.

Our first result is of negative type.

Theorem 3.1 *Let $M = [0,1]^3$ be a 3-dimensional cube. There exists a diffeomorphism $g \in D(M)$, such that for any smooth path $g_t \in D$, connecting Id and g, there exists another path $\tilde{g}_t$, connecting Id and g, such that $J\{\tilde{g}_t\}_0^1 < J\{g_t\}_0^1$. Thus, the minimal action cannot be achieved.*

We shall discuss the proof of this theorem a little later.

If the classical solution does not exist, we can look for a generalized one. The appropriate notion of *generalized flow* was introduced by Brenier [2]. In the classical fluid description fluid particles are labelled by the points of the flow domain (for example, by the positions of the particles at some time). Brenier proposed to decouple the set of particles from the set of points. Namely, he defines the set of particles as a probability space Ω with σ-algebra $\mathcal{F}$ of measurable sets and a positive measure μ, such that $\mu(\Omega) = 1$. The *generalized flow* $\mathbf{G}$ is a collection of continuous trajectories $x(\omega, t) \in M$ of the particles $\omega \in \Omega$, $0 \leq t \leq 1$.

The generalized flow $\mathbf{G}$ is called *incompressible*, if for any t and any Lebesgue measurable set $A \subset M$, we have

$$\mu\{\omega | x(\omega, t) \in A\} = \text{mes}(A) \; ; \tag{15}$$

We say that the generalized flow $\mathbf{G}$ connects Id and a given diffeomorphism $g \in D$, if for almost all $\omega \in \Omega$

$$x(\omega, 1) = g(x(\omega, 0)) \; ; \tag{16}$$

We say that the generalized flow $\mathbf{G}$ has finite action, if

$$J\{\mathbf{G}\} = \int_{\Omega} \int_0^1 \frac{1}{2} |\dot{x}(\omega, t)|^2 \, dt \; \mu(d\omega) < \infty \; . \tag{17}$$

Now we can consider, instead of the original problem, the following 'relaxed' one: given a diffeomorphism $g \in D$, find an incompressible generalized flow $\mathbf{G}$ connecting Id and g with minimal action $J\{\mathbf{G}\}$. The basic result of Brenier is the following

Theorem 3.2 *For any $g \in D$ (and, in fact, for any measure-preserving transformation g) there exists an incompressible generalized flow D, connecting Id and g, with a minimal action $J\{\mathbf{G}\}$ among all such generalized flows.*

There are different examples showing that generalized solutions may be not at all classical ones; they might look like many different fluids moving in the same volume with different velocities in different directions; the number of such fluids may be even infinite.

The proof of Theorem 3.2 follows the classical scheme. For any generalized flow $\mathbf{G}$ we can define the distribution of trajectories, i.e. the measure $\nu_{\mathbf{G}}\{dx(\cdot)\}$ in the space $X = C(0,1;M)$ of continuous trajectories in the flow domain M. If $\mathbf{G}_1, \mathbf{G}_2, \ldots$ is a sequence of generalized flows, such that $J\{\mathbf{G}_i\} \searrow \inf$, then the measures $\nu_{\mathbf{G}_i}$ form a tight (relatively weakly compact) sequence, and thus we can extract a subsequence $\nu_{\mathbf{G}_{i_k}}$ weakly converging to some measure ν. The functional $J(x(\cdot)) = \int_0^1 \frac{1}{2}|\dot{x}(t)|^2 dt$ is lower semi-continuous; this implies that

$$J(\nu) = \int_X J(x(\cdot))\nu\{dx(\cdot)\} \leq \liminf \int_X J(x(\cdot))\nu_{i_k}\{dx(\cdot)\} \,. \tag{18}$$

However, we can regard any generalized flow $\mathbf{G}$ as a measure in the space X; this change of the particle space does not affect any observable property of the flow. Thus, $\mathbf{G} = (X, \mathcal{F}, \nu, i)$ is a generalized flow, where $\mathcal{F}$ is the Borel σ-algebra in X, and i is the tautological map that labels every element of X, i.e. a trajectory in M, with that same trajectory. It is incompressible and satisfies the boundary condition $x(1) = g(x(0))$ for ν-a.a. trajectories $x(t)$, because these properties are stable under weak convergence of measures. This generalized flow has least action among all incompressible generalized flows connecting Id and g.

Generalized flows are a powerful and flexible tool for investigating the geometry of the group D itself, in particular for obtaining accurate estimates of diameter and degree of regularity of D (see Sect. 2), and for the proof of Theorem 3.1. The main role is played by the following approximation theorem [9].

Theorem 3.3 *Let $\mathbf{G} = (\Omega, \mathcal{F}, \mu, x(\omega,t))$ be a generalized flow in a cube $M = (0,1)^n$, $n \geq 3$, connecting Id and a diffeomorphism $g \in D$. Then there exists a sequence of smooth flows $g_t^{(k)}$, connecting Id and g, such that $\nu_{g_t^{(k)}} \to \nu$ weakly, and $J\{g_t^{(k)}\} \to J(\mathbf{G})$, as $k \to \infty$, where the measures $\nu_{g_t^{(k)}}$ and ν are distributions in $X = C(0,1;M)$ of trajectories of smooth flows $g_t^{(k)}$ and of the generalized flow $\mathbf{G}$.*

This theorem enables us to construct example of a diffeomorphism g of a 3-dimensional cube M_3 which cannot be connected with Id by the shortest flow, as it was stated in Theorem 3.1. Let x, y, z be Cartesian coordinates in M, $0 < x, y, z < 1$. The diffeomorphism g has the form $g(x,y,z) = (h(x,y), z)$, where h is an area-preserving diffeomorphism of a unit square M_2. As we know already (Theorem 2.1), the diameter of the group $D(M_2)$ is infinite, while the diameter of the group $D(M_3)$ is finite. This implies that there exist diffeomorphisms $h \in D(M_2)$, such that $\mathrm{dist}_{D(M_3)}(Id, g) < \mathrm{dist}_{D(M_2)}(Id, h)$.

Theorem 3.4 *If $g(x,y,z) = (h(x,y),z)$ is the above diffeomorphism, and $\mathrm{dist}_{D(M_3)}(Id, g) < \mathrm{dist}_{D(M_2)}(Id, h)$, then for any smooth flow g_t, connecting Id and g in $D(M_3)$, there exists another smooth flow $\tilde{g}_t$, connecting Id and g and such that $J\{\tilde{g}_t\}_0^1 < J\{g_t\}_0^1$.*

Here is the idea of the proof. Let g_t be any smooth flow connecting Id and g. Let us write this flow in coordinates:

$$g_t(\bar{x},\bar{y},\bar{z}) = (x(\bar{x},\bar{y},\bar{z},t), y(\bar{x},\bar{y},\bar{z},t), z(\bar{x},\bar{y},\bar{z},t)) .$$

Now let us define a generalized flow $\mathbf{G}$ whose particle space Ω is a 4-dimensional cube with coordinates $\bar{x},\bar{y},\bar{z},\bar{w}$, and with Lebesgue measure μ given by $d\bar{x}d\bar{y}d\bar{z}d\bar{w}$. For any particle $\omega = (\bar{x},\bar{y},\bar{z},\bar{w}) \in \Omega$ we define its trajectory $(x(\omega,t), y(\omega,t), z(\omega,t))$ as $(x(\bar{x},\bar{y},\bar{z},t), y(\bar{x},\bar{y},\bar{z},t), \bar{w})$. It is easy to see that the defined generalized flow is incompressible, satisfies the boundary condition (16), and its action $J(\mathbf{G})$ is equal to the horizontal component of the action of the flow g_t

$$J_h\{g_t\}_0^1 = \int_0^1\int_0^1\int_0^1\int_0^1 \frac{1}{2}\Big((\dot{x}(\bar{x},\bar{y},\bar{z},t)^2 + (\dot{y}(\bar{x},\bar{y},\bar{z},t)^2\Big)\, dxdydzdt . \qquad (19)$$

Now we can observe, that if the flow g_t is really 3-dimensional, i.e. $\dot{z}(\bar{x},\bar{y},\bar{z},t)$ is not identically zero, then $J(\mathbf{G}) < J\{g_t\}_0^1 - \varepsilon$, $\varepsilon > 0$. By Theorem 3.3, we can approximate the generalized flow $\mathbf{G}$ by a smooth flow $\tilde{g}_t$, with the same boundary conditions, i.e. connecting Id and g, so that $J\{\tilde{g}_t\}_0^1 < J(\mathbf{G}) + \frac{\varepsilon}{2}$; thus, $J\{\tilde{g}_t\}_0^1 < J\{g_t\}_0^1$. On the other hand, if $\dot{z}(\bar{x},\bar{y},\bar{z},t) \equiv 0$, then the flow is foliated into flows in the planes $z = \mathrm{const}$. In this case the action of the flow g_t equals the action of the flow h_t, connecting Id and h in $D(M_2)$; by hypothesis this is larger than the infimum of actions of 3-dimensional flows connecting Id and g. This contradicts the assumption and ends the proof of Theorem 3.4.

4. Braids and Flows

1. We have seen above that if the flow domain M is a 3-dimensional cube, then the 2-point variational problem may have no classical solution. Its generalized solution is, generally, very far away from a physically reasonable flow. Namely, if $\mathbf{G}$ is a generalized flow minimizing the action functional J, then for a set of points (x,t) of positive measure there exist many fluid particles passing through this point. Thus, the velocity field is not defined (we have a multiple-valued velocity). It is difficult to imagine a fluid moving this way. Such a flow is extremely unstable, and, worse, the smallest viscosity will make this flow impossible.

But in two dimensions the situation is much better, because there is an additional structure. To see it, consider first a smooth flow g_t, connecting $g_0 = Id$ and $g_1 = g \in D$. Let $Q = \{(x,t)|x \in M, 0 \le t \le 1\}$ be a cylinder in $\mathbb{R}^3$. Trajectories of points $x = g_t(x_0)$ are lines $\gamma(x_0)$ in Q connecting points x_0 (at the bottom surface of Q) with $g(x_0)$ (at the top surface of Q), for every x_0 on M. These lines grow monotonically in Q (i.e. the t-variable grows monotonically along every curve $\gamma(x_0)$), and the lines $\gamma(x), \gamma(y)$ do not intersect for different x, y.

We call *braid* the collection of a finite number of lines γ_i, $i = 1, \dots, N$, growing monotonically in Q and connecting points $x_1, \dots, x_N$ at the bottom end at $t = 0$ with points $y_1, \dots, y_N$ at the top end of Q, where the sequence $\{y_i\}$ is obtained from the sequence $\{x_i\}$ by some permutation. We call the system of lines $\gamma(x), x \in M$ *continual braid*, or simply *braid*, while the above are referred to as *finite braids.*

Roughly speaking, the relationship between continual braids and finite braids is similar to the relation between smooth diffeomorphisms of M and permutations of finite sets. The concept of continual braids is a generalization of finite braids, which involves a number of notions. The first one is a generalization to a *smooth braid*, as described above. A smooth braid $\mathbf{B}$ is a collection of lines $\gamma(x)$ in the cylinder Q, which are trajectories of a smooth area preserving flow g_t, $0 \le t \le 1$. We say that the braid $\mathbf{B}$ connects Id with the map $g = g_1$.

Two smooth braids $\mathbf{B}_0$ and $\mathbf{B}_1$, connecting Id with g, are called *isotopic*, if there exists a family of smooth braids $\mathbf{B}_s$, $0 \le s \le 1$, connecting Id and g and depending continuously on s. The isotopy classes of braids form a group; we can define a composition of braids $\mathbf{B}_1$, $\mathbf{B}_2$ by simply concatenating these families of lines.

In some sense multiplication of continual braids is even better defined than for finite braids, because we can define composition of two finite braids only if the sets of their endpoints are the same, while for continual braids the last condition is satisfied automatically. It is easy to see that if M is a simply connected domain, then the group of isotopy classes of braids is a semidirect product of the group $D(M)$ of area-preserving diffeomorphisms and the group Z of integers. This means that any braid $\mathbf{B}_1$ can be obtained from any other braid $\mathbf{B}_2$, connecting Id with the same diffeomorphism g, by some integer number of twists. Thus, here the situation is not very interesting from the topological viewpoint.

Any smooth braid is at the same time a flow, connecting Id and g. Thus, we can define the action of the braid, $J(\mathbf{B})$, as the action of the flow g_t:

$$J(\mathbf{B}) = \int_M \int_0^1 \frac{1}{2} |\dot{g}_t(x)|^2 \, dx dt \ . \tag{20}$$

The variational problem for smooth braids may be posed as follows: let us fix a smooth braid $\mathbf{B}_0$, connecting Id and $g \in D$; we are looking for a smooth and incompressible braid $\mathbf{B}$, isotopic to $\mathbf{B}_0$ and having minimal action $J(\mathbf{B})$.

There exist no general results concerning this problem. Perhaps something interesting may occur if the reference braid $\mathbf{B}_0$ is *very* twisted: our hypothesis is that for sufficiently large number of twists the minimum always exists. This, however, has not been not proven yet.

In order to obtain general existence results, we have to further generalize the notion of braid. A *generalized braid*, or simply a braid (all other sorts of braids are characterized by some adjectives, such as smooth, finite, ...) is a particular case of generalized flow in M, defined above. Our starting point is the following

Lemma 4.1 *Suppose* $\mathbf{G}$ *is an incompressible generalized flow in a two dimensional domain* M*, connecting* Id *and* $g \in D$ *and having finite action. Then for almost all pairs* $(\omega_1, \omega_2) \in \Omega \times \Omega$ *the trajectories* $x(\omega_1, t)$ *and* $x(\omega_2, t)$ *do not intersect, i.e.* $x(\omega_1, t) \neq x(\omega_2, t)$ *for all* $t \in [0, 1]$.

As a corollary we see that for almost every N-tuple $(\omega_1, \ldots, \omega_N)$ the lines $x(\omega_1, t), \ldots, x(\omega_N, t)$ form an N-braid, connecting N points $x_1, \ldots, x_N$ with respective points $g(x_1), \ldots, g(x_N)$ (note that this is not a canonical finite braid, for which the sets of endpoints coincide: we can still define isotopy for such non-canonical braids, provided the end-points are fixed.)

Now we can define the class of generalized braids we need. Let $g \in D$; let us fix a smooth braid $\mathbf{B}_0$, connecting Id and g. Let $\mathbf{B}$ be an incompressible generalized flow, connecting Id and g; let Ω be its particle space. For any $\omega_1, \ldots, \omega_N \in \Omega$, we can define two line systems in $Q = M \times [0, 1]$: the first system consists of trajectories $x = x(\omega, t)$ (call these lines γ_i, $i = 1, \ldots, N$), and the second consists of lines of the smooth braid $\mathbf{B}_0$, whose end-points are $x_i = x(\omega_i, 0)$ (at the bottom) and $y_i = g(x_i)$ (at the top); call these lines Γ_i, $i = 1, \ldots, N$. For almost all N-tuples $(\omega_1, \ldots, \omega_N)$, the systems $\{\gamma_i\}$ and $\{\Gamma_i\}$ form N-braids, with the same sets of end-points at the bottom and at the top of Q.

Definition 4.2 *The generalized flow* $\mathbf{B}$ *is said to be a* **braid weakly isotopic to** $\mathbf{B}_0$*, if for all* $N > 0$ *and for almost all* $(\omega_1, \ldots, \omega_N) \subset \Omega^N$*, the braids* $(\gamma_1, \ldots, \gamma_N)$ *and* $(\Gamma_1, \ldots, \Gamma_N)$ *are isotopic.*

Note that we do not define isotopy of two generalized braids (or flows) directly, but only via comparison with a smooth braid. The set of isotopy classes of all braids is very large and we don't want to discuss this subject here. There are a lot of non-trivial braids weakly isotopic to a smooth one. For example, if $\mathbf{B}_0$ is a trivial braid, consisting of straight vertical lines, then we can take any generalized flow $x(\omega,t)$, however entangled, and define a new one

$$\tilde{x}(\omega,t)=\begin{cases} x(\omega,t) & 0\le t\le \frac{1}{2} \\ x(\omega,1-t) & \frac{1}{2}\le t\le 1\,. \end{cases} \qquad (21)$$

This is a braid connecting Id to itself and weakly isotopic to $\mathbf{B}_0$.

2. Now we can formulate the generlized (relaxed) variational problem: given a diffeomorphism $g\in D$ and a smooth braid $\mathbf{B}_0$ (connecting Id and g) find a braid $\mathbf{B}$ (connecting Id and g), which is weakly isotopic to $\mathbf{B}_0$ and has minimal action $J(\mathbf{B})$.

Theorem 4.3 *The generalized variational problem has a solution.*

Proof. This is the idea of the proof: let $\mathbf{B}_1,\mathbf{B}_2,\ldots$ be a sequence of braids weakly isotopic to $\mathbf{B}_0$, such that their actions $J(\mathbf{B}_i)$ tend to an infimum. Then, as above, we can extract a subsequence $\{\mathbf{B}_{i_k}\}$, which converges weakly (as a sequence of measures in $C(0,1;M)$) to some generalized flow $\mathbf{B}$, and $J(\mathbf{B})\le \inf\limits_i J(\mathbf{B}_i)$.

Lemma 4.4 *The generalized flow* $\mathbf{B}$ *is a braid, connecting* Id *and* g *and weakly isotopic to* $\mathbf{B}_0$.

In other words, the weak isotopy class of a braid is stable with respect to weak convergence. This Lemma concludes the proof.

QED

3. The braid $\mathbf{B}$ with minimal action among all braids weakly isomorphic to $\mathbf{B}_0$ (let us call it "minimal braid") may still be very entangled (locally braids may be as wild as generic generalized flows). In fact, a minimal braid is much more regular, as described in the following:

Theorem 4.5 *Suppose* $\mathbf{B}$ *is a minimal braid, regarded as generalized flow. Then there exists a vector field* $u(x,t)\in L^2$, *satisfying* $\nabla\cdot u=0,\ u_n|_{\partial M}=0$ *in a weak sense, and such that for a.a.* $\omega\in\Omega$ *we have*

$$\dot{x}(\omega,t)=u(x(\omega,t),t). \qquad (22)$$

An idea of the proof is given below; let us first introduce the following definition. Let $\mathbf{G}$ be an arbitrary incompressible generalized flow. We have:

Definition 4.6 *The vector field* $u(x,t)$ *is called* **mean velocity** *of* $\mathbf{G}$, *if for any vector field* $v(x,t)\in C_0^\infty$ *we have*

$$\int\limits_\Omega \mu(d\omega)\int\limits_0^1(\dot{x}(\omega,t),v(x(\omega,t),t))dt=\int\limits_M\int\limits_0^1(u(x,t),v(x,t))\,dxdt\,. \qquad (23)$$

From the definition we easily deduce the following properties of the field $u(x,t)$:

$$(i) \quad J(\mathbf{G}) = \int_\Omega \mu(d\omega) \int_0^1 \frac{1}{2}|\dot{x}(\omega,t)|^2 dt \geq \int_M \int_0^1 \frac{1}{2}|u(x,t)|^2\, dx dt \; ; \qquad (24)$$

equality holds only if $\mathbf{G}$ satisfies (22).

(ii) The vector field $u(x,t)$ satisfies $\nabla \cdot u = 0$, $u_n|_{\partial M} = 0$ in a weak sense, i.e. for any scalar function $\varphi(x,t) \in C^\infty$, we have $\int\int (u, \nabla\varphi)\, dx dt = 0$.

Proof. (i) is a consequence of Fubini theorem. Namely, for a.a. $(x_0, t_0) \in Q$, there is defined a conditional measure $\nu_{x_0,t_0}(d\omega)$, which is a probability measure in Ω concentrated on the set of ω, such that $x(\omega, t_0) = x_0$, and for a.a. t, $\int \nu_{x,t}\, dx = \mu$. Then for a.a. (x,t) $u(x_0,t_0) = \int_\omega \dot{x}(\omega, t_0)\nu_{x_0,t_0}(d\omega)$; and because of convexity of $|u|^2$, $|u(x_0,t_0)|^2 \leq \int_\Omega |\dot{x}(\omega,t_0)|^2 \nu_{x_0,t_0}(d\omega)$ for a.a. (x_0,t_0), and thence (24).

(ii) For any function $\varphi(x) \in C^\infty$,

$$\frac{d}{dt}\int_\Omega \varphi(x(\omega,t),t)\mu(d\omega) = \frac{d}{dt}\int_M \varphi(x)\, dx = 0 \; ; \qquad (25)$$

but for a.a. t, the left hand side is equal to $\int_\Omega (\dot{x}(\omega,t), \varphi(x(\omega,t)))\mu(d\omega) = \int_M (u, \nabla\varphi)\, dx$; and thence (ii).

QED

The key theorem is the following:

Theorem 4.7 *Let* $\mathbf{B}$ *be any braid, weakly isotopic to a smooth braid* $\mathbf{B}_0$ *and connecting* Id *and* $g \in D$. *There exists a braid* $\tilde{\mathbf{B}}$, *which is weakly isotopic to* $\mathbf{B}_0$ *that connects* Id *and* g, *and such that for a.a.* (ω,t) *we have*

$$\dot{\tilde{x}}(\omega,t) = u(\tilde{x}(\omega,t),t), \qquad (26)$$

where $u(x,t)$ *is the mean velocity of* $\mathbf{B}$ *and* $\tilde{x}(\omega,t)$ *is the mapping corresponding to* $\tilde{\mathbf{B}}$ *(regarded as a generalized flow).*

Thus, if $\mathbf{B}$ is a minimal braid, then it has a definite velocity field $u(x,t) \in L^2$, such that for a.a. (ω,t) $\dot{x}(\omega,t) = u(x(\omega,t),t)$; otherwise $\mathbf{B}$ is not minimal, for $J(\tilde{\mathbf{B}}) < J(\mathbf{B})$, and $\tilde{\mathbf{B}}$ is also an admissible braid. We may say that the minimal braid is "combed".

The proof of Theorem 4.7 is based on a detailed analysis of topology of braids, and will be published elsewhere.

The main properties of the vector field $u(x,t)$ are described by the following:

Theorem 4.8 *(i) $u(x,t)$ is a weak solution of the Euler equations; (ii) The energy $E(t) = \int_M \frac{1}{2}|u(x,t)|^2\,dx$ is constant.*

Proof. Recall that a field $u(x,t) \in L^2$ is called a weak solution of the Euler equations, if for any test-fields $v(x,t) \in C_0^\infty$, such that $\nabla \cdot v = 0$, and any scalar test-function $\varphi(x,t) \in C_0^\infty$, such that

$$\int\int [(u, \frac{\partial v}{\partial t}) + (u, \nabla v \cdot u)]\,dxdt = 0\ ; \tag{27}$$

$$\int\int (u, \nabla\varphi)\,dxdt = 0\ . \tag{28}$$

Any regular solution of the Euler equations satisfies (27) and (28), and any regular field $u(x,t)$, satisfying (27) and (28) for any v, φ, is a classical solution of the Euler equations. But there may exist non-regular (say, discontinuous) solutions of the Euler equations, and in our case we may also anticipate singularities.

So, consider a braid **B** with a definite velocity field $u(x,t)$. Let us define an infinitesimal deformation of this braid: any point $(x,t) \in Q$ is moved to a near-by point $(x + v(x,t), t)$, where $v(x,t)$ is an infinitesimal vector field, $v \in C_0^\infty$, $\nabla \cdot v = 0$. Then the variation $\delta J = \int\int (u, \delta u)\,dxdt$;

$$\delta u = \frac{\partial v}{\partial t} + \nabla v \cdot u - \nabla u \cdot v\ . \tag{29}$$

So

$$\delta J = \int\int [(u, \frac{\partial v}{\partial t}) + (u, \nabla v \cdot u) - (u, \nabla u \cdot v)]\,dxdt; \tag{30}$$

the last term is equal to $\int\int \frac{1}{2}(\nabla|u|^2, v)\,dxdt = 0$, because $v(x,t)$ is orthogonal to the gradient of any distribution. If **B** is a minimal braid, then $\delta J(\mathbf{B}) = 0$ for any v; this gives the result (27). To prove the second part, consider deformation of Q, moving every point (x,t) into $(x, t + \psi(t))$, where $\psi(t) \in C_0^\infty$ is an infinitesimal deformation in the t-direction. Then $\delta u = \psi \frac{\partial u}{\partial t} - \frac{\partial \psi}{\partial t} u$; hence

$$\delta J = \int\int (u, \delta u) dxdt = \int\int [\psi(u, \frac{\partial u}{\partial t}) - \frac{\partial \psi}{\partial t}(u,u)]\,dxdt =$$

$$= \int\int [\psi \frac{\partial}{\partial t}(\frac{1}{2}|u|^2 - \frac{\partial \psi}{\partial t}|u|^2]\,dxdt =$$

$$= \int_0^1 [\psi(t) \int_M (u, \frac{\partial u}{\partial t})\,dx - \frac{\partial \psi}{\partial t} \int_M (u,u) dx]\,dt =$$

$$= \int_0^1 \psi(t) \int_M [(u, \frac{\partial u}{\partial t}) + \frac{\partial}{\partial t}(u,u) dx] dt = 3 \int_0^1 \psi(t) \frac{d}{dt} \int_M (u,u)\,dxdt = 0$$

for any ψ, iff $\int_M (u,u)dx = \text{const}$.

QED

The nature of weak solution thus obtained is quite unclear. Possibly, it is a piecewise-smooth flow with surfaces of a tangent discontinuity; but more complicated singularities are also possible. This question requires more work.

4. The problem of the shortest braid has another physical interpretation: it is a very particular case of the problem of magnetic equilibrium in a perfectly conducting incompressible fluid. Suppose that the fluid fills a 3-dimensional domain G, and initially there is a magnetic field $H(x)$ in G. This field remains incompressible: $\nabla \cdot H = 0$. Now suppose that the fluid in G is deformed, i.e. any point $x \in G$ goes to another point $\xi(x) \in G$, where ξ is a volume-preserving diffeomorphism of G. If the fluid is perfectly conducting, then the magnetic field is transported by the fluid, so that $H(x)$ is transformed into $(\xi_* H)(x) = \xi'(\xi^{-1}(x)) \cdot H(\xi^{-1}(x))$.

The energy of magnetic field H is given by

$$W(H) = \int_G \frac{1}{2} |H(x)|^2 \, d^3x \ , \tag{31}$$

where H is measured in appropriate units. The system tends to minimize its energy; therefore, if $H_0(x)$ is a given initial field, the system relaxes to a configuration (by a diffeomorphism ξ), such that the energy of the transported field $W(\xi_* H_0)$ is minimal. This configuration, if it exists, is called a magnetic equilibrium (see the paper [4] in this volume, concerning other aspects of magnetic equilibrium).

In general, we still do not know whether for a given field H_0 magnetic equilibrium can be achieved, and what is its nature (especially, what may be said about possible singularities of the field). Recently Laurence and Stredulinsky [6] proved that generalized magnetic equilibrium exists in the 2-dimensional case, and in a wider class of problems where the field does not depend on one coordinate.

There is however another situation, where we can say something about existence of magnetic equilibrium. This is the case of a long and thin cylinder G with magnetic field directed almost parallel to its axis. So, suppose that G is a long and thin tube, filled by an incompressible and perfectly conducting fluid; it is fitted by a pair of perfectly conducting plugs, so that the magnetic lines are fixed at the ends. Initial magnetic field H_0 has nearly constant intensity, and magnetic lines are filling G, being tangent to its side boundary. Let x_1, x_2, z be coordinates in $\mathbb{R}^3$, the base of the cylinder G being a bounded 2-d domain M, and its axis being parallel to the z-axis, so that G is defined as $x \in M$, $0 \leq z \leq L$. Then the z-coordinate is growing monotonically along every field line, and thus defines a mapping h_z from $M \times \{0\}$ to $M \times \{z\}$, $0 \leq z \leq L$. The map h_L is fixed, because the plugs and the fluid are perfectly conducting and, thus, the end-points of every magnetic line are fixed. In first approximation, as $L \gg 1$, the maps h_z are incompressible, and the energy of the field is

$$W = \int_0^L \int_M \frac{1}{2}(1 + |\frac{\partial}{\partial z} h_z(x)|^2)\, dx dz \,, \tag{32}$$

which is equal (up to a constant) to the action of an incompressible flow (where z and h_z replace t and g_t respectively). This approximation is referred to as 'Parker's model' [7], and is used as a model for solar prominences. In particular, physicists conjecture that field configurations minimizing the energy should, generically, have singularities. But every singularity (say, surface of discontinuity) is a place of concentration of electric current. These concentrations ("current layers") should play a big role in dynamics and heating of solar corona and of artificial plasma. In the frame of the Parker model we can use our results concerning existence and regularity of a minimum with minimal action. We proved that there exists a minimal braid having almost everywhere a given field $H(x_1, x_2, z)$; the z-component of this field is everywhere close to 1, while x_1- and x_2-components are *a priori* L^2-functions.

Our hypothesis is that the field $H(x, z)$ is smooth for almost all (x, z). Among possible singularities of H are, first of all, surfaces of tangent discontinuity, i.e. surfaces of discontinuity, such that the field H is tangent to them (it is easy to see that the surface of discontinuity cannot be transversal to H). Then, these surfaces may branch, forming a stratified set. It would be interesting to know whether this singular set has a finite number of strata, or whether there is a countable set of them, or more complicated singularities.

Acknowledgements

I am very grateful to the Isaac Newton Institute for the hospitality, and to the participants of the Program on Geometry and Topology of Fluid Flows for many interesting discussions.

References

1. Arnold, V. & Khesin, B. (1998) *Topological Methods in Hydrodynamics.* Applied Mathematical Sciences Series **125**, Springer Verlag, New York.
2. Brenier, Y. (1989) The least action principle and the related concept of generalized flows for incompressible perfect fluid. *J. Amer. Math. Soc.* **2**(2), 225–255.

3. Ebin, D. & Marsden, J. (1979) Groups of diffeomorphisms and the motion of an incompressible fluid. *Ann. of Math.* **92**(2), 102–163.
4. Khesin, B. (2001) Topology bounds the energy. In *An Introduction to the Geometry and Topology of Fluid Flows* (ed. R.L. Ricca), this volume. NATO-ASI Series: Mathematics, Kluwer, Dordrecht, The Netherlands.
5. Eliashberg, Y. & Ratiu, T. (1991) The diameter of the symplectomorphism group is infinite. *Invent. Math.* **103**(2), 327–340.
6. Laurence, P. & Stredulinsky, E. (2000) Two-dimensional magnetohydrodynamic equilibria with prescribed topology. *Comm. Pure Appl. Math.* **53**(9), 1177–1200.
7. Parker, E. N. (1994) *Spontaneous Current Sheets in Magnetic Fields.* International Series on Astronomy and Astrophysics, Oxford University Press, New York.
8. Shnirelman, A. (1985) The geometry of the group of diffeomorphisms and the dynamics of an ideal incompressible fluid. *Math. USSR Sbornik* **56**(1), 79–105.
9. Shnirelman A. (1994) Generalized fluid flows, their approximation and applications. *Geom. Funct. Anal.* **4**(5), 586–620.
10. Shnirelman, A. (1993) Attainable diffeomorphisms. *Geom. Funct. Anal.* **3**(3), 279–294.

VARIATIONAL PRINCIPLES, GEOMETRY AND TOPOLOGY OF LAGRANGIAN-AVERAGED FLUID DYNAMICS

DARRYL D. HOLM
Theoretical Division and Center for Nonlinear Studies
Los Alamos National Laboratory, MS B284
Los Alamos, NM 87545, USA
E-mail: *dholm@lanl.gov*

Abstract. The Lagrangian average (LA) of the ideal fluid equations preserves their transport structure. This transport structure is responsible for the Kelvin circulation theorem of the LA flow and, hence, for its convection of potential vorticity and its conservation of helicity. Lagrangian averaging also preserves the Euler-Poincaré (EP) variational framework that implies the LA fluid equations. This is expressed in the Lagrangian-averaged Euler-Poincaré (LAEP) theorem proven here and illustrated for the Lagrangian average Euler (LAE) equations.

1. Introduction

Decomposition of multiscale problems & scale-up. In turbulence, in climate modeling and in other multiscale fluids problems, a major challenge is "scale-up." This is the challenge of deriving models that correctly capture the mean, or large scale flow – including the influence on it of the rapid, or small scale dynamics. In classical mechanics this sort of problem has been approached by choosing a proper "slow + fast" decomposition and deriving evolution equations for the slow mean quantities by using, say, the standard method of averages. For nondissipative systems in classical mechanics that arise from Hamilton's variational principle, the method of averages may extend to the averaged Lagrangian method, under certain conditions.

Eulerian vs Lagrangian means. In meteorology and oceanography, the averaging approach has a venerable history and many facets. Often this averaging is applied in the geosciences in combination with additional approxi-

R.L. Ricca (ed.), An Introduction to the Geometry and Topology of Fluid Flows, 271–291.

mations involving force balances (for example, geostrophic and hydrostatic balances). It is also sometimes discussed as an initialization procedure that seeks a nearby invariant "slow manifold." Moreover, in meteorology and oceanography, the averaging may be performed in either the Eulerian, or the Lagrangian description. The relation between averaged quantities in the Eulerian and Lagrangian descriptions is one of the classical problems of fluid dynamics.

Generalized Lagrangian mean (GLM). The GLM equations of Andrews & McIntyre [1] systematize the approach to Lagrangian fluid modeling by introducing a slow + fast decomposition of the Lagrangian particle trajectory in general form. In these equations, the Lagrangian mean of a fluid quantity evaluated at the mean particle position is related to its Eulerian mean, evaluated at the displaced fluctuating position. The GLM equations are expressed directly in the Eulerian representation. The Lagrangian mean has the advantage of preserving the fundamental transport structure of fluid dynamics. In particular, the Lagrangian mean commutes with the scalar advection operator and it preserves the Kelvin circulation property of the fluid motion equation.

Compatibility of averaging and reduction of Lagrangians for mechanics on Lie groups. In making slow + fast decompositions and constructing averaged Lagrangians for fluid dynamics, care must generally be taken to see that the averaging and reduction procedures do not interfere with each other. Reduction in the fluid context refers to symmetry reduction of the action principle by the subgroup of the diffeomorphisms that takes the Lagrangian representation to the Eulerian representation of the flow field. The theory for this yields the Euler-Poincaré (EP) equations [12, 13, 15].

Lagrangian averaged Euler-Poincaré (LAEP) equations. The compatibility requirement between averaging and reduction is handled automatically in the Lagrangian averaging (LA) approach. The Lagrangian mean of the action principle for fluids does not interfere with its reduction to the Eulerian representation, since the averaging process is performed at *fixed Lagrangian coordinate.* Thus, the process of taking the Lagrangian mean is compatible with reduction by the particle-relabeling group of symmetries for Eulerian fluid dynamics. In this paper, we perform this reduction of the action principle and thereby place the LA fluid equations such as GLM into the EP framework. In doing this, we demonstrate the variational reduction property of the Lagrangian mean. This is encapsulated in the LAEP Theorem:

Theorem 1.1 (Lagrangian Averaged Euler-Poincaré) *Lagrangian averaging preserves the variational structure of the Euler-Poincaré framework.*

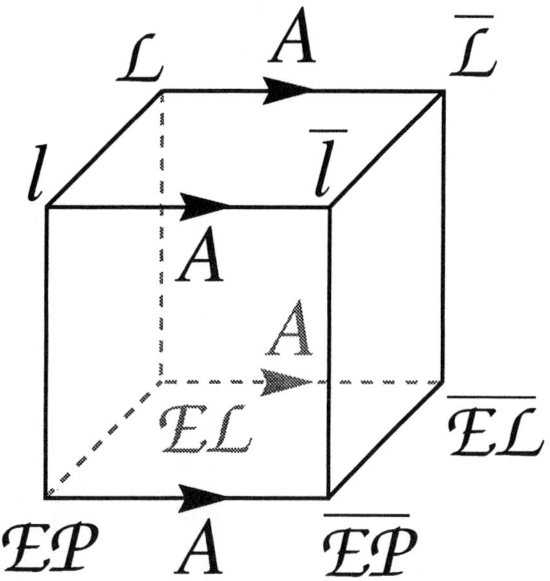

Figure 1. The Averaged EP theorem produces a cube consisting of four equivalence relations on each of its left and right faces, and four commuting diagrams (one on each of its four remaining faces).

According to this theorem, the Lagrangian mean's preservation of the fundamental transport structure of fluid dynamics also extends to preserving the EP variational structure of these equations. This preservation of structure may be *visualized as a cube.* As we shall explain, the LAEP theorem produces a cube shown in Figure 1 consisting of four equivalence relations on each of its left and right faces, and four commuting diagrams (one on each of its four remaining faces).

Euler-Lagrange-Poincaré (ELP) cube. The front and back faces of the ELP cube live in the Eulerian (spatial) and Lagrangian (material) pictures of fluid dynamics, respectively. The top face contains four variational principles at its corners and the bottom face contains their corresponding equations of motion. The horizontal edges represent Lagrangian averaging and are directed from the left to the right. The left face contains the four equivalence relations of the Euler-Poincaré Theorem on its corners and the right face contains the corresponding averaged equivalence relations. Thus, the left and right faces of the ELP cube are equivalence relations, and its front, back, top and bottom faces are commuting diagrams.

The back face of the ELP cube displays the LA preservation of variational structure in the Lagrangian fluid picture. Hamilton's principle with L yields the Euler-Lagrange equations EL in this picture, and Lagrangian averaging A preserves this relation. Namely, Hamilton's principle with the averaged Lagrangian $\bar{L}$ yields the averaged Euler-Lagrange equations $\overline{EL}$. This pair of Hamilton's principles and Euler-Lagrange equations has its counterpart in the Eulerian picture of fluid dynamics - on the front face of the ELP cube – whose variational relations are also exactly preserved by the LA process. The bottom front edge of the cube represents the GLM equations of Andrews & McIntyre [1]. Thus, the GLM equations represent a foundational result for the present theory. The six faces of the ELP cube represent six interlocking equivalence relations and commutative diagrams that enable modeling and Lagrangian averaging to be performed equivalently either at the level of the equations, as in Andrews & McIntyre [1], or at the level of Hamilton's principle. At the level of Hamilton's principle, powerful theorems from other mean field theories are available. An example is Noether's theorem, which relates symmetries of Hamilton's principle to conservation laws of the equations of motion. Fluid conservation laws include mass, momentum and energy, as well as local conservation of potential vorticity. The latter yields the Casimirs of the corresponding Lie-Poisson Hamiltonian formulation of ideal fluid dynamics and is due to the symmetry of relabeling diffeomorphisms admitted by Hamilton's principle for fluid dynamics [2, 14]. In certain cases, the fluid vorticity winding number (called helicity – a topological quantity) is also conserved. Lagrangian averaging preserves all of these conservation laws.[1] Thus, the LA Hamilton's principle yields the LA fluid equations in either the Lagrangian, or the Eulerian fluid picture, and one may transform interchangeably along the edges of the cube in search of physical and mathematical insight.

Remark 1.2 [Eulerian mean] Of course, the preservation of variational structure resulting in the interlocking commuting relationships and conservation laws of the LAEP Theorem is not possible with the Eulerian mean, because the Eulerian mean does *not* preserve the transport structure of fluid mechanics.

Remark 1.3 [Balanced approaches] The LAEP Theorem puts the approach using averaged Hamilton's principles and the method of Lagrangian averaged equations onto equal footing. This is quite a bonus for both approaches to modeling fluids. According to the LAEP Theorem, the averaged Hamilton's principle produces dynamics that is guaranteed to be verified directly by averaging the original equations, and the Lagrangian averaged equations inherit the conservation laws that are available from the symmetries of Hamilton's principle.

[1]We note that the conserved mean topological quantity resulting after Lagrangian averaging is the helicity of the mean fluid vorticity, not the mean of the original helicity.

Outline of the paper. In section 2, we begin by briefly reviewing the mathematical setting of the EP theorem from Holm, Marsden & Ratiu [12, 13]. We state the EP theorem and discuss a few of its implications for continuum mechanics in vector notation. We also sketch its proof, in preparation for proving the corresponding results for the Lagrangian Averaged Euler-Poincaré (LAEP) theorem presented in section 3. Finally, in section 4, we illustrate the LAEP theorem by applying it to incompressible ideal fluids. We also mention recent progress toward closure of these equations as models of fluid turbulence.

2. The Euler-Poincaré Theorem for Fluids with Advected Properties

2.1. MATHEMATICAL SETTING AND STATEMENT OF THE EP THEOREM

The assumptions of the Euler-Poincaré theorem from Holm, Marsden & Ratiu [12] are briefly listed below.

- There is a *right* representation of Lie group G on the vector space V and G acts in the natural way on the *right* on $TG \times V^*$: $(v_g, a)h = (v_g h, ah)$.
- Assume that the function $L : TG \times V^* \to \mathbb{R}$ is right G–invariant.
- In particular, if $a_0 \in V^*$, define the Lagrangian $L_{a_0} : TG \to \mathbb{R}$ by $L_{a_0}(v_g) = L(v_g, a_0)$. Then L_{a_0} is right invariant under the lift to TG of the right action of G_{a_0} on G, where G_{a_0} is the isotropy group of a_0.
- Right G–invariance of L permits one to define $\ell : \mathfrak{g} \times V^* \to \mathbb{R}$ by

$$\ell(v_g g^{-1}, a_0 g^{-1}) = L(v_g, a_0). \tag{1}$$

  Conversely, this relation defines for any $\ell : \mathfrak{g} \times V^* \to \mathbb{R}$ a right G–invariant function $L : TG \times V^* \to \mathbb{R}$.
- For a curve $g(t) \in G$, let

$$u(t) \equiv \dot{g}(t)g(t)^{-1} \in TG/G \cong \mathfrak{g} \tag{2}$$

  and define the curve $a(t)$ as the unique solution of the linear differential equation with time dependent coefficients

$$\dot{a}(t) = -a(t)u(t) \tag{3}$$

  where the action of $u \in \mathfrak{g}$ on the initial condition $a(0) = a_0 \in V^*$ is denoted by concatenation from the right. The solution of (3) can be written as the *advective transport relation*,

$$a(t) = a_0 g(t)^{-1}. \tag{4}$$

Theorem 2.1 (EP Theorem) *The following are equivalent:*

i *Hamilton's variational principle*

$$\delta \int_{t_1}^{t_2} L_{a_0}(g(t), \dot{g}(t))dt = 0 \tag{5}$$

holds, for variations $\delta g(t)$ of $g(t)$ vanishing at the endpoints.

ii *$g(t)$ satisfies the Euler–Lagrange equations for L_{a_0} on G.*

iii *The constrained variational principle*

$$\delta \int_{t_1}^{t_2} \ell\,(u(t), a(t))dt = 0 \tag{6}$$

holds on $\mathbf{g} \times V^*$, using variations of the form

$$\delta u = \frac{\partial \eta}{\partial t} + \mathrm{ad}_u\, \eta, \quad \delta a = -a\, \eta, \tag{7}$$

where $\eta(t) \in \mathbf{g}$ vanishes at the endpoints.

iv *The Euler–Poincaré equations hold on $\mathbf{g} \times V^*$*

$$\frac{\partial}{\partial t}\frac{\delta \ell}{\delta u} = -\mathrm{ad}^*_u \frac{\delta \ell}{\delta u} + \frac{\delta \ell}{\delta a} \diamond a. \tag{8}$$

2.2. DISCUSSION OF THE EP EQUATIONS IN VECTOR NOTATION

When mass is the *only* advected quantity, the EP motion equation (8) and the advection relation (3) for mass conservation are written as

$$\Big(\frac{\partial}{\partial t} + \mathrm{ad}^*_u\Big)\frac{\delta \ell}{\delta u} - \frac{\delta \ell}{\delta D} \diamond D = 0\,, \quad \text{and} \quad \frac{\partial D}{\partial t} = -\,\mathcal{L}_u D\,. \tag{9}$$

Here $\mathcal{L}_u$ denotes the Lie derivative with respect to velocity u, and the operations ad^* and $\diamond$ are defined using the L_2 pairing $\langle f, g\rangle = \int fg\, d^3x$. The ad^* operation is defined as (minus) the L_2 dual of the Lie algebra operation, ad, for vector fields, namely

$$-\Big\langle \mathrm{ad}^*_u \frac{\delta \ell}{\delta u}, \eta\Big\rangle = \Big\langle \frac{\delta \ell}{\delta u}, \mathrm{ad}_u\, \eta\Big\rangle\,. \tag{10}$$

In vector notation $\mathrm{ad}_u\, \eta$ is expressed as

$$\mathrm{ad}_u\, \eta = \eta u - u \eta = -\,[u, \eta] = -\,\mathrm{ad}_\eta\, u = u \cdot \nabla \eta - \eta \cdot \nabla u\,. \tag{11}$$

The diamond operation $\diamond$ is defined as (minus) the L_2 dual of the Lie derivative, namely,

$$-\Big\langle \frac{\delta \ell}{\delta a} \diamond a, \eta\Big\rangle = \Big\langle \frac{\delta \ell}{\delta a}, \mathcal{L}_\eta a\Big\rangle\,. \tag{12}$$

Here a and $\delta\ell/\delta a$ are dual tensors and $(\delta\ell/\delta a)\diamond a$ is a one-form density (dual to vector fields under L_2 pairing). In vector notation the diamond operation $\diamond$ for the example of the density D becomes

$$-\Big\langle \frac{\delta\ell}{\delta D}\diamond D,\eta\Big\rangle = \Big\langle \frac{\delta\ell}{\delta D},\mathcal{L}_\eta D\Big\rangle = \Big\langle \frac{\delta\ell}{\delta D},\mathrm{div}(D\eta)\Big\rangle = -\Big\langle D\nabla\frac{\delta\ell}{\delta D},\eta\Big\rangle . \tag{13}$$

Thus, the EP motion equation (8) may be written in Cartesian components as

$$\frac{\partial}{\partial t}\frac{\delta\ell}{\delta u^i} + \underbrace{\frac{\partial}{\partial x^j}\Big(\frac{\delta\ell}{\delta u^i}u^j\Big) + \frac{\delta\ell}{\delta u^j}\frac{\partial u^j}{\partial x^i}}_{\equiv\left(\mathrm{ad}^*_u\frac{\delta\ell}{\delta u}\right)_i} - \underbrace{D\frac{\partial}{\partial x^i}\Big(\frac{\delta\ell}{\delta D}\Big)}_{\equiv\left(\frac{\delta\ell}{\delta D}\diamond D\right)_i} = 0\,, \tag{14}$$

and the advection relation for the mass density $D \in V^*$ satisfies,

$$\Big(\frac{\partial}{\partial t}+\mathcal{L}_u\Big)\Big(D\,d^3x\Big) = 0\,, \quad \text{or} \quad \frac{\partial D}{\partial t} = -\,\mathrm{div}(D\mathbf{u})\,. \tag{15}$$

Remark 2.2

- In passing from coordinate-free forms to their component expressions, we shall write tensors in a Cartesian basis. For example, we shall include the volume form in the mass density by denoting it as Dd^3x.
- The EP motion equation and mass advection equation may also be written equivalently using Lie derivative notation as

$$\Big(\frac{\partial}{\partial t}+\mathcal{L}_u\Big)\frac{\delta\ell}{\delta u} - \frac{\delta\ell}{\delta D}\diamond D = 0\,, \quad \text{and} \quad \Big(\frac{\partial}{\partial t}+\mathcal{L}_u\Big)D = 0\,. \tag{16}$$

  The equivalence here of $\mathcal{L}_u$ and ad^*_u arises because $\delta\ell/\delta u$ is a one-form density and the equality $\mathrm{ad}^*_u\mu = \mathcal{L}_u\mu$ holds for any one-form density μ.
- In the Lie derivative notation, one proves the *Kelvin-Noether circulation theorem* immediately as a corollary, by

$$\frac{d}{dt}\oint_{c(u)}\frac{1}{D}\frac{\delta\ell}{\delta u} = \oint_{c(u)}\Big(\frac{\partial}{\partial t}+\mathcal{L}_u\Big)\frac{1}{D}\frac{\delta\ell}{\delta u} = \oint_{c(u)}\frac{1}{D}\frac{\delta\ell}{\delta D}\diamond D\,, \tag{17}$$

  for any closed curve $c(u)$ that moves with the fluid. In vector notation, this is seen as

$$\frac{d}{dt}\oint_{c(u)}\frac{1}{D}\frac{\delta\ell}{\delta\mathbf{u}}\cdot d\mathbf{x} = \oint_{c(u)}\nabla\frac{\delta\ell}{\delta D}\cdot d\mathbf{x} = 0\,. \tag{18}$$

- *Helicity conservation.* In Lie derivative notation, one may rewrite the Kelvin circulation theorem as

$$\Big(\frac{\partial}{\partial t}+\mathcal{L}_u\Big)v + dp = 0\,,$$

where

$$v = \mathbf{v} \cdot d\mathbf{x} = \frac{1}{D} \frac{\delta \ell}{\delta \mathbf{u}} \cdot d\mathbf{x}\,, \quad \text{and} \quad p = -\frac{\delta \ell}{\delta D}\,.$$

Therefore,

$$\Big(\frac{\partial}{\partial t} + \mathcal{L}_u\Big)(v \wedge dv) = -d\,(p\,dv)\,,$$

where $\wedge$ is the exterior product of differential forms. In vector notation, this is the helicity equation,

$$\frac{\partial}{\partial t}(\mathbf{v} \cdot \operatorname{curl} \mathbf{v}) + \operatorname{div}\left(\mathbf{u}\,(\mathbf{v} \cdot \operatorname{curl} \mathbf{v}) + p \operatorname{curl} \mathbf{v}\right) = 0\,. \tag{19}$$

Consequently, for homogeneous boundary conditions one finds conservation of helicity $\Lambda = \int v \wedge dv = \int \mathbf{v} \cdot \operatorname{curl} \mathbf{v}\, d^3x$. The helicity Λ is a topological quantity that measures the linkage number of the lines of curl $\mathbf{v}$, the fluid vortex lines in this case.

2.3. PROOF OF THE EP THEOREM

Proof. The equivalence of **i** and **ii** holds for any configuration manifold and so, in particular, it holds in this case.

The following string of equalities shows that **iii** is equivalent to **iv**.

$$\begin{aligned}
0 &= \delta \int_{t_1}^{t_2} \ell(u, a)\, dt = \int_{t_1}^{t_2} \Big\langle \frac{\delta \ell}{\delta u}, \delta u \Big\rangle + \Big\langle \frac{\delta \ell}{\delta a}, \delta a \Big\rangle\, dt \\
&= \int_{t_1}^{t_2} \Big\langle \frac{\delta \ell}{\delta u}, \frac{\partial \eta}{\partial t} + \mathrm{ad}_u\, \eta \Big\rangle - \Big\langle \frac{\delta \ell}{\delta a}, \mathcal{L}_\eta\, a \Big\rangle\, dt \\
&= -\int_{t_1}^{t_2} \Big\langle \frac{\partial}{\partial t} \frac{\delta \ell}{\delta u} + \mathrm{ad}^*_u \frac{\delta \ell}{\delta u} - \frac{\delta \ell}{\delta a} \diamond a\,, \eta \Big\rangle\, dt\,.
\end{aligned} \tag{20}$$

Finally we show that **i** and **iii** are equivalent. First note that the G–invariance of $L : TG \times V^* \to \mathbb{R}$ and the definition of $a(t) = a_0 g(t)^{-1}$ imply that the integrands in (5) and (6) are equal. Moreover, all variations $\delta g(t) \in TG$ of $g(t)$ with fixed endpoints induce and are induced by variations $\delta u(t) \in \mathbf{g}$ of $u(t)$ of the form $\delta u = \partial \eta / \partial t + \mathrm{ad}_u\, \eta$ with $\eta(t) \in \mathbf{g}$ vanishing at the endpoints. The relation between $\delta g(t)$ and $\eta(t)$ is given by $\eta(t) = \delta g(t) g(t)^{-1}$. This is the proof first given in [12].

QED

2.4. EULER'S EQUATION FOR AN INCOMPRESSIBLE FLUID

For an incompressible fluid, the EP theorem 2.1 yields Euler's equations as

$$\frac{\partial}{\partial t}\frac{\delta \ell}{\delta \mathbf{u}} = -\mathrm{ad}^*_{\mathbf{u}}\frac{\delta \ell}{\delta \mathbf{u}} + \frac{\delta \ell}{\delta D} \diamond D\,, \tag{21}$$

for the reduced Lagrangian

$$\ell = \int \frac{1}{2}D|\mathbf{u}|^2 - p(D-1)\,d^3\mathbf{u}\,. \tag{22}$$

Here the pressure p is a Lagrange multiplier that imposes incompressibility. The variational derivatives of this Lagrangian are given by

$$\delta\ell = \int D\mathbf{u}\cdot\delta\mathbf{u} + \Big(\frac{1}{2}|\mathbf{u}|^2 - p\Big)\delta D - (D-1)\delta p\; d^3\mathbf{u}\,. \tag{23}$$

The expected Euler equation for incompressible fluids is found upon setting $D = 1$ in equation (21) as

$$\frac{\partial}{\partial t}\mathbf{u} + \mathbf{u}\cdot\nabla\mathbf{u} + \nabla p = 0\,. \tag{24}$$

The auxiliary advection relation for the mass density D is the continuity equation

$$\frac{\partial D}{\partial t} = -\,\mathrm{div}(D\mathbf{u})\,, \tag{25}$$

which, as usual, ensures incompressibility via the constraint $D = 1$.

3. Lagrangian Averaged Euler-Poincaré (LAEP) Theory

We shall place the GLM (Generalized Lagrangian Mean) theory of Andrews & McIntyre [1] into the Euler-Poincaré framework discussed in the previous section.

3.1. GLM THEORY FROM A GEOMETRIC VIEWPOINT

In an ensemble of fluid flows, the Lagrangian fluid parcels that correspond to the same initial coordinate $\mathbf{x}_0$ in each realization have an ensemble average position as a function of time given by $\mathbf{x}(\mathbf{x}_0, t)$. This Lagrangian mean trajectory may be assigned (lifted) to a time dependent curve $\tilde{g}(t) \in G$ in the diffeomorphism group as

$$\mathbf{x}(\mathbf{x}_0, t) = \tilde{g}(t)\cdot\mathbf{x}_0 \quad \text{(Lagrangian mean position)}\,, \tag{26}$$

via the action of the diffeomorphism group G on the vector space of reference positions $\mathbf{x}_0 \in \mathbb{R}^3$. Relative to this time dependent mean position, the current position $\mathbf{x}^{\xi}(\mathbf{x}_0, t)$ of a fluid parcel possessing the Lagrangian label $\mathbf{x}_0$ in the reference configuration for a given realization ξ in the ensemble of fluid flows may be expressed as a composition of functions,

$$\mathbf{x}^{\xi}(\mathbf{x}_0, t) = \Xi\,(\mathbf{x}(\mathbf{x}_0, t), t) \equiv \Xi\,(t) \cdot (\tilde{g}(t) \cdot \mathbf{x}_0) \quad \text{(current position)}\,. \tag{27}$$

The mean of this ensemble of trajectories taken at fixed Lagrangian coordinate $\mathbf{x}_0$ is denoted as

$$\mathbf{x}(\mathbf{x}_0, t) = \overline{\mathbf{x}^{\xi}}\,(\mathbf{x}_0, t)\,. \tag{28}$$

As with any averaging method, once the GLM average is taken, information is lost and one cannot go back to the exact solution from the mean solution. In particular, different current configurations $\mathbf{x}^{\xi}\,(\mathbf{x}_0, t)$ may have the same Lagrangian mean $\overline{\mathbf{x}^{\xi}}\,(\mathbf{x}_0, t)$. However, this is of no concern in GLM theory: one may still assume that a *known current configuration in a given realization* of the ensemble may be related uniquely to the GL-mean configuration at any time by the smooth invertible map $\Xi(t)$. Viewed this way, the GLM solution is simply a Lagrangian reference configuration that changes with time according to the mean flow governed by the GLM equations and this mean flow itself is generated as a diffeomorphism from a fixed reference configuration.

This idea may also be applied in treating the average of a single solution. For example, if instead of an ensemble average we average over the phase of a single solution, say, then we think of the GLM solution as a phase average that projects onto the slow part of a solution that admits a decomposition into its mean and fluctuating parts, as in the WKB method. This phase-average projection is not invertible, and we cannot reconstruct the fast phase from the WKB amplitude alone. However, it is standard in the WKB method to use the mean solution as a reference configuration in which to specify the decomposition of the exact solution.

This idea lifts to the diffeomorphism group as the *factorization* of the Lagrange-to-Euler map as a product of diffeomorphisms,

$$g(t) = \Xi\,(t) \cdot \tilde{g}(t)\,. \tag{29}$$

This is illustrated in Figure 2.

Here the first factor $\tilde{g}(t)$ is defined by the choice of which averaging process shall be applied to the ensemble of fluid flows. This averaging process may be lifted to the diffeomorphism group action and denoted as

$$\overline{\mathbf{x}^{\xi}}(\mathbf{x}_0, t) = \bar{g}(t) \cdot \mathbf{x}_0 = \overline{\Xi\,(t) \cdot \tilde{g}(t)} \cdot \mathbf{x}_0 = \tilde{g}(t) \cdot \mathbf{x}_0 = \mathbf{x}(\mathbf{x}_0, t)\,, \tag{30}$$

or, formally,

$$\bar{g}(t) = \overline{\Xi\,(t) \cdot \tilde{g}(t)} = \tilde{g}(t)\,. \tag{31}$$

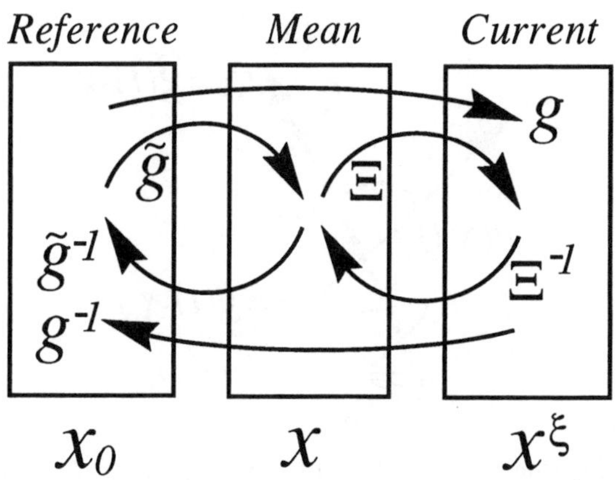

Figure 2. GLM theory factorizes the Lagrange to Euler map at a given time by first mapping the reference configuration to the mean position, then mapping that to the current position.

GLM theory selects averages that satisfy the Reynolds projection property,

$$\overline{\overline{\mathbf{x}^{\xi}}}(\mathbf{x}_0,t) = \overline{\mathbf{x}^{\xi}}\,(\mathbf{x}_0,t) = \mathbf{x}(\mathbf{x}_0,t)\,. \tag{32}$$

This property lifts to the action of the diffeomorphism group as

$$\bar{\tilde{g}}(t)\cdot\mathbf{x}_0 = \tilde{g}(t)\cdot\mathbf{x}_0\,. \tag{33}$$

We shall also write this formula as $\bar{\tilde{g}}(t) = \tilde{g}(t)$, without distinguishing between the average over the ensemble of Lagrangian trajectories and the diffeomorphism group action to which the ensemble mean motion corresponds. The ensemble average, for example, does satisfy the Reynolds projection property.

Remark 3.1 Thus, GLM theory first averages the Lagrangian trajectories over an ensemble of fluid flows. The evolution of the Lagrangian trajectories of these flows corresponds to the actions on a given reference configuration of a set of curves in the diffeomorphism group parameterized by time. GLM theory relates the action of the diffeomorphism group for each realization $g(t)$ to the action $\tilde{g}(t)$ for the ensemble-average result, by assuming that $g(t)\cdot\tilde{g}^{-1}(t) = \Xi\,(t)$ is also a diffeomorphism. This accomplishes the GLM factorization $g(t) = \Xi\,(t)\cdot\tilde{g}(t)$ for each member of the ensemble, modulo the flows that leave the given reference configuration invariant.

The composition of maps $g(t) = \Xi\,(t)\cdot\tilde{g}(t)$ yields via the chain rule the following *velocity relation*,

$$\dot{\mathbf{x}}^{\xi}(\mathbf{x}_0,t) = \frac{\partial}{\partial t}\Big|_{x}\Xi\,(\mathbf{x},t) + \frac{\partial\Xi}{\partial\mathbf{x}}\cdot\dot{\mathbf{x}}\,, \tag{34}$$

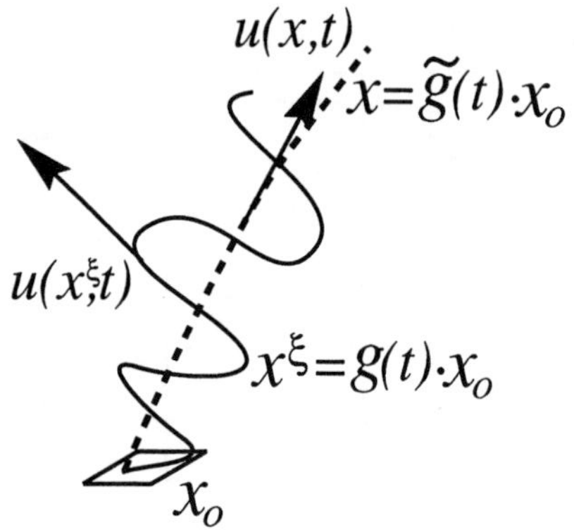

Figure 3. The GLM velocities $\mathbf{u}(\mathbf{x}^\xi, t)$ and $\bar{\mathbf{u}}^L(\mathbf{x}, t)$ are tangent to the current and mean trajectories, $\mathbf{x}^\xi$ and $\mathbf{x}$, respectively.

or, equivalently,

$$\dot{g}(t) \cdot \mathbf{x}_0 = \dot{\Xi}(t) \cdot \mathbf{x} + T\Xi \cdot (\dot{\tilde{g}}(t) \cdot \mathbf{x}_0)\,. \tag{35}$$

By invertibility, $\mathbf{x}_0 = g^{-1}(t) \cdot \mathbf{x}^\xi = \tilde{g}^{-1}(t) \cdot \mathbf{x}$. Consequently, one may define the fluid parcel velocity at the current position in terms of a vector field evaluated at the mean position as,

$$\mathbf{u}(\mathbf{x}^\xi, t) = \dot{g} \cdot g^{-1}(t) \cdot \mathbf{x}^\xi = \dot{g} \cdot \tilde{g}^{-1}(t) \cdot \mathbf{x} = \mathbf{u}^\xi(\mathbf{x}, t)\,.$$

Hence, by using the velocity relation (35) one finds,

$$\begin{aligned} \mathbf{u}^\xi(\mathbf{x}, t) &= \dot{\Xi}(t) \cdot \mathbf{x} + T\Xi \cdot (\dot{\tilde{g}}\tilde{g}^{-1}(t) \cdot \mathbf{x}) \\ &\equiv \frac{\partial}{\partial t}\Big|_x \Xi(\mathbf{x}, t) + \frac{\partial \Xi}{\partial \mathbf{x}} \cdot \bar{\mathbf{u}}^L(\mathbf{x}, t)\,. \end{aligned} \tag{36}$$

Here the *Lagrangian mean velocity* $\bar{\mathbf{u}}^L$ is defined as

$$\bar{\mathbf{u}}^L(\mathbf{x}, t) \equiv \overline{\mathbf{u}^\xi(\mathbf{x}, t)} = \overline{\dot{g}\tilde{g}^{-1}(t)} \cdot \mathbf{x} = \dot{\tilde{g}}(t)\tilde{g}(t)^{-1} \cdot \mathbf{x} = \dot{\tilde{g}}(t) \cdot \mathbf{x}_0 = \dot{\mathbf{x}}(\mathbf{x}_0, t)\,. \tag{37}$$

In the third equality we used the projection property of the averaging process and assumed that $\overline{\dot{g}} = \dot{\overline{g}} = \dot{\tilde{g}}$, so that

$$\bar{\mathbf{u}}^L(\mathbf{x}, t) = \dot{\tilde{g}}(t)\tilde{g}(t)^{-1} \cdot \mathbf{x} \equiv \tilde{\mathbf{u}}(\mathbf{x}, t)\,.$$

Thus, the Lagrangian mean velocity vector satisfies $\bar{\mathbf{u}}^L = \tilde{\mathbf{u}}$, so $\bar{\mathbf{u}}^L$ is tangent to the mean motion associated with $\tilde{g}(t)$ This is shown in Figure 3.

Hence, we may write equation (36) in terms of the *mean material time derivative* D^L/Dt as

$$\mathbf{u}^\xi(\mathbf{x},t) = \Big(\frac{\partial}{\partial t} + \bar{\mathbf{u}}^L\cdot\nabla\Big)\,\Xi(\mathbf{x},t) \equiv \frac{D^L}{Dt}\,\Xi\,(\mathbf{x},t)\,. \tag{38}$$

Likewise, for any fluid quantity χ one may define χ^ξ as the composition of functions

$$\chi^\xi(\mathbf{x},t) = \chi(\mathbf{x}^\xi,t) = \chi(\Xi\,(\mathbf{x},t),t)\,.$$

Taking the mean material time derivative and using the definition of D^L/Dt in equation (38) yields the *advective derivative relation*,

$$\begin{aligned}\frac{D^L}{Dt}\chi^\xi &= \Big(\frac{\partial\chi}{\partial t}\Big)^\xi + T\chi\cdot\frac{D^L}{Dt}\Xi\,(\mathbf{x},t)\\ &= \Big(\frac{\partial\chi}{\partial t} + T\chi\cdot\mathbf{u}\Big)^\xi \equiv \Big(\frac{D\chi}{Dt}\Big)^\xi\,.\end{aligned} \tag{39}$$

As in equation (37) for the velocity, the *Lagrangian mean* $\bar\chi^L$ of any other fluid quantity χ is defined as

$$\bar\chi^L(\mathbf{x},t) \equiv \overline{\chi^\xi(\mathbf{x},t)} = \overline{\chi(\mathbf{x}^\xi,t)} = \overline{\chi(\,\Xi\,(t)\cdot\mathbf{x},t)} = \overline{\chi(\,g(t)\cdot\mathbf{x}_0,t)}\,.$$

Taking the Lagrangian mean of equation (39) and once again using its projection property yields

$$\dot{\bar\chi}^L = \frac{D^L}{Dt}\bar\chi^L = \overline{\Big(\frac{D\chi}{Dt}\Big)}^L = \bar{\dot\chi}^L\,, \quad\text{so that}\quad \frac{D^L}{Dt}\chi^\ell = \Big(\frac{D\chi}{Dt}\Big)^\ell\,,$$

where $\chi^\ell = \chi^\xi - \bar\chi^L$ is the Lagrangian disturbance of χ satisfying $\overline{\chi^\ell} = 0$.

Remark 3.2 The Lagrangian mean commutes with the material derivative. Hence, the advective derivative relation (39) decomposes additively, as

$$\frac{D^L}{Dt}\Big(\chi^L + \chi^\ell\Big) = \overline{\Big(\frac{D\chi}{Dt}\Big)}^L + \Big(\frac{D\chi}{Dt}\Big)^\ell\,. \tag{40}$$

3.2. MEAN ADVECTED QUANTITIES AND THEIR TRANSFORMATIONS

Advective transport by $g(t)$ and $\tilde g(t)$ is defined by

$$a(\mathbf{x}^\xi,t) = a_0\cdot g^{-1}(t) \quad\text{and}\quad \tilde a(\mathbf{x},t) = a_0\cdot\tilde g^{-1}(t)\,,$$

where $a_0 = a(\mathbf{x}_0,0) = \tilde a(\mathbf{x}_0,0)$, with $a,\tilde a \in V^*$ and the factorization $g(t) = \Xi\,(t)\cdot\tilde g(t)$ implies

$$\tilde a(\mathbf{x},t) = a\cdot\,\Xi\,(\mathbf{x},t)\,.$$

Note that the right side of this equation is potentially rapidly varying, but the left side is a mean advected quantity.

Since a and $\tilde{a}$ refer to the same initial conditions, a_0, one may define a *transformation factor* $\mathcal{F}(\mathbf{x},t)$ as

$$a_0 \cdot \tilde{g}^{-1}(t) = \tilde{a}(\mathbf{x},t) = a \cdot \Xi\,(\mathbf{x},t) = a(\mathbf{x}^\xi,t) \equiv \mathcal{F}(\mathbf{x},t) \cdot a^\xi(\mathbf{x},t)\,. \qquad (41)$$

The transformation factor $\mathcal{F}(\mathbf{x},t)$ in this equation depends on the tensor type of a. For example, one computes formula (41) for an advected density as

$$\Big(D(\mathbf{x}_0) d^3\mathbf{x}_0\Big) \cdot \tilde{g}^{-1}(t) = D^\xi(\mathbf{x},t) \det(T\Xi)\, d^3\mathbf{x} = \tilde{D}(\mathbf{x},t)\, d^3\mathbf{x}\,. \qquad (42)$$

Thus, for an *advected density*, D, one finds

$$D^\xi \det(T\Xi)(\mathbf{x},t) = \tilde{D}(\mathbf{x},t)\,, \quad \mathcal{F}(\mathbf{x},t) = \det(T\Xi)\,, \quad \frac{\partial}{\partial t}\tilde{D} = -\operatorname{div}(\tilde{D}\tilde{\mathbf{u}})\,. \qquad (43)$$

For an *advected scalar function*, s, one finds

$$s^\xi(\mathbf{x},t) = \tilde{s}(\mathbf{x},t) = \bar{s}^L(\mathbf{x},t)\,, \quad \mathcal{F} = 1\,, \quad \frac{\partial}{\partial t}\tilde{s} = -\,\tilde{\mathbf{u}} \cdot \nabla \tilde{s}\,. \qquad (44)$$

For an *advected vector field*, $\mathbf{B}$, one finds

$$K^i_j\, B^{\xi\, j}(\mathbf{x},t) = \tilde{B}^i(\mathbf{x},t)\,, \quad \text{and} \quad K^i_j = \det(T\Xi\,)\,(T\Xi^{-1})^i_j\,. \qquad (45)$$

Thus, in the case of an advected vector field, one has

$$\mathsf{K} \cdot \mathbf{B}^\xi(\mathbf{x},t) = \tilde{\mathbf{B}}(\mathbf{x},t)\,, \quad \text{with} \quad \mathcal{F} = \mathsf{K} \equiv \det(T\Xi\,)\, T\Xi^{-1}\,, \qquad (46)$$

and an advection relation (e.g., a frozen-in magnetic field) given by

$$\frac{\partial}{\partial t}\tilde{\mathbf{B}} = -\,\tilde{\mathbf{u}} \cdot \nabla \tilde{\mathbf{B}} + \tilde{\mathbf{B}} \cdot \nabla \tilde{\mathbf{u}}\,. \qquad (47)$$

Finally, for an *advected symmetric tensor* S one finds

$$(T\Xi^T \cdot S^\xi \cdot T\Xi\,)_{ij} = \tilde{S}_{ij}\,, \qquad (48)$$

whose advection relation is obtained as in the other cases.

In each case, the corresponding transformation factor appears in a *variational relation* for an advected quantity, expressed via equation (41) as

$$\delta a^\xi = \delta\,(\mathcal{F}^{-1} \cdot \tilde{a}) = \mathcal{F}^{-1} \cdot \delta\tilde{a} + (\delta\mathcal{F}^{-1}) \cdot \tilde{a}\,. \qquad (49)$$

This formula will be instrumental in establishing the main result.

3.3. LAGRANGIAN AVERAGED EULER-POINCARÉ (LAEP) THEOREM

Let the following list of assumptions hold [12, 13].

- There is a *right* representation of Lie group G on the vector space V and G acts in the natural way on the *right* on $TG \times V^*$: $(v_g, a)h = (v_g h, ah)$.
- The function $L : TG \times V^* \to \mathbb{R}$ is right G–invariant.
- In particular, if $a_0 \in V^*$, define the Lagrangian $L_{a_0} : TG \to \mathbb{R}$ by $L_{a_0}(v_g) = L(v_g, a_0)$. Then L_{a_0} is right invariant under the lift to TG of the right action of G_{a_0} on G, where G_{a_0} is the isotropy group of a_0.
- Right G–invariance of L permits one to define $\ell : \mathbf{g} \times V^* \to \mathbb{R}$ by $\ell(v_g g^{-1}, a_0 g^{-1}) = L(v_g, a_0)$. Conversely, this relation defines for any $\ell : \mathbf{g} \times V^* \to \mathbb{R}$ a right G–invariant function $L : TG \times V^* \to \mathbb{R}$.
- For a curve $g(t) \in G$, let $u(t) \equiv \dot{g}(t) g(t)^{-1} \in TG/G \cong \mathbf{g}$ and define the curve $a(t)$ as the unique solution of the linear differential equation with time dependent coefficients $\dot{a}(t) = -a(t)u(t)$ where the action of $u \in \mathbf{g}$ on the initial condition $a(0) = a_0 \in V^*$ is denoted by concatenation from the right. This solution can be written as the *advective transport relation*, $a(t) = a_0 g(t)^{-1}$.
- The GLM factorization holds, $g(t) = \Xi(t) \cdot \tilde{g}(t)$, with

$$
\bar{g}(t) = \overline{\Xi(t) \cdot \tilde{g}(t)} = \tilde{g}(t) \text{ and } \bar{\tilde{g}}(t) = \tilde{g}(t) \ .
$$

Theorem 3.3 (LAEP Theorem) *The following are equivalent:*

$\bar{\text{i}}$ *The averaged Hamilton's principle holds*

$$
\delta \int_{t_1}^{t_2} \overline{L_{a_0}(g(t), \dot{g}(t))} \, dt = 0 \tag{50}
$$

for variations $\delta g(t)$ of $g(t)$ vanishing at the endpoints.

$\overline{\text{ii}}$ *The averaged Euler–Lagrange equations for $\bar{L}_{a_0}$ are satisfied on $T^*\tilde{G}$,*

$$
\overline{\frac{\delta L_{a_0}}{\delta g} \cdot T\Xi} - \overline{\frac{d}{dt} \frac{\delta L_{a_0}}{\delta \dot{g}} \cdot T\Xi} = 0 \tag{51}
$$

$\overline{\text{iii}}$ *The averaged constrained variational principle*

$$
\delta \int_{t_1}^{t_2} \overline{\ell\,(u(t), a(t))} dt = 0 \tag{52}
$$

holds, using the chain-rule induced variations

$$
\begin{aligned}
\delta u &= (\partial_t + \mathrm{ad}_u)\eta' + (T\Xi \cdot (\partial_t + \mathrm{ad}_{\tilde{u}})\tilde{\eta}\,)\,\Xi^{-1} \\
\delta a &= -\,a\,\eta = \delta(\mathcal{F}^{-1} \cdot \tilde{a})\,\Xi^{-1} \\
&= -\,a\,\eta' - (\mathcal{F}^{-1} \cdot (\tilde{a}\,\tilde{\eta}))\,\Xi^{-1}\,,
\end{aligned} \tag{53}
$$

where Lie derivatives of advected quantities by the vector fields $\eta'(t) \equiv \delta\Xi\,\Xi^{-1}$, $\tilde\eta(t) \equiv \delta\tilde g\,\tilde g^{-1}$ *and*

$$\eta \equiv \delta g\, g^{-1} = \eta' + (T\Xi \cdot \tilde\eta)\Xi^{-1}, \tag{54}$$

are indicated by concatenation on the right and these three vector fields all vanish at the endpoints.

iv *The Euler–Poincaré (EP) equation holds on* $\mathbf{g} \times V^*$

$$\Big(\frac{\partial}{\partial t} + \mathrm{ad}^*_u\Big)\frac{\delta \ell}{\delta u} = \frac{\delta \ell}{\delta a} \diamond a\,, \tag{55}$$

and the Lagrangian averaged Euler–Poincaré (LAEP) equation holds on $\tilde{\mathbf{g}} \times \tilde V^*$

$$\Big(\frac{\partial}{\partial t} + \mathrm{ad}^*_{\bar{\tilde u}}\Big)\overline{\Big(\frac{\delta \ell}{\delta u^{\xi}} \cdot T\Xi\Big)} = \overline{\Big(\frac{\delta \ell}{\delta a^{\xi}} \cdot \mathcal{F}^{-1}\Big)} \diamond \tilde a\,. \tag{56}$$

Notation. As discussed earlier, the operations ad* and $\diamond$ in equations (55) and (56) are defined by using the L_2 pairing $\langle f, g\rangle = \int fg\, d^3x$. The ad* operation is defined as (minus) the L_2 dual of the Lie algebra operation, ad, or commutator, $\mathrm{ad}_u\,\eta = -\,[u,\eta]$, for vector fields, $-\,\langle \mathrm{ad}^*_u\,\mu, \eta\rangle \equiv \langle \mu, \mathrm{ad}_u\,\eta\rangle$. The diamond operation $\diamond$ is defined as (minus) the L_2 dual of the Lie derivative, namely, $\langle b \diamond a\,,\,\eta\rangle \equiv -\langle b\,,\,\mathcal{L}_\eta a\rangle = -\langle b\,,\,a\,\eta\rangle$, where $\mathcal{L}_\eta a$ denotes the Lie derivative with respect to vector field η of the tensor a, and a and b are dual tensors.

Proof. To compute the averaged Euler-Lagrange equation (51), we use the following *variational relation* obtained from the composition of maps $g(t) = \Xi(t) \cdot \tilde g(t)$, cf. the velocity relation (35),

$$\delta g(t) = \delta\Xi(t) \cdot \tilde g(t) + T\Xi(t) \cdot \delta\tilde g(t)\,. \tag{57}$$

Hence, we find

$$\begin{aligned}
0 &= \delta \int_{t_1}^{t_2} \overline{L_{a_0}(g(t), \dot g(t))}\, dt \\
&= \int_{t_1}^{t_2} \Big(\overline{\frac{\delta L_{a_0}}{\delta g} \cdot \delta g} + \overline{\frac{\delta L_{a_0}}{\delta \dot g} \cdot \delta \dot g} \Big)\, dt \\
&= \int_{t_1}^{t_2} \overline{\Big(\Big(\frac{\delta L_{a_0}}{\delta g} - \frac{d}{dt}\frac{\delta L_{a_0}}{\delta \dot g}\Big) \cdot \delta\,\Xi(t)\Big) \cdot \tilde g}\, dt \\
&\quad + \int_{t_1}^{t_2} \Big(\overline{\frac{\delta L_{a_0}}{\delta g} \cdot T\Xi} - \overline{\frac{d}{dt}\frac{\delta L_{a_0}}{\delta \dot g} \cdot T\Xi} \Big) \cdot \delta\tilde g\, dt\,.
\end{aligned} \tag{58}$$

In the last equality, the first integral vanishes for any $\delta\,\Xi$, thus ensuring that the Euler-Lagrange equations are satisfied *before* averaging is applied.

Vanishing of the second integral for arbitrary $\delta\tilde{g}$ then yields the averaged Euler-Lagrange equations (51) and, thus, the equivalence of $\bar{\textbf{i}}$ and $\overline{\textbf{ii}}$.

The equivalence of $\overline{\textbf{iii}}$ and $\overline{\textbf{iv}}$ in the LAEP Theorem now follows by substituting the variations (53) defined using the chain rule into (52) and integrating by parts,

$$
\begin{aligned}
0 &= \delta \int_{t_1}^{t_2} \overline{\ell(u,a)}\, dt = \int_{t_1}^{t_2} \overline{\Big\langle \frac{\delta \ell}{\delta u}\,,\, \delta u \Big\rangle} + \overline{\Big\langle \frac{\delta \ell}{\delta a}\,,\, \delta a \Big\rangle}\, dt \\
&= - \int_{t_1}^{t_2} \overline{\Big\langle (\partial_t + \mathrm{ad}^*_u) \frac{\delta \ell}{\delta u} - \frac{\delta \ell}{\delta a} \diamond a \,,\, \eta' \Big\rangle}\, dt \\
&\quad - \int_{t_1}^{t_2} \Big\langle \Big(\frac{\partial}{\partial t} + \mathrm{ad}^*_{\tilde{u}}\Big) \overline{\Big(\frac{\delta \ell}{\delta u^\xi} \cdot T\Xi\Big)} \qquad\qquad (59) \\
&\qquad - \overline{\Big(\frac{\delta \ell}{\delta a^\xi} \cdot \mathcal{F}^{-1}\Big)} \diamond \tilde{a} \,,\, \tilde{\eta} \Big\rangle\, dt\,. \qquad\qquad (60)
\end{aligned}
$$

Thus, the independent variations η' in (59) and $\tilde{\eta}$ in (60) result in the EP motion equation (55) and the LAEP motion equation (56), respectively.

Finally we show that $\bar{\textbf{i}}$ and $\overline{\textbf{iii}}$ are equivalent. First note that the G–invariance of $L : TG \times V^* \to \mathbb{R}$ and the definition of $a(t) = a_0 g(t)^{-1}$ imply that the integrands in (50) and (52) are equal. In fact, this holds both before and after averaging. Moreover, all variations $\delta g(t) \in TG$ of $g(t)$ with fixed endpoints induce and are induced by variations $\delta u(t) \in \mathbf{g}$ of $u(t)$ of the form $\delta u = \partial\eta/\partial t + \mathrm{ad}_u\, \eta$ with $\eta(t) \in \mathbf{g}$ vanishing at the endpoints. The relation between $\delta g(t)$ and $\eta(t)$ is given by $\eta(t) = \delta g(t) g(t)^{-1}$. The corresponding statements also hold for the prime- and tilde-variables in the variational relations (53) that are used in the calculation of the other equivalences. This finishes the proof of the LAEP Theorem.

QED

Lie derivative vs ad.* The equality $\mathrm{ad}^*_u \mu = \mathcal{L}_u \mu$ holds for any one-form density μ (such as $\mu = \delta\ell/\delta u$, the variational derivative). Thus, the LAEP motion equation (56) may be written equivalently using Lie derivatives as

$$
\Big(\frac{\partial}{\partial t} + \mathcal{L}_{\tilde{u}}\Big) \overline{\Big(\frac{\delta \ell}{\delta u^\xi} \cdot T\Xi\Big)} = \overline{\Big(\frac{\delta \ell}{\delta a^\xi} \cdot \mathcal{F}^{-1}\Big)} \diamond \tilde{a}\,.
$$

In this notation, the advection of mass takes the form $(\partial_t + \mathcal{L}_{\tilde{u}})\tilde{D} = 0$ and one immediately obtains the

Corollary 3.4 (LA Kelvin-Noether Circulation Theorem) *The one form* $\tilde{v} \equiv \overline{\Big(\frac{\delta \ell}{\delta u^\xi} \cdot T\Xi\Big)} / \tilde{D}$ *satisfies*

$$
\frac{d}{dt} \oint_{c(\tilde{u})} \tilde{v} = \oint_{c(\tilde{u})} \frac{1}{\tilde{D}}\, \overline{\Big(\frac{\delta \ell}{\delta a^\xi} \cdot \mathcal{F}^{-1}\Big)} \diamond \tilde{a}\,,
$$

for any closed curve $c(\tilde{u})$ following the LA fluid motion.

The same result is obtained by applying LA directly to Kelvin's Theorem for the EP motion equation (55).

4. Application of the LAEP Theorem to Incompressible Fluids

4.1. THE LAGRANGIAN AVERAGED EULER (LAE) EQUATIONS

The Lagrangian averaged Euler (LAE) equations are derived from the LAEP Theorem 3.3 as follows. The corresponding averaged Lagrangian in the *material description* is given by

$$\bar{L} = \int D_0\, d^3\mathbf{x}_0 \left[\frac{1}{2}\, \overline{|\dot{\mathbf{x}}^\xi|^2} + \overline{p^\xi \left(\det \frac{\partial \mathbf{x}^\xi}{\partial \mathbf{x}_0} - 1 \right)} \right] . \tag{61}$$

Therefore, the reduced averaged Lagrangian in the *spatial picture* becomes

$$\bar{\ell} = \int d^3\mathbf{x} \left[\frac{1}{2}\, \tilde{D}\, \overline{|\mathbf{u}^\xi|^2} + \overline{p^\xi \left(\det T\Xi\, - \tilde{D} \right)} \right] , \tag{62}$$

where we have used equation (42) in the change of variables. The necessary variations of this Lagrangian are given by (dropping the $\delta\,\Xi-$ terms)

$$\begin{aligned} \delta\bar{\ell} \;=\; \int d^3\mathbf{x} \Big[& \tilde{D}\, \overline{\mathbf{u}^\xi \cdot T\Xi} \cdot \delta\tilde{\mathbf{u}} + \Big(\frac{1}{2}\, \overline{|\mathbf{u}^\xi|^2} - \bar{p}^L \Big) \delta\tilde{D} \\ & + \delta\bar{p}^L \Big(\overline{\det T\Xi} - \tilde{D} \Big) + \overline{\delta p^\ell \det T\Xi}\ \Big] . \end{aligned} \tag{63}$$

Here we substituted the pressure decomposition $p^\xi = \bar{p}^L + p^\ell$ with $\overline{p^\xi} = \bar{p}^L$ and used the projection property of the Lagrangian average. Thus, the pressure constraint implies that the mean advected density is related to the mean fluid trajectory by

$$\tilde{D} = \overline{\det T\Xi}\, . \tag{64}$$

The advected mean mass density $\tilde{D}$ satisfies the corresponding mean continuity equation

$$\frac{\partial \tilde{D}}{\partial t} = -\operatorname{div}(\tilde{D}\tilde{\mathbf{u}})\, . \tag{65}$$

Consequently, the LAE fluid velocity in general has a divergence,

$$\operatorname{div} \tilde{\mathbf{u}} = -\frac{1}{\tilde{D}} \Big(\frac{\partial}{\partial t} + \bar{\mathbf{u}}^L \cdot \nabla \Big) \tilde{D} \neq 0\, , \tag{66}$$

as was first noticed in Andrews & McIntyre [1]. In principle, one may restrict $g(t)$ and both its factors $\Xi(t)$ and $\tilde{g}(t)$ to the space of volume-preserving diffeomorphisms, for which $\det T\Xi \equiv 1$. However, some LA processes may not respect this restriction and, in general, $\operatorname{div}\tilde{\mathbf{u}} \neq 0$. In [17, 16], though, $\operatorname{div}\tilde{\mathbf{u}} = 0$ is accomplished.

The Lagrangian disturbance of the pressure p^ℓ imposes the constraint

$$\overline{\delta p^\ell \det T\Xi} = 0\,. \tag{67}$$

This constraint also arises in the self-consistent theory of wave-mean flow interaction dynamics in Gjaja & Holm [9].

The LAE equations may now be written in LAEP form (56) in components as

$$\frac{\partial}{\partial t}\tilde{v}_i + \tilde{u}^j\frac{\partial}{\partial x^j}\tilde{v}_i + \tilde{v}_j\frac{\partial}{\partial x^i}\tilde{u}^j + \frac{\partial}{\partial x^i}\tilde{\pi} = 0\,, \tag{68}$$

$$\tilde{v}_i = \frac{1}{\tilde{D}}\frac{\delta \ell}{\delta \tilde{u}^i} = \overline{u^\xi_j (T\Xi)^j_i}\,, \quad \tilde{\pi} = -\frac{\delta \ell}{\delta \tilde{D}} = -\frac{1}{2}\overline{|\mathbf{u}^\xi|^2} + \bar{p}^L\,, \tag{69}$$

When $T\Xi = Id + \nabla\xi$, one finds

$$\tilde{\mathbf{v}} = \overline{\mathbf{u}^\xi} + \overline{\frac{D^L}{Dt}\xi_j\nabla\xi^j} \equiv \bar{\mathbf{u}}^L - \bar{\mathbf{p}}\,. \tag{70}$$

The term $\bar{\mathbf{p}}$ is called the pseudomomentum in Andrews & McIntyre [1]. See, e.g., [11] for a recent discussion and more details.

Remark 4.1 [Momentum balance] The EP theory of Holm, Marsden and Ratiu [12] implies momentum balance in this case in the form,

$$\frac{\partial}{\partial t}(\tilde{D}\tilde{v}_i) + \frac{\partial}{\partial x^j}\left(\tilde{D}\tilde{v}_i\tilde{u}^j + \bar{p}^L\delta^j_i\right) = \frac{\tilde{D}}{2}\frac{\partial\,\overline{|\mathbf{u}^\xi|^2}}{\partial x^i}\bigg|_{exp}\,, \tag{71}$$

where subscript *exp* refers to the *explicit* spatial dependence arising from the Ξ–terms in $\overline{|\mathbf{u}^\xi|^2} = \overline{D^L\Xi/Dt^2}$ obtained from equation (36).

Remark 4.2 [Lagrangian Average Conservation laws] From the viewpoint of the LAEP theorem, the Kelvin circulation theorem and its associated conservation of potential vorticity for LA flows both emerge because reduction of Hamilton's principle by its relabeling symmetries in passing from the material to the spatial picture of continuum mechanics is compatible with Lagrangian averaging, which takes place at *fixed* fluid labels.

LA also preserves the kinematic symmetries of Hamilton's principle, so, conservation laws for momentum and energy for the LA dynamics are also guaranteed by Noether's theorem for the averaged variational principle when it admits space and time translations.

4.2. RECENT PROGRESS TOWARD CLOSURE

Of course, the LAE equations (68) are not yet closed. As indicated in their momentum balance relation (71), they depend on the unspecified Lagrangian statistical properties appearing as the $\Xi-$terms in the definitions of $\tilde{\mathbf{v}}$ and $\tilde{\pi}$. Until these properties are modeled or prescribed, the LAE equations are incomplete.

Progress in formulating and analyzing a closed system of fluid equations related to the LAE equations has recently been made in the EP context. Such closed model LAE equations were first obtained in Holm, Marsden and Ratiu [12, 13]. For more discussion of this type of equation and its recent developments as a turbulence model, see papers by Chen *et al.* [3, 4, 5, 6], Shkoller [19], Foias *et al.* [7, 8], Marsden, Ratiu & Shkoller [17] and Marsden and Shkoller [16]. A self-consistent variant of the LAE closure was also introduced in Gjaja and Holm [9] in the Lagrangian fluid specification, see Holm [10, 11] for further discussion of that approach.

Remark 4.3 [Transport structure] Although the Lagrangian averaged Euler (LAE) equations are not yet closed, their transport structure may still be discussed because they are derived in the Lagrangian averaged Euler-Poincaré (LAEP) context, which preserves the transport structure of ideal fluids. Thus, as in equations (18) and (19), we have

$$\frac{d}{dt}\oint_{c(\bar{\mathbf{u}})} \tilde{\mathbf{v}}\cdot d\mathbf{x} = 0\,, \quad \text{(LAE Kelvin theorem)}\,, \tag{72}$$

and

$$\frac{d}{dt}\int \tilde{\mathbf{v}}\cdot \mathrm{curl}\,\tilde{\mathbf{v}}\,d^3x = 0 \quad \text{(LAE Helicity conservation)}\,. \tag{73}$$

The LAEP approach is also versatile enough to derive LA equations for *compressible* fluid motion. This was already shown in the original GLM theory [1]. For brevity, we only remark that the LAEP approach preserves helicity conservation for isentropic compressible flows, as well, and it preserves magnetic helicity and cross-helicity conservation when applied to ideal magnetohydrodynamics (MHD). See Moffatt [18] for an early discussion of the topological interpretation of helicity in ideal fluids. For more details concerning the Euler-Poincaré approach to Lagrangian averaging in ideal MHD, see [11].

Acknowledgements

I am grateful for stimulating discussions of this topic with P. Constantin, G. Eyink, U. Frisch, J. Marsden, M. E. McIntyre, I. Mezic, S. Shkoller and A. Weinstein. Some of these discussions took place at Cambridge University while the author was a visiting professor at the Isaac Newton Institute for Mathematical Science. This work was supported by the U.S. Department of Energy under contracts W-7405-ENG-36 and the Applied Mathematical Sciences Program KC-07-01-01.

References

1. Andrews, D.G. & McIntyre, M.E. (1978) An exact theory of nonlinear waves on a Lagrangian-mean flow. *J. Fluid Mech.* **89**, 609–646.
2. Arnold, V.I. (1966) Sur la géometrie differentielle des groupes de Lie de dimension infinie et ses applications à l'hydrodynamique des fluides parfaits. *Ann. Inst. Fourier (Grenoble)* **16**, 319–361.
3. Chen, S.Y., Foias, C., Holm, D.D., Olson, E.J., Titi, E.S. & Wynne, S. (1998) The Camassa-Holm equations as a closure model for turbulent channel and pipe flows. *Phys. Rev. Lett.* **81**, 5338–5341.
4. Chen, S.Y., Foias, C., Holm, D.D., Olson, E.J., Titi, E.S. & Wynne, S. (1999) The Camassa-Holm equations and turbulence in pipes and channels. *Physica D* **133**, 49–65.
5. Chen, S.Y., Foias, C., Holm, D.D., Olson, E.J., Titi, E.S. & Wynne, S. (1999) A connection between the Camassa-Holm equations and turbulence in pipes and channels. *Phys. Fluids* **11**, 2343–2353.
6. Chen, S.Y., Holm, D.D., Margolin, L.G. & Zhang, R. (1999) Direct numerical simulations of the Navier-Stokes alpha model. *Physica D* **133**, 66–83.
7. Foias, C., Holm, D.D. & Titi, E.S. (1999) The Three Dimensional Viscous Camassa–Holm Equations, and Their Relation to the Navier–Stokes Equations and Turbulence Theory. *J. Diff. Eq.*, to appear.
8. Foias, C., Holm, D.D. & Titi, E.S. (2001) The Navier-Stokes-alpha model of fluid turbulence. *Physica D*, to appear.
9. Gjaja, I. & Holm, D.D. (1996) Self-consistent wave-mean flow interaction dynamics and its Hamiltonian formulation for a rotating stratified incompressible fluid. *Physica D* **98**, 343–378.
10. Holm, D.D. (1999) Fluctuation effects on 3D Lagrangian mean and Eulerian mean fluid motion. *Physica D* **133**, 215–269.
11. Holm, D.D. (2001) Averaged Lagrangians and the mean dynamical effects of fluctuations in continuum mechanics. *Physica D*, to appear.
12. Holm, D.D., Marsden, J.E. & Ratiu, T.S. (1998) The Euler–Poincaré equations and semidirect products with applications to continuum theories. *Adv. in Math.* **137**, 1–81.
13. Holm, D.D., Marsden, J.E. & Ratiu, T.S. (1998) Euler–Poincaré models of ideal fluids with nonlinear dispersion. *Phys. Rev. Lett.* **80**, 4173–4177.
14. Holm, D.D., Marsden, J.E., Ratiu, T.S. & Weinstein, A. (1985) Nonlinear stability of fluid and plasma equilibria. *Physics Reports* **123**, 1–116.
15. Marsden, J.E. & Ratiu, T.S. (1999) *Introduction to Mechanics and Symmetry.* Springer, New York.
16. Marsden, J.E.& Shkoller, S. (2001) The anisotropic averaged Euler equations. *J. Rat. Mech. Anal.*, to appear.
17. Marsden, J.E., Ratiu, T.S. & Shkoller, S. (2001) The geometry and analysis of the averaged Euler equations and a new diffeomorphism group. *Geom. Funct. Anal.*, to appear.
18. Moffatt, H.K. (1969) The degree of knottedness of tangled vortex lines. *J. Fluid Mech.* **35**, 117–129.
19. Shkoller, S. (1998) Geometry and curvature of diffeomorphism groups with H_1 metric and mean hydrodynamics. *J. Funct. Anal.* **160**, 337–365.

IV. RECONNECTIONS AND SINGULARITIES

THE GEOMETRY OF RECONNECTION

GUNNAR HORNIG
Topological Fluid Dynamics (NB 7/31)
Department of Physics and Astronomy
44780 Ruhr-Universität Bochum, Germany

E-mail: *gh@tp4.ruhr-uni-bochum.de*

Abstract. Reconnection is an important process of structure formation in fluid dynamics. It occurs in the form of vortex reconnection in hydrodynamics as well as in the form of magnetic reconnection in plasmas. In both cases a divergence-free field, the magnetic or the vorticity field, respectively, alter the topology of their field lines. A first definition of the process in two dimensions is given, which is wide enough to cover both forms of reconnection, and which is independent of the underlying dissipation process. It is also demonstrated that the definition can distinguish reconnection from other non-ideal processes. With the help of simple two-dimensional examples, one for magnetic reconnection and one for vortex reconnection, similarities and differences of both types of reconnection are discussed. The transition to three-dimensional configurations shows to require a more general framework, which is found in the covariant generalization of flux conservation.

1. Introduction

Reconnection is a fundamental process in fluid systems by which a divergence-free field changes its topology. The divergence-free field may for example be the vorticity field of a hydrodynamical system or the magnetic field in a plasma. A first impression of the nature of the process is provided by Figure 1.

Aside of hydrodynamics (HD) and magnetohydrodynamics (MHD) the notion of reconnection exists also in other fields of physics such as in superfluids [10], cosmic strings [12] and liquid crystals [3]. Processes similar to reconnection are also known in knot-theory as a *surgery of framed knots* [8],

R.L. Ricca (ed.), An Introduction to the Geometry and Topology of Fluid Flows, 295–312.

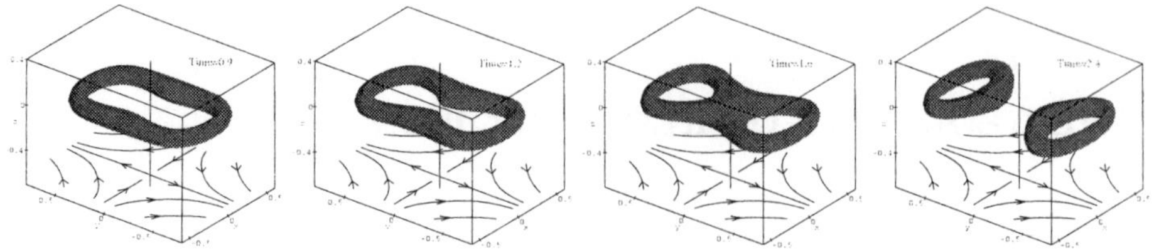

Figure 1. An example of a simple two-dimensional reconnection process. Exemplary for the evolution of the field a flux tube is followed in time. The fluid velocity is shown on the bottom of the boxes.

and in enzymology, where topoisomerase acts like reconnection on strands of the DNA [13]. However, except of vortex and magnetic reconnection all other forms of reconnection refer to singular string-like objects (cosmic strings, knots, quantized vortices, DNA) and are not based on continuous fields such as vorticity fields or magnetic fields in fluids. Therefore they are mentioned here only for completeness and are disregarded in the following. The interested reader is referred to the selected references given above.

With respect to the development of theoretical concepts vorticity or magnetic reconnection are well ahead of reconnection in other fields of physics. This is also reflected by the enormous number of more than thousand publication referring to magnetic reconnection or synonymous notions such as *merging.* Curiously vorticity reconnection (sometimes also called *cross-linking* or *cut-and-connect*) did not get that much attention although the phenomenon must have been known already for a long time in form of reconnection of smoke rings.

Among the numerous aspects under which reconnection is investigated in the literature the following exposition will focus on the topological and geometrical aspects. There are two primary reasons for this restriction. First, these are the main aspects of this book, and second, these aspects are amenable to a rigorous and very general mathematical analysis, which treats magnetic and vortex reconnection on the same basis. Especially we will prescribe macroscopic transport coefficients such as viscosity or resistivity without discussing their physical origin or quantifying them. This allows us to answer questions of how reconnection changes the topology of a magnetic or vorticity field, but not why and when it occurs. With respect to these and many other physically very important and partly still unsolved questions, we refer the reader to the numerous publications, an entry to which is given for instance by [11] and [2] for the case of magnetic reconnection and by [9] for vorticity reconnection.

The process of reconnection is remarkable in that it occurs even for very low dissipation, that is high Reynolds numbers or magnetic Reynolds numbers, respectively. For example in the plasma of the corona, the upper atmosphere of the Sun, reconnection occurs in an environment of magnetic Reynolds numbers as high as $10^8 - 10^{12}$. Usually in these regimes the idealization of infinite Reynolds numbers is a good approximation in all other respects. Therefore it is tempting to neglect dissipation completely and use the ideal equations, either the Euler-equation for HD or the ideal Ohm's law for MHD applications. Both equations, however, preserve the topology and do not allow for reconnection. The solution to this seemingly paradoxical situation is the fact that plasmas for example can develop self-amplifying processes such as the formation of thin current sheets, which can locally reduce the characteristic length scales and thereby the Reynolds number by several orders of magnitude. Hence, in such current sheets, dissipation is not negligible and reconnection is enabled. Since these sheets are very thin the dissipation of energy in these sheets compared to the magnetic energy stored in the configuration is rather small and localized. Thus the restructuring of magnetic flux occurs as if there would be no dissipation. This is the reason why reconnection can be considered as a flux-conserving process under non-continuous flows. But before we explain this in more detail we will give a short survey of the basic equations of MHD and HD in Section 2 and state the conditions for which they yield a flux-conserving equation. A main problem in the analysis of reconnection is to distinguish it from other dissipative processes. Therefore Section 3 contains a discussion of so called slippage solutions and a subsection on weak forms of flux conservation, before we come to the most simple two-dimensional reconnection processes in Section 4. The last section, Section 5 is devoted to the more complicated three-dimensional reconnection.

2. Flux-Conservation in Hydrodynamics and Magnetohydrodynamics

A divergenc-free field $\mathbf{B}$ embedded in a fluid moving with velocity $\mathbf{v}$ is said to be *flux-conserving* if the *flux*,

$$\int_{C^2(t)} \mathbf{B} \cdot d\mathbf{a} \ ,$$

is constant in time for every co-moving two-dimensional surface C^2. Here *co-moving* means that the surface can be considered as consisting of fluid elements. This is equivalent to an evolution of $\mathbf{B}$ according to

$$\begin{aligned} \partial_t \mathbf{B} - \nabla \times (\mathbf{v} \times \mathbf{B}) = 0 \qquad (1) \\ \Rightarrow \int_{C^2(t)} \mathbf{B} \cdot d\mathbf{a} = \text{constant.} \end{aligned}$$

The conservation of flux is only one of a series of *topological conservation laws.* Another one is the conservation of line integrals,

$$\partial_t \mathbf{A} + \nabla(\mathbf{v}\cdot\mathbf{A}) - \mathbf{v}\times\nabla\times\mathbf{A} = 0 \qquad (2)$$
$$\Rightarrow \int_{C^1(t)} \mathbf{A}\cdot d\mathbf{l} = \text{constant},$$

which is closely related to the flux conservation, since the curl of (2) yields a flux-conserving equation for $\mathbf{B} = \nabla \times \mathbf{A}$.

In the following we show how an equation of type (1) emerges from both magnetohydrodynamics and hydrodynamics.

2.1. MAGNETOHYDRODYNAMICS (MHD)

The set of equations of MHD,

$$\partial_t \rho + \nabla\cdot(\rho\mathbf{v}) = 0 \,, \qquad (3)$$
$$\rho\, \partial_t\mathbf{v} + \rho\mathbf{v}\cdot\nabla\mathbf{v} = -\nabla p + \mathbf{J}\times\mathbf{B} \,, \qquad (4)$$
$$\mathbf{E} + \mathbf{v}\times\mathbf{B} = \eta\mathbf{J} + \mathbf{T} \,, \qquad (5)$$
$$\nabla\times\mathbf{E} + \partial_t\mathbf{B} = 0 \,, \qquad (6)$$
$$\nabla\cdot\mathbf{B} = 0 \,, \qquad (7)$$
$$\nabla\times\mathbf{B} = \mu\mathbf{J} \,, \qquad (8)$$

consists of a continuity equation (3) for the plasma density ρ, a momentum balance (4) for the plasma velocity $\mathbf{v}$ and the Ohm's law (5) for the magnetic and electric field. They are supplemented by (6)–(8), i.e. Faraday's law, the solenoidal condition for $\mathbf{B}$ and Maxwell's equation without displacement current. The Ohm's law is the result of various approximations. Depending on the plasma under consideration, aside from the resistive term $\eta\mathbf{J}$, other terms such as the Hall term or the pressure term, could be significant. We collectively denote these terms by $\mathbf{T}$. Equations (5)–(8) can be simplified by taking the curl of (5), using (6) to substitute for the curl of $\mathbf{E}$ and finally using (8) to substitute for $\mathbf{J}$. Thus we end up with only one equation, the *induction equation* of MHD,

$$\partial_t\mathbf{B} - \nabla\times(\mathbf{v}\times\mathbf{B}) = -\nabla\times(\hat{\eta}\,\nabla\times\mathbf{B} + \mathbf{T}) \qquad (\hat{\eta} = \eta/\mu) \,, \qquad (9)$$

to which (7) is only an initial condition for the magnetic field. Now the whole system of equations (3), (4) and (9) does not contain the electric field anymore, which justifies the name magnetohydrodynamics. The relative magnitude of the second term on the l.h.s. of (9) in comparison with the first term on the r.h.s. is given by the *magnetic Reynolds number*

$$Re_m = \frac{\|\nabla\times(\mathbf{v}\times\mathbf{B})\|_{\text{char}}}{\|\nabla\times(\hat{\eta}\nabla\times\mathbf{B})\|_{\text{char}}} = \frac{vL}{\hat{\eta}} \,,$$

where we used characteristic values for the velocity v, and derivatives $\partial_x \sim 1/L$. For large values of Re_m which are especially obtained in the case of large length scales L, e.g. for astrophysical plasmas, we can neglect the resistive term in (9). With a similar consideration we can disregard the second term on the r.h.s. so that we end up with

$$\partial_t \mathbf{B} - \nabla \times (\mathbf{v} \times \mathbf{B}) = 0 \ , \tag{10}$$

which is equivalent to (1). Thus the magnetic flux through any comoving surface is conserved, a consequence that is commonly called *Alfvén's theorem.* Equation (10) is sometimes called *ideal induction equation*, since it can be derived from the *ideal Ohm's law* $\mathbf{E} + \mathbf{v} \times \mathbf{B} = 0$, which we would have obtained in the limit of large Reynolds numbers directly from Ohm's law (5).

Exercise 2.1 By substituting $\mathbf{E} = -\partial_t \mathbf{A} - \nabla\phi$ in the ideal Ohm's law show that we can obtain a conservation law for the vector potential $\mathbf{A}$ of the same type as (2) if we choose an appropriate gauge $\mathbf{A} \to \mathbf{A} + \nabla\chi$, $\phi \to \phi - \partial_t \chi$. In this case what kind of conservation law is valid for the helicity density $\mathbf{A} \cdot \mathbf{B}$?

2.2. HYDRODYNAMICS (HD)

An approximation similar to the one which leads to the ideal induction equation can be applied in hydrodynamics. Here the set of equations consists of the continuity equation (11) and the momentum balance (12) for a viscous fluid (*Navier-Stokes equation*):

$$\partial_t \rho + \nabla \cdot (\rho \mathbf{v}) = 0 \ , \tag{11}$$

$$\rho \partial_t \mathbf{v} + \rho \mathbf{v} \cdot \nabla \mathbf{v} = -\nabla p + \mathbf{F} \ , \tag{12}$$

$$\text{where} \quad \mathbf{F} = \partial_k \left(\nu(\partial_k v^i + \partial_i v^k - 2/3 \delta_{ik} \partial_l v^l) + \partial_i \zeta \partial_l v^l \right) \vec{e}_i \ .$$

The comparatively complicated viscous term on the r.h.s. of (12) gets much simpler if we consider only incompressible fluids with constant viscosity, for which we have

$$\Rightarrow \ \partial_t \mathbf{v} + \rho \mathbf{v} \cdot \nabla \mathbf{v} = -\nabla p + \nu \Delta \mathbf{v} \quad \text{with } \nu = \text{ constant}, \ \nabla \cdot \mathbf{v} = 0 \ . \tag{13}$$

By taking the curl of the Navier-Stokes equation we have an equation for the vorticity $\mathbf{w} = \nabla \times \mathbf{v}$:

$$\partial_t \mathbf{w} - \nabla \times (\mathbf{v} \times \mathbf{w}) = \nabla \times \left(1/\rho \nabla p + \nabla v^2/2 + 1/\rho \mathbf{F} \right) \ . \tag{14}$$

Note that an HD approximation which results in a conservation law of the form of (1) requires not only large Reynolds numbers $Re = vL/\nu$ (so as to disregard the viscous term), but also an *isentropic* flow $p = p(\rho)$ (so that $\nabla \times (1/\rho \nabla p)$ disappears). These assumptions give

$$\partial_t \mathbf{w} - \nabla \times (\mathbf{v} \times \mathbf{w}) = 0 , \tag{15}$$

and the conservation of vorticity which is the so called *Kelvin's Theorem.*

Question 2.2 Can the Navier-Stokes equation, analogous to Ohm's law, be re-written to take the same form of equation (2)?

3. Dynamics of Divergence-Free Fields

In the previous section we saw that MHD as well as HD yield under suitable approximation a conservation equation for the flux of a divergence-free field. Since the conservation of flux also implies a conservation of field lines, the magnetic field lines or vorticity field lines are conserved as well. This feature is sometimes referred to as *frozen-in* field lines. These properties help a lot to envisage possible evolutions of magnetic or vorticity fields. We can think of an evolution of these fields as if all field lines are ropes (tied together at null points of the field) which are drifting with the flow of the fluid. One can hardly underestimate the importance of this picture in MHD or HD. Many models are conceived on the basis of these comparatively simple and descriptive conception. However, the concept of frozen-in field lines and flux would not have been so successful if it would be valid *only* in the limit of infinite Reynolds numbers. Its success results from the fact that it is often tolerant with respect to deviations from the limit of vanishing dissipation. Moreover, one can extend the notion of flux-conservation by allowing flux conservation not only with respect to the fluid flow $\mathbf{v}$ but with respect to arbitrary flows.

3.1. FLUX-CONSERVING FLOWS

Consider the evolution of a divergence-free field $\mathbf{G}$ which instead of being governed by an ideal flux-conserving equation, is now disturbed by a non-ideal term $\nabla \times \mathbf{N}$,

$$\partial_t \mathbf{G} - \nabla \times (\mathbf{v} \times \mathbf{G}) = -\nabla \times \mathbf{N} . \tag{16}$$

Note that an arbitrary evolution of $\mathbf{G}$ can be cast in this form as long as $\mathbf{N}$ is left unspecified. We now ask ourselves under which conditions the evolution of $\mathbf{G}$ can still be considered flux-conserving *in presence* of the non-ideal term $\nabla \times \mathbf{N}$? That is, under which conditions a velocity field $\mathbf{u}$ can take the place of $\mathbf{v}$ to satisfy

$$\partial_t \mathbf{G} - \nabla \times (\mathbf{u} \times \mathbf{G}) = 0, \qquad \mathbf{u} = \mathbf{v} + \Delta\mathbf{v} \ . \tag{17}$$

Thus we look for a solution $(\Delta\mathbf{v}, \Phi)$ of the equation

$$\Delta\mathbf{v} \times \mathbf{G} = -\mathbf{N} + \nabla\Phi \ .$$

A necessary condition to satisfy is

$$\mathbf{G} \cdot \mathbf{N} = \mathbf{G} \cdot \nabla\Phi \ . \tag{18}$$

For $\mathbf{G} \neq 0$ this condition is also sufficient and the solution is given by

$$\Delta\mathbf{v} = (\mathbf{N} - \nabla\Phi)_\perp \times \mathbf{G}/G^2 \ , \tag{19}$$

where $\perp$ denotes the component perpendicular to $\mathbf{G}$ and $G = \|\mathbf{G}\|$.

A wide class of configurations ($\mathbf{G}$,$\mathbf{v}$,$\mathbf{N}$), which satisfy (18), are represented by two-dimensional (2D) configurations, i.e. solutions for which $\mathbf{G}$ and $\mathbf{v}$ depend only on two space coordinates and only have components in these two coordinate directions, e.g. $\mathbf{G} = G_x(x,y,t)\mathbf{e}_x + G_y(x,y,t)\mathbf{e}_y$ in Euclidian coordinates. In this case the two-dimensionality in (16) requires $\nabla \times \mathbf{N} \cdot \mathbf{e}_z = 0$ and thus $\mathbf{N} = N_z(x,y)\mathbf{e}_z + \nabla\Phi_N(x,y)$. Therefore $\Phi = -\Phi_N$ satisfies (18), and (19) yields

$$\Delta\mathbf{v} = N_z(x,y)\, \mathbf{e}_z \times \mathbf{G}/G^2 \ . \tag{20}$$

If we require that $\Delta\mathbf{v}$ vanishes on a boundary where $\mathbf{N}$ vanishes the velocity $\Delta\mathbf{v}$ is *uniquely* determined up to a component parallel to $\mathbf{G}$.

Whenever a solution for $\Delta\mathbf{v}$, and therefore for $\mathbf{u}$ in (17), exists the field lines of $\mathbf{G}$ will move with a velocity $\mathbf{u}$ different from the fluid velocity $\mathbf{v}$. An observer will therefore see a slippage of the field lines with respect to the fluid elements. The slippage velocity $\Delta\mathbf{v}$ is proportional to the non-ideal term $\mathbf{N}$ and can be chosen to be zero when $\mathbf{N}$ tends to zero. This is the reason why the concept of frozen-in flux works in so many cases *in spite of the fact that is derived from a highly idealized situation.* In most cases a small deviation in terms of a small $\mathbf{N}$ will lead only to a small (and hardly noticeable) difference between the velocity of the field lines and the fluid velocity. This result is the basis for the following definition.

Definition 3.1 *Let* $\mathbf{G}$ *be a divergence-free field in a domain* D *of a fluid with velocity* $\mathbf{v}$. *If the flux-conserving equation (1) does not hold for* $\mathbf{v}$, *but for a different velocity field* $\mathbf{u}$ *(with* $\mathbf{u}_{\partial D} = \mathbf{v}_{\partial D}$*), we call this a* **flux-conserving flow** *or a* **slippage solution**. *The velocity* $\mathbf{u}_\perp - \mathbf{v}_\perp$ *is the* **slippage velocity**.

Note that a flux-conserving flow is, in general, not unique and there is no need to have a unique minimal slippage velocity (in the sense of maximal norm), since the problem of finding a $\Delta\mathbf{v}$-solution involves the existence of a global potential Φ. $\mathbf{u}$ and $\mathbf{v}$ must satisfy the same boundary condition if we want to ensure that existence of solutions for two adjacent domains (D_1 and D_2) implies existence of solution for the union $D_1 \cup D_2$.

Remark 3.2 There is no need for $\mathbf{u}$ to be necessarily a particle velocity. If $||G||^2$ is sufficiently small in (19), then $\mathbf{u}$ becomes greater than the speed of light, and this without any violation of physical laws.

Exercise 3.3 The Ohm law, with Hall and pressure terms included, but no resistivity, is given by the equation

$$\mathbf{E} + \mathbf{v} \times \mathbf{B} = (e\,n)^{-1}(\mathbf{J} \times \mathbf{B} - \nabla P) \ ,$$

where e is the electron charge and n the density. State the conditions under which flux-conserving solutions exist. In this example what is the meaning of $\mathbf{u}$?

Exercise 3.4 Show that the total helicity

$$\int_V \mathbf{A} \cdot \mathbf{G} \; d^3\mathbf{x} \ , \qquad \mathbf{n} \cdot \mathbf{G}|_{\partial V} = 0 \ ,$$

is independent of the choice of the vector potential $\mathbf{A}$ of $\mathbf{G}$ (if integrated over a volume where $\mathbf{G}$ is tangent to the boundary) and that it is constant in time for every flux-conserving flow (see [7] for more information about helicity).

3.2. A WEAK FORM OF FLUX CONSERVATION

Although we can often find flux-conserving flows, there are many cases where non-ideal effects cannot be abosrbed in a slippage velocity. This is especially true for reconnections, although reconnections are, in a certain sense, almost flux-conserving. To understand this one has to understand the difference between dissipation of energy and 'dissipation' of flux of a divergence-free field.

To appreciate this consider an idealized field in a toroidal flux tube FT that is $\mathbf{G} = 0$ everywhere outside the flux tube. If the whole field is dissipated according to

$$\partial_t \mathbf{G} = \eta \Delta \mathbf{G}, \tag{21}$$

then both total energy $(1/2)\int_{FT} G^2 \, d^3\mathbf{x}$ and flux $\int_C \mathbf{G} \cdot d\mathbf{A}$ decrease with t. Evidently, in this case there is no flux-conserving flow for $\mathbf{G}(\mathbf{x}, t)$.

The situation changes if, for example, η, and therefore the dissipation in (21), is non-zero only in one half of the torus. Again the total energy of the field decreases, but this time the flux of the tube has to stay constant. There is no dissipation of flux whenever at least on one cross-section the dissipation is zero. In other words the dissipation of flux requires a dissipative process in the whole volume of the flux tube, whereas dissipation of energy is a local process. Note, however, that this *weak* form of flux conservation does not imply existence of the *strong* form of flow. For instance, if the **G**-field is twisted in the dissipative half torus of the previous example, the twist will vanish for $t \to \infty$ since $\nabla \times \mathbf{G}$ tends to zero. This will change the connectivity of field lines within the flux tube and is in contradiction to the existence of a flux-conserving flow.

The weak form of flux conservation is found in fluids with very low Reynolds numbers that can develop significant dissipation only in very localized regions, as for instance in current sheets or vortex sheets. Usually these regions are too thin to contain closed flux tubes; hence, all the flux within the dissipation region is connected to the dissipation-free environment. Reconnection, for instance, is a form of weak flux conservation which does not allow a slippage solution.

4. 2D Reconnection

The so called two-dimensional reconnection is a model for a real reconnection process in three-dimensional (3D) space which is approximately invariant in one space direction (here the z-direction). Thus all quantities depend only on two space coordinates (x and y) and time. Moreover we require that **G** and **v** have no z-component. The domain D of the (x, y) coordinates is a compact subset of $\mathbb{R}^2$ with a boundary ∂D where the non-ideal term **N** vanishes. Thus we have the same situation which lead us to the solution of equation (20). The existence of such solutions is questionable at points where $\mathbf{G} = 0$. In the generic situation where **N** is non-zero at null **G**-points we shall have a $\Delta\mathbf{v}$ singularity. The type of this singularity depends on the local structure of **G** at the null point. Since there are only two structurally stable types of nulls for two-dimensional fields, a hyperbolic null (X-point) and an elliptic null (O-point), we get two typical singularities of $\Delta\mathbf{v}$ at null points of **G**.

An elliptic null point of **G** generates a $\Delta\mathbf{v}$ which points radially inwards or outwards depending on whether **N** and $\nabla\times\mathbf{G}$ are parallel or anti-parallel (see Figure 2). For example for $\mathbf{N} = [0, 0, 1]$ and $\mathbf{G} = [-y, x, 0]$, **N** and $\nabla \times \mathbf{G}$ are aligned; thus

$$\Delta\mathbf{v} = \mathbf{N} \times \mathbf{G}/G^2 = (x^2 + y^2)^{-1}[-x, -y, 0]\,.$$

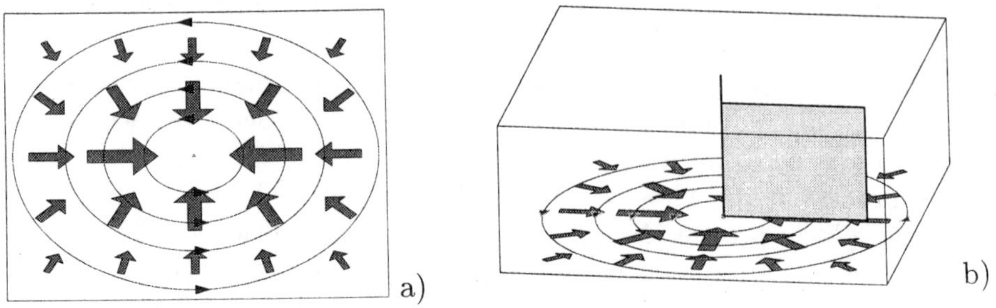

Figure 2. a) Local solution for $\Delta\mathbf{v}$ at an elliptic null of $\mathbf{G}$ for a non-vanishing non-ideal term $\mathbf{N}$. The direction of the arrows depends on the orientation of $\mathbf{N}$ with respect to the circulation of $\mathbf{G}$. b) The surface of integration used to determine the rate of flux annihilation.

It has a singularity of type $1/\sqrt{(x^2+y^2)}$ due to the linearity of a generic, structurally stable null of $\mathbf{G}$. These solutions must be interpreted as either a destruction of flux, if $\Delta\mathbf{v}$ points towards the null, or a creation of flux if $\Delta\mathbf{v}$ is directed away from the null. The rate at which flux is created or annihilated at the null is equal to the change of flux per time through a certain surface. This surface (see Figure 2b) is chosen to be tangent to the field lines of $\mathbf{u}$ and comoving with the flow of $\mathbf{u}$ except for one boundary which is co-moving with the null. Since this surface is comoving except for one boundary its flux can change only by a transport of $\mathbf{G}$-flux across this boundary, i.e. across the null line. The transport velocity of the flux across this boundary is the limit of $\mathbf{u}(\mathbf{x}) - \tilde{\mathbf{v}}$ (where $\tilde{\mathbf{v}}$ is the velocity of the null) for $\mathbf{x}$ tending to $\mathbf{x}_0$, the site of the null. Thus we have for the rate of flux creation

$$\begin{aligned} \frac{d}{dt}\int_C \mathbf{G}\cdot d\mathbf{A} &= \lim_{\mathbf{x}\to\mathbf{x}_0}\int_z (\mathbf{u}(\mathbf{x}) - \tilde{\mathbf{v}})\times\mathbf{G}\cdot d\mathbf{l} \\ &= \lim_{\mathbf{x}\to\mathbf{x}_0}\int_z \Delta\mathbf{v}(\mathbf{x})\times\mathbf{G}\cdot d\mathbf{l} = -N_z(\mathbf{x}_0,t)\Delta z\ . \end{aligned} \tag{22}$$

or simply $N_z(\mathbf{x}_0,t)$, normalizing per unit length in z direction.

For the other type of null, the hyperbolic null point or X-point, the same consideration leads to a pattern as shown in Figure 3a. The singularity again generically scales as $1/\sqrt{(x^2+y^2)}$, but is now of X-type (stagnation flow). The analysis of the loss of flux from the previous case holds here for the inflow direction, while for the outflow direction we have a creation of flux at the same rate, as can be seen from (22). Thus the flux appears to be transported in a finite time from both sides onto the null line (along the inflow direction), and then reappears along the outflow direction but now connecting different parts of the incoming strands of $\mathbf{G}$-field lines. This

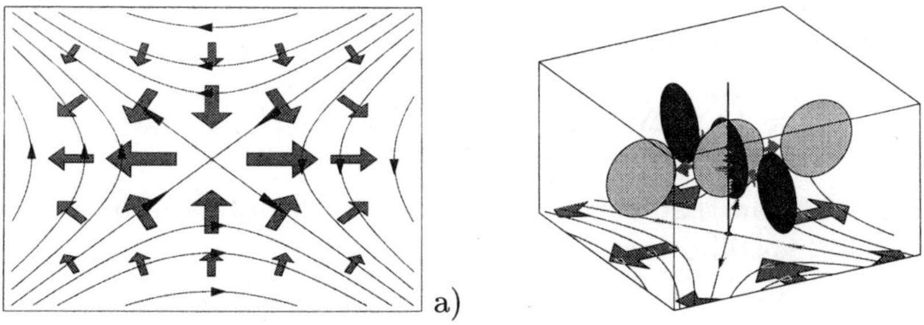

a) b)

Figure 3. a) Local solution for $\Delta\mathbf{v}$ at a hyperbolic null of $\mathbf{G}$ for a non-vanishing non-ideal term $\mathbf{N}$. The direction of the arrows depends on the orientation of $\mathbf{N}$ with respect to the orientation of $\mathbf{G}$. b) The evolution of two circular cross sections of $\mathbf{G}$-flux tubes, which are transported in the singular $\Delta\mathbf{v}$-flow.

leads to the impression that incoming flux is cut along the null line and thereupon newly glued together when leaving in the outflow direction, as shown in Figure 1. In Figure 3b the evolution of the cross sections of this tube in the vicinity of the null point is shown. This feature justifies the name *reconnection.*

The rate at which flux is cut and reconnected along the null line, is called the *reconnection rate*, and is essentially given by equation (22). Since in the process of reconnection flux is simultaneously lost in one direction and created along the other direction there is no use in giving the reconnection rate a sign, hence,

$$\frac{d}{dt}\Phi_{rec} = \|N_z(\mathbf{x}_0, t)\|\Delta z \ . \tag{23}$$

One should be aware that in early models of magnetic reconnection the term reconnection rate was often associated with the Alfvénic Mach number of the plasma velocity in the inflow region. Also note that the reconnection rate as given by Eq. (23) still holds if the null point has the X-point topology but is non-linear in x and y, for example $\mathbf{G} = [y, x^3, 0]$. Examples of MHD solutions with null points of this kind can be found in [2]. These considerations lead us to the following statement about reconnection (which we can take as a definition of reconnection).

Proposition 4.1 *A two-dimensional solenoidal field* $\mathbf{G}$ *in a fluid domain* D *with velocity* $\mathbf{v}$ *exhibits reconnection if there exists a solution* $\mathbf{u}$ *of (17) such that* $\mathbf{u} = \mathbf{v}$ *on the boundary of* D*, with a singularity at an X-null point of* $\mathbf{G}$ *such that the reconnection rate given by 23 is finite.*

Although most models of reconnection are stationary there is no doubt that in reality reconnection will be a time dependent process. It will start with the occurrence of $\mathbf{N} \neq 0$ at the null of $\mathbf{G}$ at time t_i and it will end as soon as $\mathbf{N}$ is again zero at $t = t_e$. Integrating over the whole process yields the total reconnected flux:

$$\Phi_{rec} = \int_{t_i}^{t_e} \int_z \Delta\mathbf{v} \times \mathbf{G} \cdot d\mathbf{l}\, dt = -\Delta z \int_{t_i}^{t_e} N_z(\mathbf{x}_0, t) dt \,.$$

Exercise 4.2 Show that under 2D reconnection the total helicity (see previous exercise) is still conserved.

4.1. MHD EXAMPLE

Consider the following local stationary solution of the set of MHD equations (3)–(8) with $\mathbf{T} = 0$ in the vicinity of a hyperbolic null point of a 2D magnetic field.

$$\begin{aligned}
\rho &= \text{constant}, && (24)\\
\mathbf{B} &= [y, kx, 0], \quad k > 1 && (25)\\
\mathbf{v} &= [-x, y, 0], && (26)\\
\mathbf{E} &= [0, 0, E_z], && (27)\\
\hat{\eta} &= (E_z - kx^2 - y^2)/(k-1), && (28)\\
p &= p0 - 1/2(\rho + k^2 - k)x^2 - 1/2(\rho + 1 - k)y^2 \,. && (29)
\end{aligned}$$

All quantities are normalized by their characteristic values and the solution holds only in the region near the null where resistivity η and pressure are positive. There are rather elaborated analytical and numerical solutions, but for the sake of illustration this example is sufficient. Note that the non-ideal term is represented by a resistive term $\mathbf{N} = \eta\mathbf{J}$, as in most models of magnetic reconnection. (However, there are plasmas where the resistive term is not important and other terms such as the electron inertia take over the rôle of the non-ideal term. This shows the advantage of not relating $\mathbf{N}$ to any particular dissipation mechanism.)

If we want to answer the question whether reconnection occurs in our example we first have to solve equation (17) for $\mathbf{u}$ by using for $\mathbf{G}$ the magnetic field $\mathbf{B}$ (from 25), which yields

$$\mathbf{u} = (\mathbf{E} - \eta\mathbf{J}) \times \mathbf{B}/B^2 = (k^2x^2 + y^2)^{-1}[-kE_z x, E_z y, 0] \,. \qquad (30)$$

Therefore $\mathbf{u}$ shows a singularity at the X-point of $\mathbf{B}$ which also leads to a finite reconnection rate of $E_z = N_z$ per unit length in z-direction, and thus satisfies the definition of reconnection. Note that due to the steady state of the system $\mathbf{u} \equiv 0$ would have also satisfied (17). However, this solution does not satisfy the condition $\mathbf{u} = \mathbf{v}$ on the boundary and is excluded by 4.1. (The boundary in our example is given by the implicit equation $\eta(x, y) = 0$.)

This demonstrates that the boundary condition must be included in 4.1 to avoid trivial solutions.

4.2. HD EXAMPLE

Although HD and MHD share the same type of equation for the evolution of the vorticity field and the magnetic field, there is no one-to-one correspondence between models of magnetic reconnection and vorticity reconnection. The primary reason is that in HD a solution of the vorticity equation has also to satisfy $\mathbf{G} = \mathbf{w} = \nabla \times \mathbf{v}$. Such a restriction is absent in MHD. Moreover, apart from solutions with constant viscosity, the standard expressions for the non-ideal term $\mathbf{N}$ is more complicated in HD. These two reasons complicate the analysis of reconnection in HD enormously, which explains why there are much fewer models or investigations of reconnection in HD compared to reconnection in plasmas. Another reason is that usually the Reynolds numbers of hydrodynamics are much lower than those of astrophysical flows plasmas. Thus the reconnection processes are much less violent and often masked by an omnipresent dissipation.

We start by constructing a stationary two-dimensional solution for vortex reconnection, a solution where all quantities depend only on two space coordinates (x, y) and the vorticity vectors lie in the x-y-plane. Using the analogy with magnetic reconnection we assume that vorticity has the form $\mathbf{w} = [y, kx, 0]$ as in the previous example of magnetic reconnection. However, the flow velocity $\mathbf{v} = [-x, y, 0]$ does not satisfy $\mathbf{w} = \nabla \times \mathbf{v}$. It must be modified with an additional z-component $v_z(x, y, t)$ so that $\mathbf{v}$ does not depend on z and

$$\mathbf{w} = \nabla v_z(x, y, t) \times \mathbf{e}_z \ .$$

From our discussion of reconnection in two-dimensional systems we know that reconnection can only occur if the non-ideal term (in the HD case the parenthesis on the r.h.s. of (14)) has a non-vanishing z-component at the X-point. Note that due to the two-dimensionality of the system the only term which can provide this component is the viscous term. Therefore the time-derivative of $\mathbf{v}$ in (12) has to balance this z-component of the viscous term. For a stationary reconnection process (with respect to $\mathbf{w}$) $\partial_t \mathbf{v}$ has to be constant in time, hence $v_z \sim t$.

By taking these conditions into account a full solution is given by

$$\begin{aligned}
\mathbf{v} &= [-x, y, y^2/2 - kx^2/2 - tE_z] \ , && (31)\\
\nu &= E_z/(l-1) - k/(3\,k-1)\,x^2 - 1/(k-3)\,y^2 \ , && (32)\\
p &= p_0 + (k+1)/(6\,k-2)\,x^2 - (k+1)/(2\,k-6)\,y^2 \ . && (33)
\end{aligned}$$

This solution satisfies (12) with a non-ideal term for viscous, incompressible flows in a region near the reconnection axis where $p > 0$, $\nu > 0$ and

$k > 3$. Equations (31)–(33) must be considered a solution only locally, in the neighborhood of the reconnection line, because of the simple polynomial ansatz for the vorticity: it cannot reproduce all the complex features of vortex reconnection such as 'bridging', secondary reconnection, etc. This simple system, however, captures several basic properties of vortex reconnection. First, although we started with the most elementary reconnection process, the process is not stationary anymore with respect to $\mathbf{v}$, since the v_z component grows linearly in time. Thus there is no true stationary vortex reconnection. Moreover, the existence of a non-vanishing v_z component breaks the two-dimensionality, which was present in the example of magnetic reconnection. This complicates the flow structure of the process enormously. But despite of the complexity, the reconnection process of the vorticity is comparatively simple. So both examples satisfy equation (17) with the singular two-dimensional flow $\mathbf{u} = (k^2x^2 + y^2)^{-1}[-kE_z x, E_z y, 0]$.

Since we know the transport velocity of the vorticity we can calculate the time necessary for two flux tubes to reconnect. This reconnection time is given by the time which is required for an arbitrary point x_0 on the x-axis to reach the z-axis under the flow $\mathbf{u}$. Hence

$$\int_{x_0}^{0} dt = \int_{x_0}^{0} 1/u^x \, dx = -\int_{x_0}^{0} kx/E_z \, dx = \frac{x_0^2 k}{2E_z} \, .$$

This result is in agreement with numerical results [4], which show a quadratic dependence of the reconnection time on the distance between the flux tubes. Note that this is not an accidental feature of our model, but is due to the generic structure of the reconnecting flows in space-time.

In analogy with magnetic reconnection, the reconnected flux is given by the integral along the reconnection line. The reconnected flux and rate are therefore given by

$$\Phi_{rec} = -\iint \partial_t \mathbf{v} \cdot d\mathbf{l} \, dt \, , \qquad \frac{d\Phi_{rec}}{dt} = -\int \partial_t \mathbf{v} \cdot d\mathbf{l} \, . \tag{34}$$

5. 3D Reconnection

Many physical phenomena show marked differences in the transition from two to three dimensions and so does reconnection. However, the crucial difference occurs already in the transition from a field $\mathbf{G}$ with only two components to a field with all three components, even if these depend only on two space coordinates. For instance, in the example of magnetic reconnection (see equation (29)) we can add a constant component b_z in the z-direction of the magnetic field and modify the electric field to $\mathbf{E} = [-yb_z, -xb_z, E_z]$. This does not affect the stationarity of the solution nor the structure of the

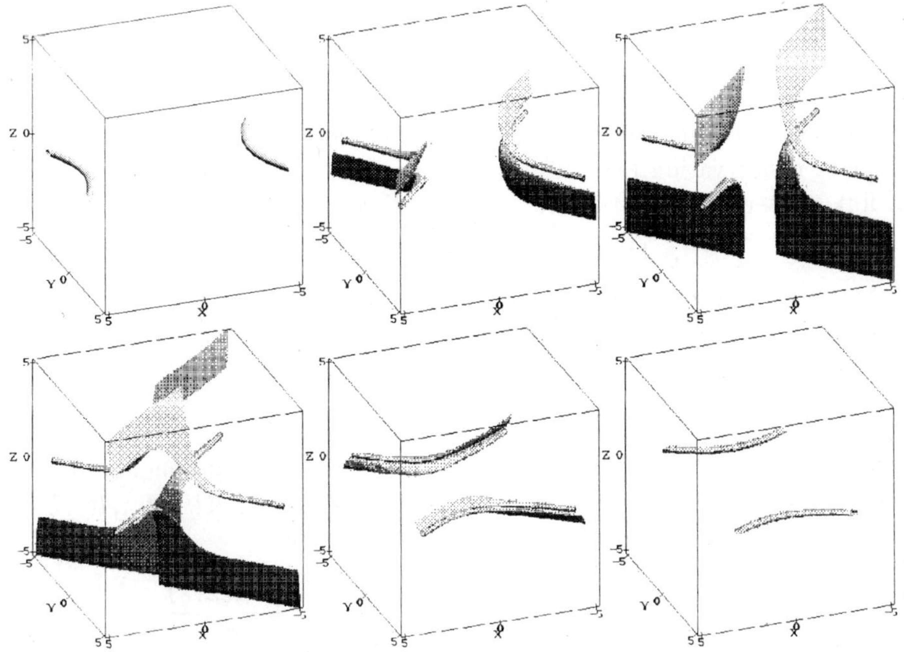

Figure 4. Reconnection in the case of a non-vanishing **G**-field. The non-ideal term **N** is localized near the z-axis.

plasma flow $\mathbf{v}$ or the non-ideal term $\eta \mathbf{J}$. However, no flux-conserving flows, neither singular nor smooth, can exist since the necessary condition (18) is not satisfied. Correspondingly, the evolution of flux tubes in this case is completely different from the 2D reconnection. This is shown in Figure 4, where two flux tubes coming from a dissipation-free external region collide into a non-ideal region: the magnetic field has a small constant component b_z. The flux tubes are integrated at each time starting from their cross sections (in the dissipation-free external region), co-moving with $\mathbf{v}$. The sequence of images show that the flux tubes loose their identity, that is they split, as soon as they enter the non-ideal region. Thereupon, the strands of the different ends anchored in the ideal external region, flip around each other and eventually merge as they leave the non-ideal region differently connected. Thus with respect to the initial and final state of the flux tubes the process is similar as in the two-dimensional case; but contrary to the 2D case the whole evolution is smooth and there is no instant where a cutting process or something similar takes place.

Since this pattern of reconnection occurs for arbitrarily small b_z it should not be considered as a new type of reconnection, but as a generic one, whereas the 2D case with $b_z = 0$ is an exceptional and structurally unstable one. Thus we need to incorporate this process into our notion of reconnection if we want to apply this notion to the real, three-dimensional case.

The key to describe both types of reconnection on the same mathematical basis is to generalize the notion of flux conservation. First note that the flux conservation (1) is the non-relativistic limit of the covariant form of flux conservation in 4-space with coordinates $x^0 = ct, x^1, x^2, x^3$:

$$\epsilon^{\alpha\beta\gamma\delta}\partial_\alpha U^\nu G_{\nu\beta} = 0 \tag{35}$$

$$\Leftrightarrow \left\{ \begin{array}{l} \partial_0(U^0\mathbf{G}_E + \mathbf{U}\times\mathbf{G}_B) + \nabla(\mathbf{G}_E\cdot\mathbf{U}) = 0 \\ U^0\partial_0\mathbf{G}_B - \nabla\times(\mathbf{U}\times\mathbf{G}_B) - \nabla U^0\times\mathbf{G}_E = 0 \end{array} \right.$$

$$\Leftrightarrow L_U\omega_G^2 = 0\ , \tag{36}$$

where $G_{\alpha\beta}$ denotes the field tensor, with the electric components $\mathbf{G}_E$ and the magnetic components $\mathbf{G}_B$. The four-velocity $U^{(4)} = (U^0, U^1, U^2, U^3) = (U^0, \mathbf{U})$ is related to the usual velocity in space by

$$\mathbf{u} := \frac{\mathbf{U}}{U^0} = \frac{d\mathbf{F}_G/ds}{dF_G^0/ds}, \quad \text{where} \quad U^{(4)} = \frac{dF_G(x,s)}{ds}\ , \tag{37}$$

and F_G is the flow of $U^{(4)}$ in 4-space. Note that the parameter is now s instead of t, but they are related by $U^0 = d(ct)/ds$. Here it is important to note that equations (35) are covariant with respect to *any* coordinate transformation in 4-space, in particular they are not restricted to a Minkowski transformation; also $U^{(4)}$ is not normalized to 1 or c.

For the MHD case the field tensor is the electromagnetic field and $\mathbf{G}_E$ and $\mathbf{G}_B$ are the usual electric and magnetic field. For the hydrodynamic case $G_{\alpha\beta}$ is the vorticity tensor, i.e. $\mathbf{G}_E = -\partial_0\mathbf{v} - \nabla v^0$ and $\mathbf{G}_B = \nabla\times\mathbf{v}$.

$L_U\omega_G^2 = 0$ in (36) is the Lie-derivative of the 2-form of the G field with respect to the flow $U^{(4)}$. The Lie-derivative theorem [1] then implies that the flux integrated over a two-dimensional co-moving surface C is constant with respect to the parameter s which describes the $U^{(4)}$-flow in space-time. Equation (36) yields

$$\int_C G dA = \int_C G_{\mu\nu} dx^\mu dx^\nu = \text{constant.} \tag{38}$$

If the time component U^0 of the 4-flow $U^{(4)}$ is constant, e.g. $U^0 \equiv 1$, the general parameter of the flow s coincides with the time t, and the covariant form of the flux conservation yields the ordinary flux conservation for the magnetic part. From (1), with $U^0 = $ constant, we have

$$\int_C \mathbf{G}_B \cdot d\mathbf{A} = \text{constant.}$$

The advantage of the covariant formulation is that now the singularity in $\mathbf{u}$, which we encountered in reconnection, can be represented by a smooth $U^{(4)}$-flow with a null at the former singularity. For instance, for the singular flow (30) a smooth 4-velocity is given by

$$U^{(4)} = [U^0, U^x, U^y, U^z] = [(k^2x^2 + y^2), -kE_z x, E_z y, 0] \,. \tag{39}$$

More about the relation between singular flows in 3-space and their smooth representation in 4-space can be learned if we integrate the system (35):

$$U^\nu G_{\nu\beta} = \partial_\beta \Phi \;\Leftrightarrow\; \left\{ \begin{array}{l} \mathbf{G}_E \cdot \mathbf{U} = -\partial_0 \Phi \\ U^0 \mathbf{G}_E + \mathbf{U} \times \mathbf{G}_B = \nabla \Phi \,. \end{array} \right. \tag{40}$$

From this it is obvious that whenever we can set $\Phi \equiv 0$, the time component of the equation is a trivial implication of the equation for the space components. Thus fields $\mathbf{G}_E, \mathbf{G}_B$ which allow for a certain $U^{(4)}$-flow also satisfy

$$\mathbf{G}_E + \mathbf{u} \times \mathbf{G}_B = 0 \,,$$

with the corresponding 3-vector flow $\mathbf{u} = \mathbf{U}/U^0$. This in turn also implies (17), i.e. the conservation of $\mathbf{G}_B$-flux under $\mathbf{u}$. This is a *a posteriori* justification of our interpretation that the flux is conserved under a singular velocity field in the 2D reconnection case.

The 3D reconnection case, however, requires a non-vanishing Φ because in this case $\mathbf{G}_E \cdot \mathbf{G_B} = 1/U^0 \mathbf{G}_B \cdot \nabla\Phi \neq 0$ as in the example at the beginning of this section. Note that for this example (i.e. $\mathbf{E} = [-yb_z, -xb_z, E_z]$, $\mathbf{B} = [y, kx, b_z]$ $(k > 0)$) a solution for the covariant equations (35) is

$$U^{(4)} = (y^2 + kx^2 + 1)^{(-1)}[(y^2 + kx^2), -x, y, 0]; \quad \Phi = -b_z xy \,.$$

This flow is smooth and its 3-flow $\mathbf{u} = \mathbf{U}/U^0$ shows the same singularity as in the 2D example. However, due to a non-vanishing Φ the covariant flux conservation does not imply the existence of the ordinary $\mathbf{G}_B$-flux conservation nor the conservation of $\mathbf{G}_B$ field lines. This is the reason why we see the strange splitting and flipping of flux tubes in Figure 4.

This example shows that there exists a more general kind of flux conservation which captures also the case of reconnection in a non-vanishing $\mathbf{G}_B$-field. Correspondingly it is possible to give a definition of reconnection on the basis of the covariant flux conservation in place of (17) (see [7]) analogous to our statement for the 2D case. This however requires additional considerations which are beyond the scope of this article.

Exercise 5.1 Show by means of the example given above that the total helicity is in general not conserved under 3D reconnection.

6. Conclusions

The process of reconnection is an important process capable of altering the topology of a divergence-free field. In most applications of fluid dynamics reconnection is driven by a dissipative term, e.g. a resistive or viscous term. It is therefore important to distinguish reconnection from other processes which involve dissipation. It turns out that reconnection is special in the sense that the flux of a field is not reduced but is only restructured. Therefore the effect of reconnection can be modeled with the help of flux-conserving flows with singularities of a specific type. This helps to understand the effect of reconnection on the structure of a field even if the details of the underlying dissipation mechanism are unknown as it is often the case in magnetic reconnection, and it also reveals the similarities of magnetic and vortex reconnection.

Acknowledgements

I would like to acknowledge the hospitality of the Isaac Newton Institute as well as financial support from the Volkswagen Foundation.

References

1. Abraham, R., Marsden, J.E. & Ratiu, T. (1983) *Manifolds Tensor Analysis and Applications.* Applied Mathematical Sciences **75**, Springer-Verlag.
2. Biskamp, D. (2000) *Magnetic Reconnection in Plasmas.* Cambridge University Press, Cambridge.
3. Chuang, I., Durrer, R., Turok, N., & Yurke, B. (1991) Cosmology in the laboratory: defect dynamics in liquid crystals. *Science* **251**, 1336–1342.
4. De Waele, A.T.A.M. & Aarts, R.G.K.M. (1994) Route to vortex reconnection. *Phys. Rev. Lett.* **72**, 482–485.
5. Hornig, G. & Schindler, K. (1996) Magnetic topology and the problem of its invariant definition. *Phys. Plasmas* **3**, 781–791
6. Hornig, G. (1997) The covariant transport of electromagnetic fields and its relation to magnetohydrodynamics. *Phys. Plasmas* **4**, 646–654.
7. Hornig, G. (1999) The evolution of magnetic helicity under reconnection. In: *Magnetic Helicity* (ed. A.A. Pevtsov *et al.*), pp. 157-165. American Geophysical Union, Washington.
8. Kauffman, L.H. (1991) *Knots and Physics.* Series on Knots and Everything **1**, World Scientific.
9. Kida, S. & Takaoka, M. (1994) Vortex Reconnection. *Annu. Rev. Fluid Mech.* **26**, 169–189.
10. Koplik, J. & Levine, H. (1993) Vortex reconnection in superfluid helium. *Phys. Rev. Letters* **71**, 1375–1378.
11. Priest, E. & Forbes, T. (2000) *Magnetic Reconnection.* Cambridge University Press, Cambridge.
12. Shellard, E.P.S. (1987) Cosmic string interactions. *Nucl. Phys. B* **282**, 624–656.
13. Sumners D. (1990) Untangling DNA. *Math. Intell.* **12**(3), 71–80.

EULER SINGULARITIES FROM THE LAGRANGIAN VIEWPOINT

STEPHEN CHILDRESS
Courant Institute of Mathematical Sciences
251 Mercer Street, New York, NY 10012, USA
E-mail: *childres@cims.nyu.edu*

Abstract. In the analysis of singular Euler flows, real or imagined, Lagrangian variables offer an attractive option, although it is well known that in many flow problems Lagrangian methods can be intractable. In other examples, Lagrangian variables are found to simplify the representation. We examine here some aspects of Lagrangian analysis of fluid flows, then focus on simple singular solutions of Euler's equations having infinite kinetic energy. We use these solutions to explore the different forms taken by these flows in Eulerian and Lagrangian formulations. We then apply a Lagrangian construction to a set of singular flows in two dimensions, as examples of a general method. Finally, we comment on the analogous but far more difficult application of the method to singular Euler flows in three dimensions.

1. Introduction

In the sixth edition of Lamb's *Hydrodynamics*, there are two entries under "Lagrangian", both referring to forms taken by the governing equations, not to solutions of flow problems using these coordinates. This underscores the great differences that can exist between the complexity of Eulerian and Lagrangian fluid descriptions. Indeed, most *steady* three-dimensional fluid flows, including ones with simple Eulerian form, have regions of chaotic Lagrangian structure.

In fact, Lamb does give a beautiful example of time-dependent flow in Lagrangian coordinates, namely the family of exact nonlinear gravity waves discovered by Gertsner, and independently by Rankine. We shall begin our

R.L. Ricca (ed.), An Introduction to the Geometry and Topology of Fluid Flows, 313–328.

discussion by looking in detail at Gertsner's waves.

The Lagrangian coordinates of a velocity field $\mathbf{u}(\mathbf{x}, t)$ defined in $\mathbb{R}^N$ are defined as solution of the system of ordinary differential equations

$$\frac{d\mathbf{x}}{dt} = \mathbf{u}(\mathbf{x}(t), t), \tag{1}$$

with initial conditions

$$\mathbf{x}(0) = \mathbf{x}_0(\mathbf{a}). \tag{2}$$

Here the map $\mathcal{L}$,

$$\mathbf{a} \to \mathbf{x}_0, \tag{3}$$

is a one-to-one map of $\mathbb{R}^N$ onto itself. The components of the vector $\mathbf{a}$ are called the Lagrangian coordinates of the flow $\mathbf{u}$. These coordinates may thus be considered to be labels uniquely identifying fluid particles. Frequently the map $\mathcal{L}$ is taken to be the identity, in which case the label is the initial position of the particle in question.

Gertsner's waves are given by the following Lagrangian coordinates in two dimensions:

$$x_1 = a_1 + \frac{1}{k} e^{ka_2} \sin k(a_1 + ct), \; x_2 = a_2 - \frac{1}{k} e^{ka_2} \cos k(a_1 + ct). \tag{4}$$

Here c is a constant to be chosen below, and $\mathbf{x}_0 = (a_1 + \frac{1}{k} e^{ka_2} \sin ka_1, \; a_2 - \frac{1}{k} e^{ka_2} \cos ka_1)$. Then $\det(d\mathbf{x}_0/d\mathbf{a}) = 1 - e^{2ka_2}$. Thus these coordinates are defined for k, a_2 nonzero.

We now show that these waves satisfy Euler's equations for a gravity wave with free surface a_2 =constant, provided that $c^2 = g/k$. Now Euler's equations in Lagrangian coordinates (with gravity acting in the direction $-\mathbf{i}_2$), take the form

$$\frac{\partial^2 x_i}{\partial t^2} + \frac{\partial(\frac{p}{\rho} + g x_2)}{\partial a_j} \frac{\partial a_j}{\partial x_i} = 0, \quad \frac{\partial}{\partial t} \det(d\mathbf{x}/d\mathbf{a}) = 0. \tag{5}$$

The second equation, which is the Lagrangian form of the incompressibility condition, expresses the conservation of volume needed to keep density constant following a fluid particle. It is satisfied by (4), as noted above. The first of (5) expresses the conservation of momentum, and balances acceleration with pressure and gravity forces. All time derivatives are taken with fixed $\mathbf{a}$.

The momentum balance is satisfied by (4) provided that

$$\frac{\partial(\frac{p}{\rho} + g x_2)}{\partial a_i} = -\frac{\partial^2 x_j}{\partial t^2} \frac{\partial x_j}{\partial a_i}$$

$$= [c^2 k e^{ka_2} \sin k(a_1 + ct),\ c^2 k e^{2ka_2} - c^2 k e^{ka_2} \cos k(a_1 + ct)]. \tag{6}$$

Thus if $c^2 = g/k$ we have

$$\frac{p}{\rho} = -ga_2 + \frac{c^2}{2} e^{2ka_2} + const, \tag{7}$$

so a free surface, on which p must be constant, may be taken as any curve $a_2 =$ constant.

If we compute the velocity $\mathbf{u}$, and then the matrix of velocity derivatives $D_{ij} = \frac{\partial u_i}{\partial x_j}$ from (4), there results

$$D_{ij} = \frac{cke^{ka_2}}{1 - e^{2ka_2}} \begin{pmatrix} -\sin k(a_1 + ct) & \cos k(a_1 + ct) + e^{ka_2} \\ \cos k(a_1 + ct) - e^{ka_2} & \sin k(a_1 + ct) \end{pmatrix}. \tag{8}$$

We then have

$$\frac{\partial \mathbf{D}}{\partial t} + \mathbf{D}^2 = \frac{cke^{ka_2}}{1 - e^{2ka_2}} \begin{pmatrix} e^{ka_2} - \cos k(a_1 + ct) & -\sin k(a_1 + ct) \\ -\sin k(a_1 + ct) & e^{ka_2} + \cos k(a_1 + ct) \end{pmatrix}. \tag{9}$$

A tedious but straightforward calculation verifies that the matrix on the right of (9) is in fact the Eulerian matrix $-\frac{\partial^2 (p/\rho + gy)}{\partial x_i \partial x_j} = -\frac{\partial^2 (p/\rho)}{\partial x_i \partial x_j}$, expressed in Lagrangian variables, so we have the Lagrangian relation

$$\frac{\partial \mathbf{D}}{\partial t} + \mathbf{D}^2 + \mathbf{\Pi} = 0, \quad \Pi_{ij}(\mathbf{a}, t) = \frac{\partial^2 (p/\rho)}{\partial x_i \partial x_j}. \tag{10}$$

This equation can be deduced directly from the Euler form of the momentum equation, by taking the gradient and expressing the result in Lagrangian form.

A second relation we can check is obtained by taking the $\mathbf{a}$ derivative of (1) to get an evolution equation for the Jacobian $J_{ij} = \frac{\partial x_i}{\partial a_j}$,

$$\frac{\partial \mathbf{J}}{\partial t} = \mathbf{DJ}. \tag{11}$$

With

$$\mathbf{J} = \begin{pmatrix} 1 + e^{ka_2} \cos k(a_1 + ct) & e^{ka_2} \sin k(a_1 + ct) \\ e^{ka_2} \sin k(a_1 + ct) & 1 - e^{ka_2} \cos k(a_1 + ct) \end{pmatrix}, \tag{12}$$

and using (8), we easily verify (11) for this solution.

It is not easy to deduce from these arguments how Gertsner might have obtained this solution. It is likely, however, that (4) was written down for finite amplitude waves on the basis of the essentially identical Lagrangian form of small-amplitude waves, the latter being an easy consequence of the Eulerian representation. [1]

We thus have a good example of a solution in Lagrangian form with no simple Eulerian form. (Note in particular that the waves are rotational.) But unfortunately there is no suggested prescription for finding other equally interesting Euler solutions.

The following construction would in principle be possible, however. First select a symmetric matrix function $\mathbf{\Pi}(\mathbf{a},t)$. Then solve (10) for a traceless matrix function $\mathbf{D}(\mathbf{a},t)$. Next solve (11) for a matrix function $\mathbf{J}$. The latter has a constant determinant since $(\det \mathbf{J})_t = (\nabla \cdot \mathbf{u}) \det \mathbf{J}$. We must however satisfy compatibility constraints insuring that each row of $\mathbf{J}$ is a Lagrangian gradient.

That fact that this procedure works in principle may be summarized as follows:

Proposition 1.1 *Let $\mathcal{L}$ be the identity, and suppose we seek to solve the incompressible Euler equations with unit density,*

$$\mathbf{u}_t + \mathbf{u} \cdot \nabla \mathbf{u} + \nabla p = 0, \quad \nabla \cdot \mathbf{u} = 0, \tag{13}$$

Use the procedure just outlined to construct a $\mathbf{J}$ with determinant unity and initial value equal to the identity, such that the rows of $\mathbf{J}$ are gradients with respect to $\mathbf{a}$ of functions X, Y, Z of $\mathbf{a}, t$. Then, there is a pressure field p such that $\mathbf{u} = (\frac{\partial X}{\partial t}, \frac{\partial Y}{\partial t}, \frac{\partial Z}{\partial t}), p$ solves (13).

Proof. Assume that $\mathbf{D}$ and $\mathbf{J}$ are thus constructed, with $\mathbf{J}$ a Jacobian of determinant unity, and introduce X, Y, Z as above. Then X is unique since $X(\mathbf{a}, 0) = a_1$, and similarly for Y, Z. Define $\mathbf{u} = (\frac{\partial X}{\partial t}, \frac{\partial Y}{\partial t}, \frac{\partial Z}{\partial t})$. From $\det(\mathbf{J}) = 1$ and the identity $(\det \mathbf{J})_t = (\nabla \cdot \mathbf{u}) \det \mathbf{J}$ we see that $\nabla \cdot \mathbf{u} = 0$. Then $(\partial u_i / \partial x_j) = \mathbf{J}_t \mathbf{J}^{-1}$, and we see from (11) that $\mathbf{D}$ is the matrix of velocity derivatives. From (10) it then follows that $\Pi_{ij} = \partial C_i / \partial x_j$ for some vector $\mathbf{C}$. Indeed $C_i = \frac{\partial u_i}{\partial t}|_{\mathbf{x}} + \mathbf{u} \cdot \nabla u_i$. But since was assumed $\mathbf{\Pi}$ to be symmetric, it must be of the form $\partial^2 p / \partial x_i \partial x_j$ for some scalar field p, unique up to an additive function $\mathbf{v}(t) \cdot \mathbf{x}$. (The last function is associated with an arbitrary linear acceleration of the coordinate frame of the Lagrangian system, which allows us to position the frame on any fluid particle.) Then $\mathbf{u}, p$ as defined solves (13).

QED

Such a procedure involves two matrix equations and very stringent compatibility conditions, and is unlikely to be workable with any generality. On

[1] We have not been able to obtain the paper by Gertsner referenced in [13]. However Rankine, see [16], indicates at the beginning of his paper that this is his approach.

the other hand it is an approach we shall consider here with the aim of seeking solutions where the matrix $\mathbf{\Pi}$ has a rather simple form. It should be noted that in two dimensions $\mathbf{D}$ can be chosen as any traceless matrix whose time derivative is symmetric, cf. (8), in which case the problem reduces to solving (11) for a compatible $\mathbf{J}$.

2. Singular Solutions

Lagrangian variables can exhibit singularities for disparate reasons:

Example 2.1 When the Eulerian flow is *steady* but singular at a point– an example is two-dimensional flow into a sink, $\mathbf{u} = r^{-2}(-x, -y)$. If (r, θ) and ρ, ϕ are polar coordinates for $\mathbf{x}$ and $\mathbf{a}$; we obtain the Lagrangian coordinates $r = (\rho^2 - 2t)^{1/2}, \theta = \phi$. (In these examples $\mathcal{L}$ is the identity.)

Example 2.2 When the Eulerian flow is steady and regular everywhere but unbounded at infinity– an example is $\mathbf{u} = (x^2, -2xy)$; here

$$\mathbf{x} = ((1/a_1 - t)^{-1}, \quad a_2 a_1^2 (1/a_1 - t)^2). \tag{14}$$

Example 2.3 When the Eulerian flow is not steady but is regular everywhere and unbounded at infinity– an example is

$$\mathbf{u} = (x, -y)(1-t)^{-1}, \quad \text{where } \mathbf{x} = (a_1(1-t)^{-1}, a_2(1-t)). \tag{15}$$

The last example is typical of a class of unphysical singular Euler flows with unbounded kinetic energy, which can therefore blow up over the entire domain at a single singularity time. The fact that the velocity derivatives blow up as the inverse time to singularity is consistent with an important estimate obtained by Beale, Kato & Majda [2]. This states that if an Euler flow loses regularity at the finite time T, then the integral with respect to time of the modulus of the supremum of the vorticity, must diverge at time T.

Example 2.3 in fact has zero vorticity, but we can easily generalize it to analogous flows with vorticity, by considering other solutions of (10),(11) such that $\mathbf{D}, \mathbf{\Pi}, \mathbf{J}$ *are functions of time only.* We shall refer to such solutions as *uniform gradient flows.* Uniform gradient flows were analyzed by Vieillefosse [19] and Cantwell [4],and have been extensively studied in connection with the evolution of small-scale structure , see [10],[11].

For the uniform gradient flows the Eulerian and Lagrangian formulations merge and can be viewed as generated by a given symmetric matrix $\mathbf{\Pi}(t)$ with the property that the trace of $\mathbf{\Pi}$ equals that of $-\mathbf{D}^2$. For example, a model adopted by Vieillefosse [19] takes $\mathbf{\Pi}$ to be a multiple of the identity consistent with the last constraint. Depending upon the choice of $\mathbf{\Pi}$ and the initial conditions, these uniform gradient flows may or may not terminate at a singularity.

Example 2.4 In view of the Beale-Kato-Majda estimate it is tempting to study a special family of singular solutions (which include the uniform gradient solutions but are not restricted to those flows) where

$$\mathbf{D} = \mathbf{D}_0(\mathbf{a})(t^* - t)^{-1}, \quad \mathbf{\Pi} = \mathbf{\Pi}_0(\mathbf{a})(t^* - t)^{-2}, \tag{16}$$

$\mathbf{D}_0, \mathbf{\Pi}_0$ being now constant matrices, and t^* is an arbitrary singularity time. We then must have

$$\mathbf{D}_0 + \mathbf{D}_0^2 + \mathbf{\Pi}_0 = 0. \tag{17}$$

Assuming a suitable solution of (17) has been found (trace $(\mathbf{D}_0) = 0$, $\mathbf{\Pi}_0$ symmetric), Lagrangian coordinates of the flow are given by

$$\mathbf{x}(\mathbf{a}, t) = \mathbf{a}e^{-\mathbf{D}_0\sigma}, \; \sigma = \ln (t^* - t), \tag{18}$$

or, if $\mathcal{L}$ is the identity, by

$$\mathbf{x}(\mathbf{a}, t) = \mathbf{a}e^{\mathbf{D}_0(\ln t^* - \sigma)}. \tag{19}$$

Example 2.5 A more general assumption than that of the last example is

$$\mathbf{D} = \mathbf{D}_0(\mathbf{a}, s)(t^* - t)^{-1}, \quad \mathbf{\Pi} = \mathbf{\Pi}_0(\mathbf{a}, s)(t^* - t)^{-2}, \; s = -\ln (t^* - t) = -\sigma. \tag{20}$$

Then we have

$$\frac{\partial \mathbf{D}_0}{\partial s} + \mathbf{D}_0 + \mathbf{D}_0^2 + \mathbf{\Pi}_0 = 0, \quad \frac{\partial \mathbf{J}}{\partial s} = \mathbf{J} \cdot \mathbf{D}_0. \tag{21}$$

We can thus entertain the idea that an Euler singularity might exist which is "unsteady" in the time variable s, the subscript zero variables having an bounded attractor or weak secondary singularity in that variable even as velocity gradients diverge with respect to t. (That is to say, exponential growth or decay with respect to σ would be already expressed in powers of $t^* - t$, and so that behavior is excluded.)The logarithmic variable σ will reappear in the explicit solutions considered in the next section.

3. Singular Flows with Stagnation-Point Similitude

Stuart [17], [18] has discussed an interesting class of singular flows in three dimensions which can be analyzed with particular success using Lagrangian methods. The velocity field considered by Stuart had the Eulerian form

$$\mathbf{u} = (f(x, t), yg(x, t), zh(x, t)). \tag{22}$$

This is a special case of what we refer to as *stagnation point similitude*, since such flows emerge from the modeling of flow near a stagnation point.

The linear growth in coordinates again places these flows in the unphysical category of infinite energy solutions. Nevertheless the structure and analysis of their solutions is provocative and suggestive. Also, it is conceivable that such solutions are "inner" approximations to physical singular flows with bounded kinetic energy.

Childress, Ierley, Spiegel & Young [6] treated a simpler class of two-dimensional flows, having a velocity field of the form

$$\mathbf{u} = (\psi_y, -\psi_x), \quad \psi = yf(x,t). \tag{23}$$

The singular structure obtained in these 2D solutions indicates the importance of unbounded energy, since it is well-known that 2D Euler flows in a bounded region, which are smooth initially, stay smooth for all time [12]. A rigorous treatment of some related two-dimensional flows is given in [8], and three-dimensional Euler and Navier-Stokes flows obtained from the stretching of vortex tubes introduces some similar ideas, see [15], [14]. Singularities of this type have been shown to arise as a means of "re-three-dimensionalization" of stretched, nearly 2D Euler flows, see [5].

3.1. THE 2D CASE

The example discussed in this sub-section, taken from [6], is of the two-dimensional form (23). The equation for f follows from substitution of (23) into the 2D vorticity equation. After one integration we obtain:

$$f_{tx} + ff_{xx} - f_x^2 = h(t), \tag{24}$$

where $h(t)$ is an arbitrary function. We localize f to the interval $0 \le x \le 2$ and require that $f(0,t) = f(2,t) = 0$. Then (24) implies $h(t) = -2\int_0^2 f_x^2 dx$. We shall, as an example, adopt the initial condition $f(x,0) = f_0(x) = x(2-x)/2$. The solution can be shown to be

$$f(x,t) = \Big[\frac{2(e^{2\tau} - e^{(2-x)\tau})}{\tau(e^{2\tau}-1)} - \frac{x}{\tau}\Big]\dot{\tau}, \tag{25}$$

where $\tau(t)$ satisfies

$$\dot{\tau} = \Big(\frac{\sinh\tau}{\tau}\Big)^2, \; \tau(0) = 0. \tag{26}$$

We are interested here mainly in the behavior near the singularity, rather than the solution method, and we refer to [6] for details. We can remark however one means of obtaining (25) is to show that f_{xx} and the Jacobian of the relevant Lagrangian variables, satisfy the same relation. This leads to an analog of the Gauss representation of vorticity in three dimensions in terms of the Jacobian. In the present case this representation may be integrated to obtain the Lagrangian coordinates.

From (26) we find that blow-up occurs when $t = t^* = \pi^2/6$, with

$$\tau = -\frac{1}{2}\ln T + \frac{1}{2}\ln(\frac{1}{2}\ln^2 T) + o(1), \tag{27}$$

as $T =\longrightarrow 0$, where

$$T = t^* - t. \tag{28}$$

From (25) and (27) we have

$$f(x,t) \approx \frac{x}{t^* - t}(\frac{1 - e^{x\tau}}{x\tau}), \tag{29}$$

for x and $t^* - t$ small. The function of $x\tau$ in the last expression effectively "resolves" the singularity formation at $x = 0$, $t = t^*$, and for small $x\tau$ we have $f \sim \frac{x}{t^* - \tau}$. On the other hand all points x in the interval become singular at the same time $t = t^*$.

We now contrast these results with the Lagrangian form of the singularity. It should be understood that the Lagrangian coordinate is here one-dimensional, and only in the "interesting" coordinate x. We are in effect labeling all fluid particles on a given line $x =$ constant by the same label, since they stay on the line even as they move vertically along it. The Jacobian $J = \partial x/\partial a$ is given by

$$J = (\sinh \tau/\tau)[\cosh \tau + (a - 1)\sinh \tau]^{-1}, \tag{30}$$

and the Lagrangian coordinate $x(a,t)$ is given by

$$x = \tau^{-1}\ln(1 + ae^{\tau}\sinh \tau). \tag{31}$$

The expression for $f_x = D(a,t)$ is then

$$D = \Big[\frac{2e^{2\tau}}{e^{2\tau} - 1}\;\frac{1}{1 + ae^{\tau}\sinh \tau} - \frac{1}{\tau}\Big]\dot{\tau}. \tag{32}$$

Near blow-up the dominant contribution to D comes close to $a = 0$, and we obtain, using (27)

$$D = \frac{1}{T}\Big[\frac{1 - 2/\ln T}{1 + a\ln^2 T/(4T)} + \frac{1}{\ln T} + o(\frac{1}{|\ln T|})\Big]. \tag{33}$$

There is again a "focusing" of the singularity near $a = 0$.

The form of (33) is interesting because, while the structure of the singularity is resolved in a, the nonlocal term $h(t)$ associated with the pressure is actually negligible. From $h(t) = -2\int_0^2 f_x^2 dx$ one can estimate directly from (32) that $h = O(T^{-2}|\ln T|^{-1})$, and so is smaller by a factor $|\ln T|^{-1}$ than $\dot{D}$ or D^2. We note that the pressure is given by

$$p = -x^2 h(t)/2 + y^2 h(t)/2 + P(x,t),\ P_{xx} = 2f_x^2. \tag{34}$$

To check whether

$$\dot{D} + D^2 \approx 0 \tag{35}$$

is the equation satisfied locally, suppose that D takes the form

$$D = A(\sigma)(\mu(\sigma,a) + T)^{-1} + B(\sigma)T^{-1},\ \sigma = \ln T. \tag{36}$$

Substituting into $\dot{D} + D^2 = h(t)$ we have, with $\Gamma = \mu + T$,

$$-A_\sigma T^{-1}\Gamma^{-1} + A\Gamma^{-2} + \mu_\sigma A\Gamma^{-2}T^{-1} - B_\sigma T^{-2}$$
$$+BT^{-2} - A^2\Gamma^{-2} - 2AB\Gamma^{-1}T^{-1} - B^2T^{-2} = h(t). \tag{37}$$

We now postulate that

$$\sigma\mu_\sigma = k(\Gamma - T), \tag{38}$$

i.e., that μ is proportional to σ^k. Comparing terms in (37) we then have

$$A(1 - k/\sigma - A)A = 0,\ A_\sigma - kA/\sigma + 2AB = 0,\ -B_\sigma + B - B^2 = hT^2. \tag{39}$$

Thus

$$A = 1 - k/\sigma,\ B = \frac{k(\sigma - k - 1)}{2\sigma(\sigma - k)}, \tag{40}$$

with h given by the last of (39). For $k = 2$ we have $\mu = \sigma^2 g(a)$, and note that these expressions agree with the local expansion (33) when $g = a/4$. In general this calculation must yield expansions of the assumed form for all possible initial conditions. We also verify the estimate on h given above, and that D may be calculated to first order by neglecting $h(t)$ entirely.

4. Further Examples in Two-Dimensions

We now consider the approach outlined in section (1) in two dimensions. We may then disregard (10) and simply take $\mathbf{D}$ to be a given traceless matrix function of $\mathbf{a}, t$, whose time derivative is symmetric. Our aim is then to solve (11) for a compatible Jacobian and thereby deduce Lagrangian variable and the flow.

To illustrate the procedure we shall restrict attention to matrices of the special form

$$\mathbf{D} = \frac{\mathbf{D}_0(\mathbf{a})}{t^*(\mathbf{a}) - t} + \mathbf{D}_1(\mathbf{a}),\quad \mathbf{D}_0 = \begin{pmatrix} \alpha & \beta \\ \beta & -\alpha \end{pmatrix}. \tag{41}$$

We first diagonalize $\mathbf{D}_0$ in the form $\mathbf{R}^{-1}\mathbf{D}_0\mathbf{R}$ where

$$\mathbf{R} = \begin{pmatrix} \cos\theta & -\sin\theta \\ \sin\theta & \cos\theta \end{pmatrix}, \tag{42}$$

where θ is a function of $\mathbf{a}$ and $(\alpha, \beta) = \lambda(\mathbf{a})(\cos 2\theta, -\sin 2\theta), \lambda > 0$. Requiring that $\mathbf{J}$ is initially the identity $\mathbf{I}$, we have the following equation for $\mathbf{K} = \mathbf{R}^{-1}\mathbf{J}$:

$$\frac{\partial \mathbf{K}}{\partial t} = \left[(t^* - t)^{-1}\begin{pmatrix} \lambda & 0 \\ 0 & -\lambda \end{pmatrix} + \begin{pmatrix} A & B \\ C & -A \end{pmatrix}\right]\mathbf{K},\ \mathbf{K}(0) = \mathbf{R}^{-1}. \tag{43}$$

Here we have defined $\mathbf{D}_1$ in terms of the functions A, B, C of $\mathbf{a}$. We now discuss a few special cases of (43).

Example 4.1 Let $A = B = C = 0$. In this case (43) may be integrated explicitly:

$$J_{11} = \frac{(1-\cos 2\theta)}{2}(1 - t/t^*)^{\lambda} + \frac{(1+\cos 2\theta)}{2}(1 - t/t^*)^{-\lambda},$$

$$J_{12} = J_{21} = \frac{\sin 2\theta}{2}[(1 - t/t^*)^{\lambda} - (1 - t/t^*)^{-\lambda}],$$

$$J_{22} = \frac{(1+\cos 2\theta)}{2}(1 - t/t^*)^{\lambda} + \frac{(1-\cos 2\theta)}{2}(1 - t/t^*)^{-\lambda}. \tag{44}$$

The compatibility conditions are

$$\frac{\partial J_{11}}{\partial a_2} = \frac{\partial J_{12}}{\partial a_1}, \quad \frac{\partial J_{21}}{\partial a_2} = \frac{\partial J_{22}}{\partial a_1}. \tag{45}$$

When these derivatives are computed from (44) the time appears in the functions

$$(1 - t/t^*)^{\pm\lambda},\ (1 - t/t^*)^{\pm\lambda - 1},\ \ln(1 - t/t^*)(1 - t/t^*)^{\pm\lambda}. \tag{46}$$

These six functions are linearly independent provided that $\lambda \neq 1/2$. In this case the compatibility equations yield twelve relations. For example the coefficients of $(1 - t/t^*)^{\pm\lambda-1}$ yield

$$(1+\cos 2\theta)\frac{\partial t*}{\partial a_2} + \sin 2\theta \frac{\partial t*}{\partial a_1}, \quad (1+\cos 2\theta)\frac{\partial t*}{\partial a_1} - \sin 2\theta\frac{\partial t*}{\partial a_2} = 0, \tag{47}$$

which together imply $\frac{\partial t*}{\partial a_1} = \frac{\partial t*}{\partial a_2} = 0$. The remaining equations show that λ and θ are also independent of $\mathbf{a}$. Thus we have a uniform gradient solution and deduce the flow in the complex form

$$u - iv = \lambda(x + iy)e^{2i\theta}(t^* - t)^{-1}. \tag{48}$$

This is the singular flow given as Example 2.3 above, rotated through an angle θ.

Example 4.2 The case $\lambda = 1/2$ can be similarly treated. From the coefficient of $(1 - t/t^*)^{-3/2}$ we again have the first of (47). The coefficients of $(1 - t/t^*)^{-1/2}$ now give

$$\cos 2\theta \frac{\partial \theta}{\partial a_1} - \sin 2\theta \frac{\partial \theta}{\partial a_2} + \frac{1}{2}\frac{\partial \ln t^*}{\partial a_2} = 0, \tag{49}$$

$$\sin 2\theta \frac{\partial \theta}{\partial a_1} + \cos 2\theta \frac{\partial \theta}{\partial a_2} + \frac{1}{2}\frac{\partial \ln t^*}{\partial a_1} = 0. \tag{50}$$

Eliminating t^* from the first of (47),(49), and (50) we obtain

$$(1 + \cos 2\theta)\frac{\partial \theta}{\partial a_1} - \sin 2\theta \frac{\partial \theta}{\partial a_2} = 0. \tag{51}$$

Thus θ is constant on the characteristic lines

$$\frac{da_2}{da_1} = -\tan\theta = \tan\psi, \tag{52}$$

where ρ, ψ are polar coordinates in the a_1, a_2-plane. Thus $\theta = -\psi$. Then, using (51), and a second equation obtain by multiplying (51) by $1 - \cos 2\theta$, in (49),(50) we obtain

$$-\frac{\partial \theta}{\partial a_1} + \frac{1}{2}\frac{\partial \ln t^*}{\partial a_2} = 0, \quad \frac{\partial \theta}{\partial a_2} + \frac{1}{2}\frac{\partial \ln t^*}{\partial a_1} = 0. \tag{53}$$

Thus θ and $\sqrt{t^*}$ are conjugate harmonic functions of a_1, a_2. Thus there is a real constant C such that

$$\sqrt{t^*}e^{-i\theta} = C(a_1 + ia_2). \tag{54}$$

Using this expression in (44), we may integrate to obtain the Lagrangian coordinates

$$x(\rho,\psi) + iy(\rho,\psi) = C^{-1}e^{i\psi}\sqrt{\rho^2 C^2 - t}. \tag{55}$$

Differentiating with respect to t, and eliminating $\mathbf{a}$ we obtain the Eulerian velocity field

$$(u, v) = -\frac{C^2}{2}\frac{(x, y)}{x^2 + y^2}. \tag{56}$$

This is a steady potential sink flow, our example 2.1 above.

Example 4.3 We now treat the case $A = C = 0$ in (43). This introduces vorticity into the flow, which is a function of $\mathbf{a}$ alone. We now simplify the calculation somewhat by not requiring that $\mathcal{L}$ be the identity. The $\mathbf{K}$ can be any solution of (43) with non-zero determinant. In the present case we obtain

$$\mathbf{K} = \begin{pmatrix} \mu B C_1 T^{1+\lambda} + C_2 T^{-\lambda} & \mu B D_1 T^{1+\lambda} + D_2 T^{-\lambda} \\ C_1 T^{\lambda} & D_1 T^{\lambda} \end{pmatrix}, \tag{57}$$

where $T = (t^* - t), \mu = -(1+2\lambda)^{-1}$. After multiplication by $\mathbf{R}$, compatibility leads after some simplification, from the coefficients of $\tan^{\pm\lambda-1}$,

$$C_i \frac{\partial t^*}{\partial a_2} = D_i \frac{\partial t^*}{\partial a_1}, \quad i = 1, 2. \tag{58}$$

This leads directly to $\det \mathbf{K} = 0$ and a degenerate problem. The only way out of this impasse is for the set $\pm\lambda, \pm\lambda - 1, 1+\lambda$ to have repeated elements, so $\lambda = \pm 1/2, -1$ We leave the cases $\lambda = \pm 1/2$ to the interested reader, and consider here only the case $\lambda = -1$.

Then we have

$$\mathbf{KR} = \begin{pmatrix} C_2 T + B C_1 & D_2 T + B D_1 \\ C_1 T^{-1} & D_1 T^{-1} \end{pmatrix}. \tag{59}$$

With $\mathbf{R}^{-1}\mathbf{KR} = \mathbf{J}$ the compatibility is easily shown to make θ independent of $\mathbf{a}$, so we may take $\mathbf{R} = \mathbf{I}$. Imposing compatibility directly on (59) leads to the following equations:

$$C_1 \frac{\partial t^*}{\partial a_2} = D_1 \frac{\partial t^*}{\partial a_1}, \quad \frac{\partial D_1}{\partial a_1} = \frac{\partial C_1}{\partial a_2}, \quad \frac{\partial D_2}{\partial a_1} = \frac{\partial C_2}{\partial a_2}, \tag{60}$$

$$\frac{\partial (C_2 t^*)}{\partial a_2} + \frac{\partial (B C_1)}{\partial a_2} = \frac{\partial (D_2 t^*)}{\partial a_1} + \frac{\partial (B D_1)}{\partial a_1}. \tag{61}$$

We see that $(C_2, D_2) = \nabla_{\mathbf{a}} F, (C_1, D_1) = G'(t^*)\nabla_{\mathbf{a}} t^*$ where G is an arbitrary function of t^*, and F an arbitrary function of $\mathbf{a}$. Using these expressions in (61) and imposing the condition $\det \mathbf{J} = 1$ we arrive at

$$\frac{\partial(F, G)}{\partial(a_1, a_2)} = 1, \quad (G')^2 \frac{\partial(B, t^*)}{\partial(a_1, a_2)} = 1. \tag{62}$$

Thus $BG' - F = h'(t^*)$ and we obtain simple expressions for the Lagrangian coordinates, but in the Lagrangian variables F, G:

$$x(F, G) = (t^*(G) - t)F + H(G), \quad y(F, G) = \int \frac{dG}{t^*(G) - t}, \tag{63}$$

where $H(G) = h(t^*(G))$. It can be seen that the Eulerian flow has the form $(u, v) = (-x f_y(y, t) + g(y, t), f(y, t))$, so these flows are closely related to the those with stagnation point similitude.

It should now be clear that the method envisioned here depends strongly on the analysis of expansions of systems of ordinary differential equations near regular singular points, see e.g. [20],[7]. Some preliminary and partial general results along these lines, still restricted to two dimensions, suggest that whenever the eigenvalues of $\mathbf{D}_0$ are $\pm\lambda$ with 2λ not an integer, then any Lagrangian expansion of $\mathbf{D} \sim \mathbf{D}_0 T^{-1} + \ldots$ will determine a uniform gradient Euler flow. The case in which $\mathbf{D}$ has eigenvalues ± 1 can be treated using "shear" and similarity transformations, see [20]. Partial results suggest flows similar to flow with stagnation point similitude.

5. Three Dimensions

In three dimensions $\mathbf{D}$ acquires a non-trivial (i.e., non-constant) skew-symmetric part, associated with vortex stretching, and this renders the problem of deducing Euler flows from Lagrangian expansions far more difficult. In the model of [19], the uniform gradient solutions are considered in the approximation where $\mathbf{\Pi}$ has the simplest form needed to insure the constraint trace($\mathbf{D}$) = 0, namely

$$\mathbf{\Pi} = -\frac{1}{3}\text{trace}(\mathbf{D}^2)\mathbf{I}. \tag{64}$$

Applied to (17), relative to axes where the symmetric part of $\mathbf{D}$ is diagonal, this approximation yields

$$\mathbf{D}_0 + \mathbf{D}_0^2 - \frac{1}{3}\text{trace}(\mathbf{D}_0^2)\mathbf{I}. = 0. \tag{65}$$

Relative to principle axes of the symmetric part we have

$$\mathbf{D}_0 = \begin{pmatrix} \lambda_1 & -\Omega_3 & \Omega_2 \\ \Omega_3 & \lambda_2 & -\Omega_1 \\ -\Omega_2 & \Omega_1 & \lambda_3 \end{pmatrix}. \tag{66}$$

Here Ω is twice the local vorticity vector, and $\sum_k \lambda_k = 0$. From the off-diagonal elements of (65) we obtain the equation $\pm(\lambda_1 - l)\Omega_1 + \Omega_2\Omega_3 = 0$ plus its cyclic permutations. Thus two components of Ω must vanish, which we take to be Ω_1, Ω_3, so that $\lambda_2 = 1$. From the diagonal elements of (65) we then have

$$\lambda_i^2 + \lambda_i - \frac{1}{3}(\sum_{k=1}^{3} \lambda_k^2 + \Omega_2^2) = 0, \; i = 1, 3. \tag{67}$$

It is easy to show that, if $\Omega_2 \neq 0$, then λ_1 and λ_3 lie on either side of unity, i.e., λ_2 is the intermediate eigenvalue of the symmetric part of $\mathbf{D}_0$. Thus, *when* (65) *is imposed the vorticity determined by* $\mathbf{D}_0$ *is aligned with the intermediate eigenvalue of the local rate of strain matrix.* This result is intriguing because of evidence that this alignment occurs in simulations of turbulence, see e.g. [1].

In analogy with the two-dimensional calculations we pose

Question 5.1 What are the possible Euler flows determined by $\mathbf{QA}_0\mathbf{Q}^T$, where $\mathbf{Q}(\mathbf{a})$ is orthogonal (the T here denoting transpose), and $\mathbf{A}_0$ has the form

$$\mathbf{A}_0 = \begin{pmatrix} c & 0 & \Omega \\ 0 & 1 & 0 \\ -\Omega & 0 & -c-1 \end{pmatrix}, \tag{68}$$

but either is otherwise an arbitrary function of $\mathbf{a}$?

Since the compatibility constraints multiply dramatically in three dimensions, some streamlining of the method we have outlined in 2D is needed. We note that (swirling) sink flows and uniform gradient flows will be included in these results and may be the only flows which can be so obtained. Developments similar to the matrix form of (36) with logarithmic time corrections in the variable σ or s again are possible and go through as in the scalar case.

More difficult is the continuation of the representation of $\mathbf{D}$ to non-singular terms, e.g. (allowing also for s-dependence):

Question 5.2 What compatible Lagrangian flows are determined by a traceless matrix satisfying (10) having the form

$$\mathbf{D} = \mathbf{Q}(\mathbf{a}, s)[\mathbf{A}_0(\mathbf{a}, s)(t^*(\mathbf{a}) - t)^{-1} + \mathbf{A}_1(\mathbf{a}, s)]\mathbf{Q}^T(\mathbf{a}, s), \tag{69}$$

where $s = -\ln(t^* - t)$?

If it is assumed $\mathbf{Q}, \mathbf{A}_0, \mathbf{A}_1$ are independent of s then (10) implies

$$\mathbf{A}_0\mathbf{A}_1 + \mathbf{A}_0\mathbf{A}_1 + \mathbf{\Pi}_1 = 0, \;\; \mathbf{A}_1^2 + \mathbf{\Pi}_2 = 0, \tag{70}$$

where $\mathbf{\Pi}_i(\mathbf{a}), i = 1, 2$ are symmetric, trace$(\mathbf{A}_1) = 0$. Relative to coordinates where $\mathbf{A}_0$ is diagonal, it is seen that $\mathbf{A}_1$ must be symmetric, so the vorticity is entirely carried by the diverging term.

Since volume is preserved at a singularity as elsewhere, any singular structure in Lagrangian form in three dimensions must represent blobs of fluid which are either flattening into a pancake or else stretching out rodlike (and perhaps simultaneously twisting). Thus both $\mathbf{J}$ and $\mathbf{J}^{-1}$ must be singular, a point that has been emphasized by Constantin, see [9]. The velocity should then diverge as well, unless the volume of the region of enhanced stretching is simultaneously shrinking to zero size as the singularity is approached. Some aspects of this behavior is present in the solutions with stagnation-point similitude, but this kind of structure, where t^* in particular has a nontrivial variation with $\mathbf{a}$, is not seen in the other examples above. In any case conservation of total kinetic energy dictates that this *must* occur in bounded flows with bounded total kinetic energy.

6. Concluding Remarks

The methods discussed in the present paper were examined with the intent of finding accessible Euler flows in Lagrangian variables which might be inaccessible by Eulerian means. Our focus has been on flows with singular structure in the Lagrangian setting. To date the limited calculations have yielded no new solutions, and the flows so obtained, including those which can be classified as examples of Euler blow-up, are unphysical in the unboundedness of the kinetic energy. The scope of Euler flows which do exhibit blowup, physical or unphysical, remains an open problem.

Although the approach we use is direct and therefore very difficult in three dimensions, there are probably substantial economies available in the algebra, which would allow a thorough analysis of some choices of $\mathbf{D}$. We have not explored the extension to complex-valued dependent and independent variables, but see [3] for the application of complex variable methods to certain Euler flows. The analysis of *near singularities*, involving a large vorticity local in space and time, can also be approached using Lagrangian variables. It may indeed be the only problem yielding physically realizable flows. One might envisage an optimization problem where maximum vorticity is sought at time 1 over a class of normalized initial velocity and vorticity fields.

Acknowledgements

I would like to kindly acknowledge the hospitality of the Isaac Newton Institute during this conference, and the financial support from the National Science Foundation under grant No. DMS 970-4546 at New York University. Some of the ideas discussed here grew out of conversations with E.A. Spiegel.

References

1. Ashurst, W.T., Kerstein, A.R., Kerr, R.M., & Gibson, C.H. (1987) Alignment of vorticity and scale gradient with strain rate in simulated Navier-Stokes turbulence. *Phys. Fluids* **30**, 2343–2353.
2. Beale, J.T., Kato, T. & Majda, A.J. (1989) Remarks on the breakdown of smooth solutions for the 3-D Euler equations. *Commun. Math. Phys.* **94**, 61–66.
3. Caflisch, R.E. (1993) Singularity formation for complex solutions of the 3D incompressible Euler equations. *Physica D* **67** , 1–18.

4. Cantwell, B.J. (1984) Exact solution of a restricted Euler equation for the velocity gradient tensor. *Phys. Fluids A* **4**, 782–793.
5. Childress, S. (1987) Nearly two-dimensional solutions of Euler's equations. *Phys. Fluids* **30**, 944–953.
6. Childress, S., Ierley, G.R., Spiegel, E.A., & Young, W.R. (1989) Blow-up of unsteady two-dimensional Euler and Navier-Stokes solutions having stagnation-point form. *J. Fluid Mech.* **203**, 1–22.
7. Coddington, E.A. & Levinson, N. (1955) *Theory of Ordinary Differential Equations.* McGraw-Hill.
8. Constantin, P. (2000) The Euler equations and nonlocal conservative Riccati equations. *Int. Math. Res. Notices* **9**, 455–465.
9. Constantin, P. (2000) Some open problems and research directions in the mathematical study of fluid dynamics. *Preprint.*
10. Dresselhaus, E. (1992) *Material Element Stretching and Alignment in Turbulence.* Ph.D. thesis, Columbia University.
11. Dresselhaus, E. & Tabor, M. 1991 The stretching and alignment of material elements in general flow fields. *J. Fluid Mech.* **236**, 415–444.
12. Kato, T. (1967) On the classical solutions of the two-dimensional non-stationary Euler equation. *Arch. Rat. Mech. Anal.* **25**, 188-200.
13. Lamb, H. (1932) *Hydrodynamics.* Dover Publications, Sixth edition.
14. Moffatt, H.K. (2000) The interaction of skewed vortex pairs: a model for blow-up of the Navier-Stokes equations. *J. Fluid Mech.* **409**, 51–68.
15. Ohkitani, S.K. & Gibbon, J.D. (2000) Numerical study of singularity formation in a class of Euler and Navier-Stokes flows. *Phys. Fluids* **12**, 3181–3194.
16. Rankine, W.J.M. (1863) On the exact form of waves near the surface of deep water. *Trans. Roy. Soc. London* **153**, 127–138.
17. Stuart, J.T. (1987) Nonlinear Euler partial differential equations: singularities in their solution. In *Proceedings in honor of C.C. Lin* (ed. D.J. Benney *et al.*), pp. 81–95. World Scientific, Singapore.
18. Stuart, J.T. (1991) The Lagrangian picture of fluid motion and its implication for flow structures. *IMA J. Appl. Math* **46**, 147–163.
19. Vieillefosse, P. (1984) Internal motions of a small element of fluid in an inviscid flow. *Physica A* **125** , 150–162.
20. Wasow, W. (1965) *Asymptotic Expansions of Ordinary Differential Equations.* Interscience.

ANALYSIS OF A CANDIDATE FLOW FOR HYDRODYNAMIC BLOWUP

RICHARD B. PELZ
Department of Mechanical and Aerospace Engineering
Rutgers University
Piscataway, New Jersey 08854-8058 USA

E-mail: *pelz@jove.rutgers.edu*

Abstract. An octahedral, vortical (divergence-free) flow is defined, and reasons are given for why it is a candidate for self-similar, point-collapse and blowup under evolution of the Euler equations. Results from a vortex filament model of such flows are then reviewed and the data subsequently analyzed. The Gauss map, a mapping of the vorticity tangent field, is examined. A Leray-Beltrami flow, defined as a force-free flow in the Leray collapse frame, is shown to develop in the inner region. Finally, the vorticity is found to scale as the inverse square of the distance from the origin, the center of the point collapse. A discussion follows on the ramification of these findings to possible blowup in the Navier-Stokes equations.

1. Introduction

Mixing topology, differential geometry and fluid mechanics may generate new approaches to the outstanding problem of existence and smoothness of incompressible flows. One of the main goals of the Newton Institute programme "Geometry and Topology of Fluid Flows" was to highlight this problem and develop methods to solve it. The recent announcement that this problem is one of seven "Millennium Prize Problems" worth one million US dollars each, may also motivate further investigations. The fact that this problem is unsolved is an embarrassment to the mathematics community as well as the physics and engineering communities who routinely use the equations of hydrodynamics for research and design.

The regularity analysis of the Navier-Stokes equation for incompress-

R.L. Ricca (ed.), An Introduction to the Geometry and Topology of Fluid Flows, 329–341.

ible flows, a parabolic PDE with quadratic nonlinearity, has been unexpectantly difficult. Likewise, the numerical analyses of this problem, while yielding hints, have not been refined enough to provide answers to such a subtle question. This has led some researchers to proclaim that $1M is not enough. Indeed, for critical problems such as this, dedicated financial support must be provided *a priori* not *a posteriori.* Funding for this will be money well spent; work on this problem will undoubtedly open new areas of mathematics, numerical analysis and physics as well as bridge seemingly separate disciplines.

The statement of the Clay Prize, written by C. Fefferman [7], is to find a proof to one of the following four problems:

A. the existence and smoothness of Navier-Stokes solutions for all initially smooth divergence-free flows in $\mathbb{R}^3$;
B. the same as A but for the three-torus T^3;
C. the breakdown of Navier-Stokes solutions in $\mathbb{R}^3$;
D. the breakdown of Navier-Stokes solutions in T^3.

The external forcing is zero and the energy of the initial flows is finite.

A main method of attack has been to examine similar but simpler model equations. Recent examples include a one-dimensional model equation [4], the one-dimensional Ginzburg-Landau equation [20], the two-dimensional geostrophic equation [5, 6], and forced two-dimensional flows [3, 13, 16].

The computational attempts at the full problem [8, 10, 1] have yielded candidate flows that have yet to be fully explored. Initial value simulations will never yield more than circumstantial evidence. The detection of the non-algebraic behavior in t for the function $[(T-t)^2+\epsilon^2]^{-\alpha/2}$, where T is the critical time and α and ϵ are constants, $\alpha > 0$ and $0 < \epsilon < T$, which differs from $(T-t)^{-\alpha}$, begins only at a time $t \sim T-\epsilon$. Integration to a time less than the critical time will, thus, never resolve whether the singularity is in real space or not.

In this paper, the highly symmetric, candidate flow [1, 17] is described in terms of its discrete group representation, and then data from a vortex filament model of this flow are analyzed for scaling and topological features.

The flow in question is invariant under the full octahedral group of symmetries, which keeps the necessary three-dimensionality of the flow, but allows a reduction by a factor of 48 in the fundamental domain. The octahedral group and octahedral flows will be discussed in the next section. Even with this reduction, the dynamic range obtained through field simulations is still too small for much quantitative analysis [1]. A greatly enhanced temporal and spatial range can be found through the use of the vortex filament model, where the smooth but compact vorticity field is approximated by a singular distribution of vorticity confined to one or more space curves.

The spatial variable becomes the parameterization of the curves making the model one-dimensional.

Filament models must be used with care. Using a singular vorticity distribution (even though it is smoothed), to investigate the question of smoothness seems contradictory: initial flows must be smooth. The shape dynamics of the vortex tubes, which may be crucial in blowup, are not contained within the vortex filament model. The filament model of the octahedral flows [17], however, has permitted the identification of scaling properties that will be sought in future field simulations.

Section 3 reviews the filament solution. The initial value problem is solved numerically, and an asymptotic problem is observed to develop in time. Near the origin, an inner region develops with a scaling that approaches the singular, self-similarity proposed by Leray [12]. Data from this simulation is then used in the analysis of subsequent sections. In Section 4, the Gauss map of the filaments is examined. The unit tangent to the filament in $\mathbb{R}^3$ defines this mapping to the sphere S^2. The "Leray-Beltrami" flow is introduced in Section 5 as a force-free flow in the Leray frame. It is shown that the Leray solution is approached in a ball of constant radius. In Section 6 it is shown that the vorticity scales as the inverse square of the radius; thus, the velocity is in L^p where $p > 3$. The ramifications of this result in light of the recent results of Nečas, Růžička & Šverák [15] and Tsai [22] on the space of Leray solutions for the Navier-Stokes equations is then discussed.

2. Octahedral Flows

Numerical evidence reported in references [1, 17, 19, 18] indicates that particular vortical flows that are invariant under the action of the full octahedral group of rotations, O_h, may evolve towards a point-wise collapse and blowup in finite time. In this section, symmetries of the octahedral group are reviewed, and octahedral flows are defined.

The full octahedral group are the 48 proper and improper rotations that leave an octahedron invariant. Centered at the origin, the octahedron is oriented so that the six vertices lie on the Cartesian axes. The four-fold rotational symmetry about each axis is denoted $\{6C_4, 3C_2\}$ (see the reference texts [11, 23]). About the four lines through the centroids of opposite faces, there is a three-fold rotational symmetry denoted $8C_3$. One set of the reflectional symmetries is about the three zero planes and is denoted $3\sigma_h$. The other set of reflections, $6\sigma_d$, are about the the six diagonal planes: $x = \pm y$, $y = \pm z$ and $z = \pm x$. There is also a two-fold rotation $6C_2'$ about these planes. These operations together with the identity E, inverse i and 14 improper rotations S make O_h a 48th order group.

Scalar fields that are (even or odd) invariant under the action of the group elements are associated with the ten irreducible representations in the O_h character table shown below.

O_h	E	$8C_3$	$6C_4$	$3C_3$	$6C_2'$	i	$8S_6$	$6S_4$	$3\sigma_h$	$6\sigma_d$
A_{1g}	1	1	1	1	1	1	1	1	1	1
A_{2g}	1	1	-1	1	-1	1	1	-1	1	-1
E_g	2	-1	0	2	0	2	-1	0	2	0
T_{1g}	3	0	1	-1	-1	3	0	1	-1	-1
T_{2g}	3	0	-1	-1	1	3	0	-1	-1	1
A_{1u}	1	1	1	1	1	-1	-1	-1	-1	-1
A_{2u}	1	1	-1	1	-1	-1	-1	1	-1	1
E_u	2	-1	0	2	0	-2	1	0	-2	0
T_{1u}	3	0	1	-1	-1	-3	0	-1	1	1
T_{2u}	3	0	-1	-1	1	-3	0	1	1	-1

TABLE 1. The character table for the full octahedral point group, O_h.

The Cartesian velocity components are associated with the degenerate representation T_{1u}. For a point collapse about the origin, however, the use of a toroidal/poloidal solenoidal decomposition is more physically meaningful. It can be shown that the poloidal velocity field is a linear combination of two scalars associated with A_{1g}, the fully symmetric representation, and A_{2g}, which is even with respect to $8C_3$ and $3\sigma_h$, but odd with respect to $6C_4$ and $6\sigma_d$. The toroidal component is a combination of fields belonging to the representation A_{1u}, even with respect to the rotational symmetries, but odd with respect to the reflections, and A_{2u}, even with respect to $8C_3$ and $6\sigma_d$ and odd with respect to $6C_4$ and $3\sigma_n$. Thus, a three-dimensional, divergence-free field with octahedral symmetry, an octahedral flow, can be defined as the set of flows with toroidal and poloidal components being suitably-behaved scalar fields associated with these four representations. Casting the dependent variables in terms of symmetric fields of a degenerate point group allows a closer association with other collapse phenomena.

The origin of the octahedral flow is a degenerate critical point. The reflectional symmetries restrict the axes to be streamlines incoming to or outgoing from the origin. They also make the Taylor expansion about the origin only odd in space. The divergence-free condition constrains the origin to be a hyperbolic point with $(\partial u/\partial x + \partial v/\partial y + \partial w/\partial z)|_0 = 0$. The rotational symmetry restricts all axes to be the same; hence, $\partial u/\partial x|_0 = \partial v/\partial y|_0 = \partial w/\partial z|_0 = 0$. The expansion about the origin of the radial velocity is

$$v_r = a_3 \frac{x^4 + y^4 + z^4 - 3(x^2y^2 + y^2z^2 + z^2x^2)}{\sqrt{x^2 + y^2 + z^2}} + O(x^5) . \qquad (1)$$

The first toroidal term is xyz.

The observation from numerical experiments [1] is that vortex dipoles centered on each axis form with induced velocity towards the origin. The first radial coefficient, a_3, is negative; hence, the six axes are incoming streamlines whereas the eight diagonals are outgoing. The configuration is observed to be stable and attracting. The dipoles accelerate towards the origin and a_3 monotonically decreases through the last integration time.

To construct such an octahedral, dipolar flow as an initial condition, a compact, positive distribution of (normal) vorticity is placed on the zero-planes, $z = 0$ for example, in such a way that the centroid is above the diagonal, $y_0 > x_0$. The dipole formed across the y axis has induced velocity towards the origin. Then let the radius of curvature of the vortex tube be large compared to the centroid distance to the origin, $\sqrt{x_0^2 + y_0^2}$. The curvature is such that the tube curves away from the origin and connects up to its image at some larger distance away.

With symmetries, the flow close to the origin is made up of three orthogonal, vortex quadrupoles symmetric about the origin and in close proximity to each other. A positive strain rate is induced in front of an advancing dipole which is orthogonal to both its induced velocity and vorticity, but parallel to the vorticity of another quadrupole. As discussed in [18] each dipole is immersed in a positive strain rate produced by the images. Restrictions on vortex-line curvature coming from self-generated axial strain rate are then lifted.

3. The Filament Simulation

In the field simulation of [1], the vorticity was concentrated in elongated tubes. Even though they were arranged as dipoles, the deformation or flattening of the cores was relatively mild. Vortex line curvatures were also mild. Because of this, the octahedral, vortical flow modeled by vortex filaments was examined in [17].

The octahedral symmetries were enforced in the filament simulations. The $3\sigma_h$ reflectional symmetries allowed an eight-fold reduction in the number of degrees of freedom, and the $8C_3$ rotational symmetries allowed a further three-fold reduction, for a total of a 24-fold reduction. Six elongated vortex rings were the initial conditions. One quarter of one closed loop formed the fundamental filament which was discretized to a polygonal curve with the position of the nodes being the dependent variables. A local

Figure 1. Initial position of the fundamental filament (left) and its position at a later time in the simulation (right). The surface is a collection of cylindrical bands at each node of the polygonal curve. The radius of the band is half of the core size and the center is at the filament. The shape of the surface near the origin is roughly conical.

core model was used to desingularize the Biot-Savart kernel: the local core cross-sectional area times the local axial strain rate remained constant. To preserve the Helmholtz condition when the filament was close to a reflectional symmetry plane, a two-length scale desingularization was used. In this way, cores did not overlap.

In Figure 1 a plot of the location and core size of the fundamental filament is given (see reference [17] for expanded views with all the images). The initial location (left) and location after some time has elapsed (right) are shown. The surface represents the core size at each point on the curve. As can be seen, a roughly conical surface develops during the simulation and is aligned in the radial direction. Close to the origin, the filament becomes toriodal and intersects the symmetry plane; this is defined as the inner region. Outside the radial region, the filament has moved in the binormal

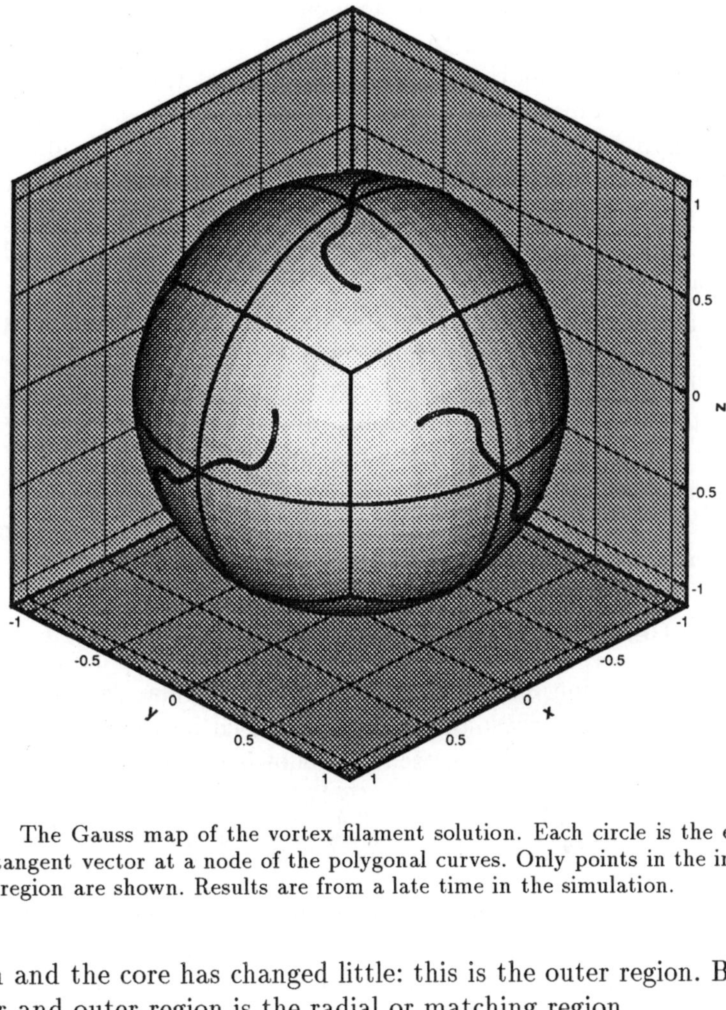

Figure 2. The Gauss map of the vortex filament solution. Each circle is the endpoint of a unit tangent vector at a node of the polygonal curves. Only points in the inner and matching region are shown. Results are from a late time in the simulation.

direction and the core has changed little: this is the outer region. Between the inner and outer region is the radial or matching region.

4. The Gauss Map

The Gauss map is the mapping of the unit tangent of a curve in $\mathbb{R}^3$ onto a curve on the unit sphere S^2. For each point on the curve, the point where the unit tangent vector with tail at the origin touches the sphere is the image. The parameterization of the curve also gives a parameterization of the image curve. The tangent of the image curve is a function of the curvature of the pre-image, curvature is torsion, etc..

Plotted in Figure 2 is the Gauss map for the filament curve at a late time in the simulation. The image of the nodes of the polygonal filament curve are plotted as small circles. The point where the filament is on the zero plane (see Figure 1) has its tangent normal to the plane, hence the image is on a pole. The fact that the tangent of the image at the poles has two components means that the inner filament at the symmetry plane has both curvature components.

Moving from the symmetry plane, the image of the fundamental filament, extends into the middle of the spherical quadrilateral. The image is straight for a while, indicating little torsion in the inner filament. Then the image develops curvature of one sign, then of the opposite sign. The torsion of the filament has a similar behavior.

Only points on the filament in the inner and matching regions are plotted. All the nodes in the matching region are mapped to the endpoint in the middle of the quadrilateral. As was apparent from Figure 1, the filament in the matching region is radial. Gauss images of the 23 reflections and rotations of the fundamental curve are also shown.

It is interesting to observe the symmetries of the Gauss curve. The $8C_3$ three-fold rotational symmetries exist as well as the $3C_2$ two-fold symmetries about the three lines through opposite vertices.

5. Leray-Beltrami Flows

Given the restriction on the dimension of the singular set [2, 21], a point collapse in space is a reasonable scenario. Based on the scale similarity of the equations, Leray [12] proposed the following transformation

$$u(x,t) = \frac{U(\xi)}{\sqrt{2a(T-t)}}, \quad \xi = \frac{x}{\sqrt{2a(T-t)}}, \quad p(x,t) = \frac{P(\xi)}{2a(T-t)}, \tag{2}$$

where p and u are the pressure and incompressible velocity fields that satisfy the Navier-Stokes equations in x and t and a is a positive constant. The Navier-Stokes equations under this transformation become

$$(U + a\xi) \times \Omega = \Pi - \nu \Delta U , \tag{3}$$

where $\Omega = \nabla \times U$, $\Pi = P + U^2/2 + a\xi \cdot U$ and $\nu > 0$. If there exists a nonzero solution U to Equation 3, then u blows up at $t = T$. In the inviscid case, the scaling can be more general, the $1/2$ power in the velocity and length can be replaced by the real number α and $1 - \alpha$, respectively.

In reference [17], the simulations showed that the flow approached the Leray scaling for the part of the filament near the origin. That is, the closest distance of the filament to the origin, which occurs on the symmetry plane, scaled as $\sqrt{T-t}$ where T was estimated. Many orders of magnitude were found easily with this one-dimensional model. Also, the minimum core size, which also occurred on the symmetry plane, had the same scaling. The reason why the core size has this behavior, was because the local core model was used and the maximum strain rate along the filament scaled as $1/(T-t)$.

The induced velocity of the filament at the point on the symmetry plane scaled as $1/\sqrt{T-t}$, thus there was an acceleration towards the origin. Dipolar induced velocity is inversely proportional to distance between vortex centroids and in the perpendicular direction. Simple induced velocity has zero acceleration. The induced velocity from the image dipoles, however, provide an inward flow tending to decrease the centroid distance of the fundamental dipole, which then allows the acceleration.

A force-free flow in the Leray frame would satisfy a modified Beltrami condition: $U + a\xi$ parallel to Ω. Let such a flow which obeys this condition be called a *Leray-Beltrami* flow. One way to show that a flow is Leray-Beltrami is to calculate the degree of alignment of the two unit vectors. Let ϕ be the angle between the vorticity vector, Ω, and the modified velocity, $U + a\xi$ at each point in the flow where both vectors are nonzero. The cosine of the angle ϕ can then be found as follows:

$$\cos\phi = \frac{\Omega \cdot (U + a\xi)}{|\Omega||U + a\xi|} \,. \tag{4}$$

If these two vectors are parallel, then this quantity will be unity.

In the filament model, the alignment angle is only defined on the filament. The results from the filament simulation are shown in Figure 3 for a number of times. Note that the abscissa is the radial location with respect to the Leray scaling.

Estimates of the critical time T and normalization constant a are found by fitting to the data at late times. The value of a is set such that at the filament point $(x_0, y_0, 0)$ on the reflectional symmetry plane, $u_r = -(a/x_0)^2\sqrt{x_0^2 + y_0^2}$.

Figure 3 shows that the region of alignment increases with time. The initial time is the shortest curve and located on the left. Six later times are also plotted from left to right. The values of $T-t$ for these seven curves are $3.0204, 1.15 \times 10^{-1}, 5.5 \times 10^{-3}, 2.7 \times 10^{-4}, 1.3 \times 10^{-5}, 6.3 \times 10^{-7}, 3.1 \times 10^{-8}$. Since the time-step is based on the inner variable and an equal number of time steps are used between outputs, the curves appear equally spaced.

The observation is that an inner region of alignment forms and increases with time. The edge of the alignment region increases by $\sqrt{T-t}$ suggesting that this region is inside a ball of constant radius in the outer variables.

The two terms in the quantity $U + a\xi$ do not scale the same with radius; the velocity U scales as ξ^{-1}. As the velocity falls off with radius and the Leray-Beltrami condition is maintained, the vorticity must align in the radial direction, which occurs in the matching region. Assuming that P

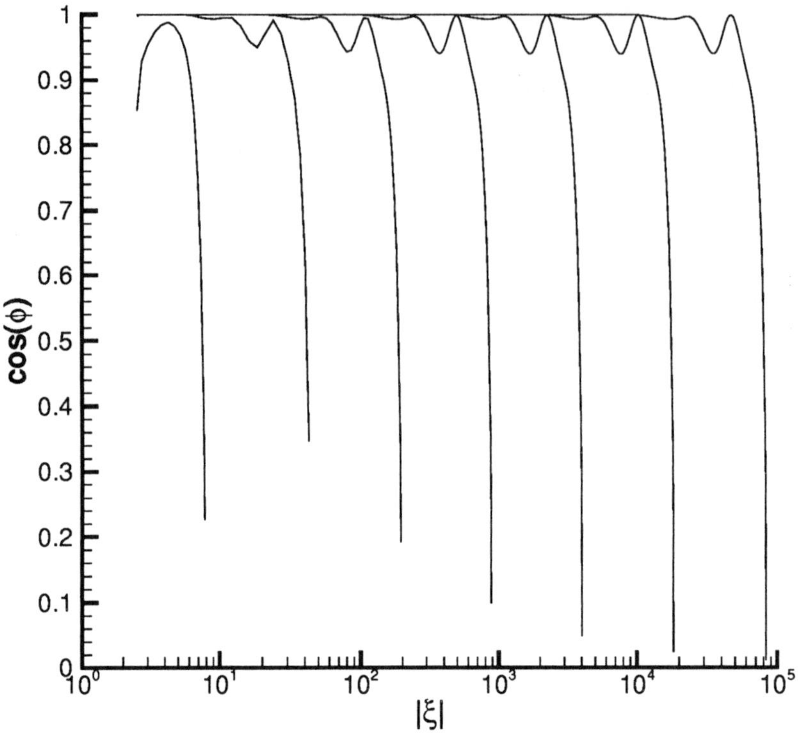

Figure 3. Plot of $\cos\phi = [\Omega \cdot (U + a\xi)]/(|\Omega||U + a\xi|)$ versus $|\xi|$ for each node on the vortex filament at times $T - t$ (displayed from left to right): 3.0204 (initial), 1.15×10^{-1}, 5.5×10^{-3}, 2.7×10^{-4}, 1.3×10^{-5}, 6.3×10^{-7} and 3.1×10^{-8}. In the inner and matching region the filament tangent is parallel to the modified velocity $U + a\xi$. The values of T and a are extrapolated from the data.

scales as $1/\xi^2$, the Leray equation in the matching region approaches $a\xi \times \Omega = 0$. For analysis of this region see reference [9].

6. Spatial Behavior of Vorticity

A plot of poloidal (radial) vorticity and toroidal vorticity versus Leray-scale radius is given in Figure 4. The data are taken at a late time in the simulation. The radial vorticity clearly scales as $1/\xi^2$ while the toroidal scales as $1/\xi^4$. Roundoff error causes the oscillatory behavior at 10^{-9}.

For a filament model, the vorticity is proportional to $1/\sigma^2$, where σ is the core size. The $1/\xi^2$ behavior of the radial vorticity is thus caused by the core scaling linearly with ξ. This supports the observation from Figure 1 that the tube was conical. For a discussion on why this scaling is to be expected, see Moffatt [14].

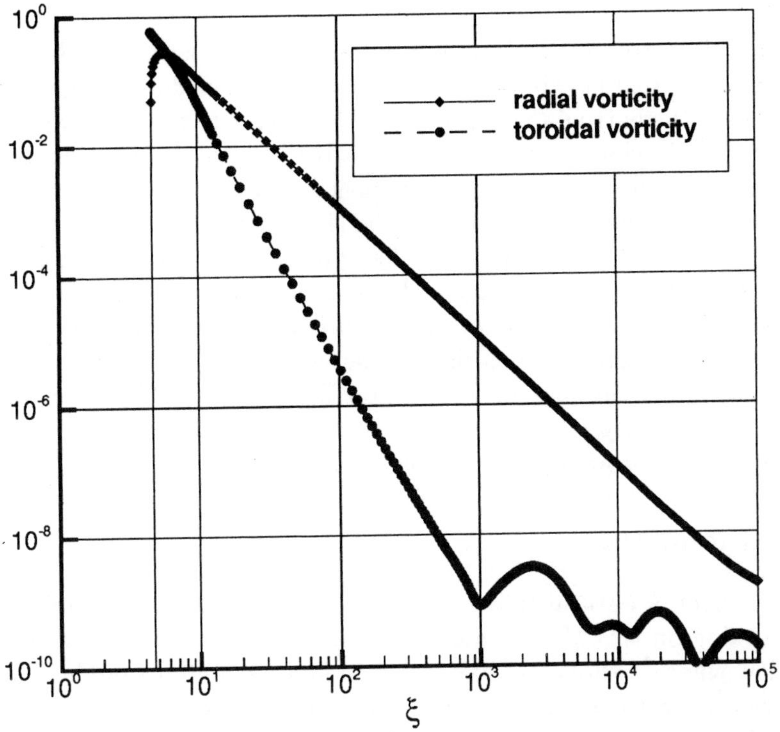

Figure 4. The plot of the radial and toroidal vorticity components versus radius for the filament solution at a late time. The vorticity is proportional to the inverse of the core size squared. The Leray scaling is used. The ranges of inner, matching and outer regions are approximately $\xi < 10, \xi = [10, 10^4], \xi > 10^4$. Roundoff behavior starts at 10^{-9}.

It is interesting to extrapolate these findings to the solution of the Euler and Navier Stokes equations. Of all the admissible scaling exponents α for the Euler equations, the filament model approached the only value, $\alpha = 1/2$, that is also admissible by the Navier-Stokes equations. The Leray solution is approached asymptotically as $t \to T$. With the scaling $U \sim 1/\xi$, U would be an element of L^p for $p > 3$. The Nečas, Růžička & Šverák [15] result for the Navier-Stokes equations is that U is trivial if $U \in L^p$ for $p = 3$, but does not limit behavior for $p > 3$. The more recent result of Tsai [22], U is trivial if $U \in L^p$ for $p > 3$ for the Navier Stokes solutions, is, however, a serious obstacle to a simple extrapolation to the Navier-Stokes equation.

Referring to Equation 3, Tsai showed that Π is a constant for a significant set of flows which satisfy the Leray equation. If, in addition, the flow satisfies the Leray-Beltrami condition, then $\nu\Delta U = 0$. If the outer boundary condition on U is zero, only the trivial solution is possible for

$nu > 0$.

On physical grounds it may be argued that such a point collapse may occur for an inviscid flow but not for a viscous one. The dipole aligned in the z direction with centroid (x_0, y_0) on the $z = 0$ plane is used as the example. The first radial velocity term, a_3, is negative if $y_0 > x_0$ (see Equation 1). The presence of viscosity will cause diffusion of ω_x across the $x = 0$ plane. It is known from the Gauss map that the (filament) dipoles have curvature in both the x and y directions at the symmetry plane $z = 0$, so ω_x is nonzero in the inner region. If ω_x scales as x near the plane, $\Omega_{\xi\xi}|_{\xi=0}$ will be nonzero, and there will be a continuous breaking of vortex lines from the dipole and reconnecting across the symmetry plane. This not only depletes the circulation of the dipole, but creates a positive a_3 from the induced velocity of the newly formed tubes. Reconnection will create the outgoing state before the incoming state has collapsed completely.

7. Final Remarks

There has been a natural split between mathematicians and computationalists working on this Millennium problem. The former have tried to prove regularity, whereas the latter have tried to find a counterexample. The former have proved theorems, but many needed to make critical assumptions that made the theorems less applicable to the physical situation. The latter have presented many plots and heuristic arguments for blowup, but no proof.

Results presented in this paper, albeit from a filament model, may provide a common problem for the two groups. The indication is that there is a well-defined, matched-asymptotic problem in which the inner region is a Leray-Beltrami flow and the matching region has radial vorticity which scales as r^{-2}. The octahedral symmetry together with the $\xi = x/\sqrt{T-t}$ similarity may make the computational problem tractable, whereas the Leray-Beltrami property and connections to group theory suggests an analytic approach is possible.

Acknowledgements

I would like to acknowledge the hospitality of the Isaac Newton Institute. Financial support from UK EPSRC (Grant GR/K99015) is also kindly acknowledged. Thanks also goes to Rutgers University for granting me a competitive leave for my stay at the Newton Institute.

Much of this work comes from discussions with Keith Moffatt, Stephen Cowley, Phil Boyland, John Greene, Rob Ghrist and Vladimir Šverák. I wish to thank them all.

References

1. Boratav, O.N. & Pelz, R.B. (1994) Direct numerical simulation of transition to turbulence from a high-symmetry initial condition. *Phys. Fluids* **6**, 2757–2784.
2. Caffarelli, L., Kohn, R.V. & Nirenberg, L. (1982) Partial regularity of suitable weak solution of the Navier-Stokes equations. *Comm. Pure Appl. Math* **35**, 771–831.
3. Constantin, P. (2000) The Euler equations and nonlocal conservative Riccati equations. *Int. Math. Res. Notices* **9**, 455–465.
4. Constantin, P, Lax, P.D. & Majda, A. (1996) A simple one-dimensional model of the three-dimensional vorticity equation. *Comm. Pure Appl. Math* **38**, 715-724.
5. Constantin, P., Majda, A., & Tabac, E. (1994) Formation of strong fronts in the 2-D quasigeostrophic thermal active scalar. *Nonlinearity* **7**, 1495-1533.
6. Cordoba, D. (1998) Non-existence of simple hyperbolic blow-up for the quasi-geostrophic equation. *Annals of Mathematics* **148**, 1135-1152.
7. Fefferman, C. (2000) Existence and smoothness of the Navier-Stokes equations. `www.claymath.org/prizeproblems/navierstokes.htm`.
8. Grauer, R., Marliani, C. & Germaschewski, K. (1998) Adaptive mesh refinement for singular solutions of the incompressible Euler equations. *Phys. Rev. Letts.* **80**(19), 4177–4180.
9. Greene, J.M. & Pelz, R.B. (2000) Stability of postulated, self-similar, hydrodynamic blowup solutions. *Phys. Rev. E* **62**, 7982.
10. Kerr, R.M. (1993) Evidence for a singularity in the three-dimensional Euler equations. *Phys. Fluids* **6**, 1725–1746.
11. Kettle, S.F.A. (1985) *Symmetry and Structure.* Wiley.
12. Leray, J. (1934) Sur le mouvement d'un liquide visqueux emplissant l'espace. *Acta Math.* **63** 193–248.
13. Malham, S.J.A. (2000) Collapse of a class of three-dimensional Euler vortices. *Proc. R. Soc. Lond.* A **456**, 2823–2833.
14. Moffatt, H.K. (2000) The interaction of skewed vortex pairs: a model for blow-up of the Navier-Stokes equations. *J. Fluid Mech.* **409**, 51–68.
15. Nečas, J., Růžička, M. & Šverák, V. (1996) On Leray's self-similar solutions of the Navier-Stokes equations. *Acta Math.* **176**, 283–294.
16. Ohkitani K. & Gibbon, J.D. (2000) Numerical study of singularity formation in a class of Euler and Navier-Stokes flows. *Phys. Fluids* **12**, 3181–3194.
17. Pelz, R.B. (1997) Locally self-similar, finite-time collapse in a high-symmetry vortex filament model. *Phys. Rev. E* **55**, 1617–1626.
18. Pelz, R.B. (2001) Symmetry and the hydrodynamic blowup problem. *J. Fluid Mech.*, in press.
19. Pelz, R.B. & Gulak, Y. (1997) Evidence for a real-time singularity in hydrodynamics from time series analysis. *Phys. Rev. Lett.* **79**(25), 4998–5001.
20. Plechac, P & Šverák, V. (2001) On self-similar singular solutions of the complex Ginzburg-Landau equation. *Comm. Pure Appl. Math.*, in press.
21. Scheffer, V. (1977) Hausdorff measure and the Navier-Stokes equations. *Comm. Math. Phys.* **55**, 97–112.
22. Tsai, T.-P. (1998) On Leray's self-similar solutions of the Navier-Stokes equations satisfying local energy estimates. *Arch. Rat. Mech. Anal.* **143**, 29–51.
23. Wherrett, B.S. (1986) *Group Theory for Atoms, Molecules and Solids.* Prentice Hall International.

SUBJECT INDEX

M

N

O

P

Q

R

S